T0343413

Recycling and Deinking of Recovered Paper

Recycling and Deinking of Recovered Paper

Pratima Bajpai

Pulp and Paper,
Consultants,
Patiala, India

AMSTERDAM • BOSTON • HEIDELBERG • LONDON • NEW YORK • OXFORD
PARIS • SAN DIEGO • SAN FRANCISCO • SINGAPORE • SYDNEY • TOKYO

Elsevier
32 Jamestown Road, London NW1 7BY
225 Wyman Street, Waltham, MA 02451, USA

First edition 2014

Notices

Knowledge and best practice in this field are constantly changing. As new research and experience broaden our understanding, changes in research methods, professional practices, or medical treatment may become necessary.

Practitioners and researchers must always rely on their own experience and knowledge in evaluating and using any information, methods, compounds, or experiments described herein.

In using such information or methods they should be mindful of their own safety and the safety of others, including parties for whom they have a professional responsibility.

To the fullest extent of the law, neither the Publisher nor the authors, contributors, or editors, assume any liability for any injury and/or damage to persons or property as a matter of products liability, negligence or otherwise, or from any use or operation of any methods, products, instructions, or ideas contained in the material herein.

British Library Cataloguing-in-Publication Data
A catalogue record for this book is available from the British Library

Library of Congress Cataloging-in-Publication Data
A catalog record for this book is available from the Library of Congress

ISBN: 978-0-12-416998-2

For information on all Elsevier publications visit our website at store.elsevier.com

This book has been manufactured using Print On Demand technology. Each copy is produced to order and is limited to black ink. The online version of this book will show color figures where appropriate.

Contents

Preface

Paper recycling has been practised for many decades in numerous countries around the world. The paper industry is among those that almost perfectly meet the expectations of society for the sustainability of their raw material base, the environmental compatibility of their processes and the recyclability of their products, and it has done so for a long time. In particular, recyclability has dramatically improved the image of the paper industry in the past few years. Recycling of paper and paper-related products is very important for sustainable economic growth. It helps save landfill space and costs, reduces the energy requirements for paper manufacturing and extends the supply of fibre. Paper recycling reduces consumption of precious natural resources (wood, water, minerals and fossil fuels). It is an important service to society and is profitable. Rigorous scientific research supports the benefits of recycled paper, and government agencies, environmental groups and many other large purchasers have adopted policies mandating its use. Recycling technologies have improved in recent years through advances in pulping, flotation deinking and cleaning/screening. This has resulted in the quality of paper made from secondary fibres approaching that of virgin paper. However, the process is much more eco-friendly than the virgin-papermaking process. By using recycled paper, companies can take a significant step towards reducing their overall environmental impacts. Most of the paper producers throughout the world are increasing their use of recycled fibres for paper products. The percentage of recycled fibre in almost every grade of paper and board, particularly in printing and writing papers, has more than doubled in the past decade; in the next 10–15 years it is expected to triple or quadruple. Earlier, paper was recycled mainly because it made economic sense to do so, whereas now it has been driven more by a collective environmental mandate of sorts. The marketplace for secondary pulps and papers continues to develop and mature. Deinked pulp has become a principal raw material for many papermaking operations around the world. Many newsprint and tissue grades commonly contain 100% deinked pulp, which is also a substantial furnish constituent of other grades, such as lightweight coated for offset, and printing and writing papers for office and home use. The scientific and technical advances in recycling and deinking are discussed in this book.

1 Introduction*

1.1 The Paper and Paperboard Industry in the Global Market

The pulp and paper industry is one of the largest industries in the world. It is dominated by North American, Northern European and East Asian countries. Latin America and Australasia also have significant pulp and paper industries. Over the next few years, it is expected that both India and China will become the key countries in the industry's growth. World production of paper and paperboard is around 390 million tonnes and is expected to reach 490 million tonnes by 2020. In 2009, total global consumption of paper was 371 million tonnes (Figure 1.1). In North America, total paper consumption declined 24% between 2006 and 2009.

Growth is speedy in Asia; it accounts for almost 40% total world paper and paperboard production whereas the European Union (EU) and North America account for about one quarter each. The profitability of pulp and paper industry has been weak on the global level in recent years. Excess capacity has led to falling product prices that have, with the impact of rising production costs, eroded the industry's profitability globally.

Consumption of paper and paperboard per person varies significantly from country to country. One person uses about 60 kg of paper a year on average; the extremes are 265 kg for each US resident and some 7 kg for each African. In the heavily populated areas of Asia, only around 40 kg of paper per person is used. This means that Asian consumption will continue to grow intensely in the coming years if developments there follow the example of the West. In Finland, consumption of paper and paperboard per person is about 194 kg. Although India's population is about 7% of the world's population, it consumes barely 2% of the global paper output with consumption per person at only 9 kg against a global average of 55 kg, 65 kg in China and 215 kg in Japan. Rapid growth in Asian paper production in recent years has increased the region's self-sufficiency, lessening the export opportunities available to both Europeans and Americans. Moreover, Asian paper has started to enter Western markets – from China in particular. Global contest has increased noticeably as the new entrants's cost level is considerably lower than in competing Western countries.

The European industry has been dismantling overcapacity by shutting down mills that do not make any profit. In all, over 5% of the production volume in Europe has been closed down in the past few years. Globally speaking, the products of the forest industry are primarily consumed in their country of production, so it can be considered a domestic-market industry. The largest trade flows are between the countries of

*Some excerpts taken from Bajpai Pratima, 'Advances in Recycling and Deinking', 2006 with kind permission from Pira International, UK.

Recycling and Deinking of Recovered Paper. DOI: http://dx.doi.org/10.1016/B978-0-12-416998-2.00001-5

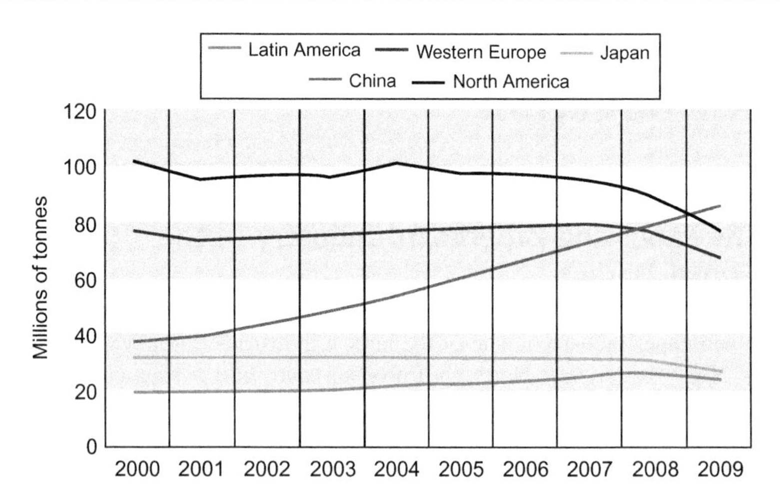

Figure 1.1 Total paper and paperboard consumption; North America versus other selected regions.
Source: Reproduced with permission from EPN *environmentalpaper.org/.../state-of-the-paper-industry-2011-full.pdf.*

Central Europe and between the Nordic countries and Central Europe. Furthermore, a lot of Canadian paper is exported to the USA. In Asia, Korea is a significant exporter. The most significant intercontinental trade flows are directed from Europe and North America to Asia, from Europe to North America as well as from North America to South America. The profitability of paper industry companies has been weak overall in the recent years. Overcapacity has led to falling prices and this, coupled with rising production costs, is eating into the sector's profitability.

1.2 General Aspects of Paper Recycling

Paper recycling in an increasingly environmentally conscious world is gaining importance (Anon, 2004a,b, 2005a,b; Bajpai, 2006; DIPP, 2011; Edinger, 2004; Francois, 2004; Friberg & Brelsford, 2002; Friberg et al., 2004; Raivio, 2006; Robbins, 2003; Rooks, 2003; Selke, 2004). Recycled fibres play a very important role today in the global paper industry as a substitute for virgin pulps. Paper recovery rates continue to increase each year in North America and Europe (with the exception of 2009–2010 in Europe owing to a dip in production during the economic downturn). In March, the American Forest & Paper Association launched its Better Practices Better Planet 2020 initiative, establishing an ambitious goal of 70% paper recovery by 2020 (the recovery rate was 63.5% in 2010) (Jourdan, 2011). Much increase in paper recovery can be attributed to the increase in easy residential and commercial recycling through single-stream recovery systems, as 87% of Americans now have access to curbside or drop-off paper recycling programmes. Figure 1.2 shows the US and Canadian recovery rates.

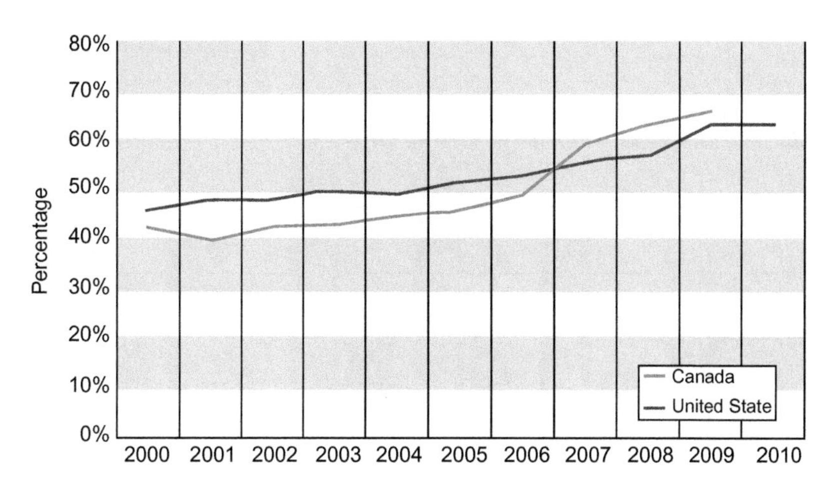

Figure 1.2 US and Canadian recovery rates.
Source: Reproduced with permission from EPN *environmentalpaper.org/.../state-of-the-paper-industry-2011-full.pdf.*

The recycling rate in Europe reached 70.4% in 2011 (Figure 1.3). The total amount of paper collected and recycled in the paper sector remains stable at 58 million tonnes, the same as in the previous years. However, this is an increase of 18 million tonnes since 1998, the base year for the first voluntary commitment the paper value chain set itself for increasing recycling in Europe. A net volume of 9.2 million tonnes (or 15.9%) of the total 58 million tonnes was imported for recycling by third countries outside the commitment region of the EU-27 plus Norway and Switzerland. Thirteen European countries exceed the 70% recycling rate, 12 European countries are below 60%. Figure 1.4 shows recycling rate in world regions in 2010. Europe is the leader in paper recycling.

Exports of recovered fibre from the USA to Asia have grown rapidly, representing a nearly three-fold increase since 2002. These exports are primarily destined for China. In 2009, approximately 36% of fibre recovered in the USA was exported to Asia.

With rapid developments in deinking processes for the reuse of secondary fibres being made, the recycling process is becoming increasingly efficient. The quality of paper made from secondary fibres is approaching that of virgin paper. Its manufacture is much more eco-friendly than that of virgin paper. The main drivers leading to increased paper recycling have traditionally been economic, that is limited availability and, hence, higher costs of virgin wood fibres. However, during the past decade, environmental and ecological concerns have become increasingly important (Stawicki & Barry, 2009). As the paper industry strives towards full sustainability, recycling becomes an increasingly important component of the supply chain.

Cost competition and the legal requirements in many countries mainly promote the use of recovered paper (Putz, 2006). The effect of environmentalists through 'green' movements and the level of acceptance in the market of paper made from recycled fibres are additional driving forces that vary by country. Recovered paper use is an environmentally friendly issue according to the recycled fibre processing

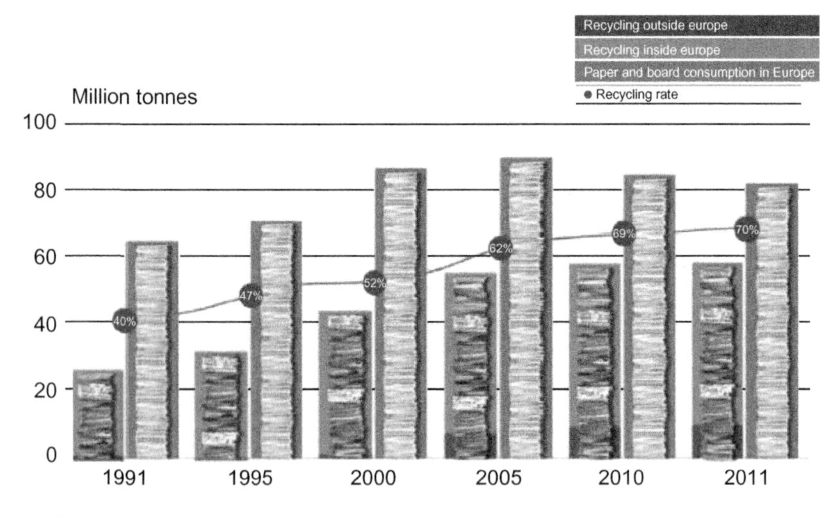

Figure 1.3 European paper recycling (1991–2011).
Source: Reproduced with permission from CEPI.

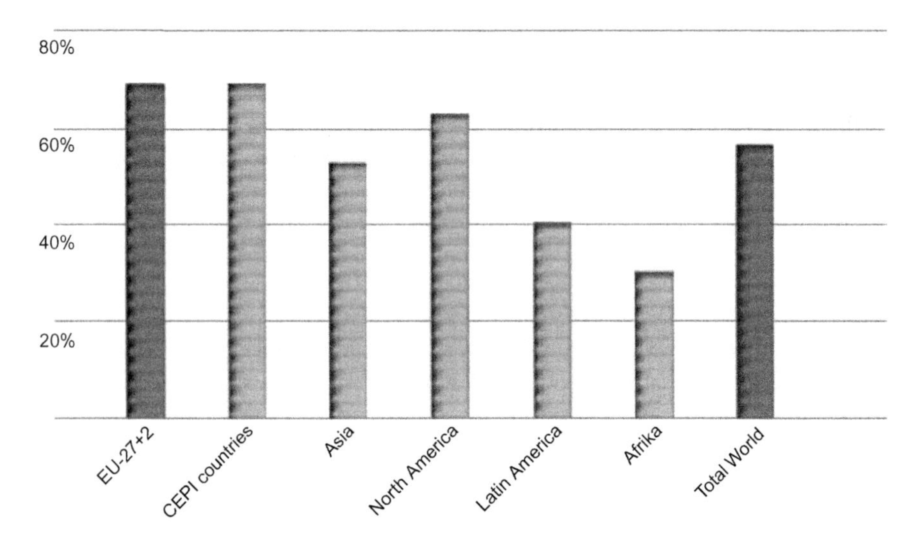

Figure 1.4 Recycling rates in world regions in 2010.
Source: Reproduced with permission from CEPI.

paper industry, environmentalists, governmental authorities and often even the marketplace. Recycling preserves forest resources and energy used for production of mechanical pulps for paper manufacturing. Moreover, recovery and recycling of used paper products avoids landfilling. The processing of recycled products requires comparatively little fresh water per tonne of paper produced. However, the solid waste rejects and sludge from recovered paper processing mills typically present a problem. The rate of formation of such residues is between 5% and 40%, depending on the recovered paper grade processed and the paper grade produced. The average rate of rejects and sludges totals about 15%, calculated from the recovered paper input on an air-dried basis. Because landfilling of organic matter has no future in several countries, most organic waste requires burning to reduce its volume. Effective, clean incineration technologies are available that control flue gas emissions, and the heat content of the residues and sludges contributes to self-supporting incineration. The final waste can be used as raw materials in other industries or can be discarded. An increasing volume of rejects and sludges can be used in brick works, the cement industry and for other purposes.

Most of the paper producers throughout the world are increasing their use of recycled fibres for paper products. The percentages of recycled fibre in almost every grade of paper and boards, and particularly in printing and writing papers, has more than doubled in the past decade. In the next 10–15 years, it is expected to triple or quadruple. Earlier, paper recycling was done mainly because it made economical sense to do so, whereas now it has been driven more by a collective environmental mandate of sorts. The marketplace for secondary pulps and papers continues to develop and mature. This huge secondary paper undertaking is primarily aimed at reducing volumes of waste paper in the world's growing landfill sites.

Paper recycling has become an increasingly important industry. Every year the percentage of paper that is recycled increases compared with the percentage that ends up in landfills. The paper recycling industry has seen marked changes over the past decade. Previously, recycled fibre was mostly used to produce products of lower quality. Today, because of new technology, recycled fibre can sometimes be used nearly interchangeably with new fibre to make even the highest quality grades of paper.

The larger quantities of waste paper available have helped to reduce the costs of recycling and provide a greater array of recycled paper and paper products. The process begins with collection, which is still one of the most expensive aspects of paper recycling. Besides collecting, the collection process involves sorting the paper into categories, baling and transporting the paper to a facility that will manufacture the waste paper into pulp. The first step at one of the repulping facilities is put the paper into large vats where it is soaked, reducing the paper into fibres. This process is known as repulping. When ink starts to separate from the fibres, chemicals are added to prevent the ink from reattaching to the paper fibres. The ink is then removed from the pulp in a deinking system, which is a series of screens that remove ink and additives. Then the pulp is cleaned several times with heat and chemicals, which removes additional ink. The pulp then enters a flotation device, where a chemical mixture containing calcium soap is introduced. Air bubbles form in this pulp. The chemical mixture causes any remaining ink to float to the surface where it can be skimmed away. After the deinking process, the pulp is ready to be manufactured into paper and related products in a similar manner to that by which paper is produced from wood pulp.

Some paper and board grades produced can use recycled fibres exclusively. This can include corrugating medium and test liner, or newsprint. Other grades use blends of recycled and virgin fibres. Europe is the largest producer of writing and printing grades whereas North America is the largest producer of newsprint (more exactly, Canada is the world's largest producer), tissue, container board and board.

According to Kenny (2005), 49% of fibrous raw material used in the paper industry is derived from recovered materials. Nearly 56% of paper products are being recycled, with a potential maximum of 81%. The remaining 19% represents unrecoverable paper and board. Sustainable development seeks to reach the maximum paper recyclability by improving knowledge of fibre flows. Recovered paper is commercially circulated throughout the world in forms of packaging and paper products. The Asian economy is rapidly expanding, with increased consumption of paper products. North American paper consumption is slowing or reversing. European consumption is growing slowly, but slightly faster in Eastern Europe. Asia is investing in increased papermaking capacity as well as importing paper products from the USA, but its reliance on recovered paper will increase. Collection of paper products is rising in Europe at a higher rate than consumption, except in Germany. Worldwide circulation of recovered paper will encourage collectors to improve quality, levelling standards and organising separate collection channels with a more consistent approach. New technology will help reduce paper loss during production and will encourage development of products based on recycled materials. The Western world can still increase collection rates, and developing economies will develop their own recovered paper sources.

Moore (2004) has reported that more than 80% of paper mills in the USA use recovered paper to make many of their products, and 200 mills in the USA use recovered paper exclusively. Of the paper currently recovered in the USA, 95% is recycled into new paper products, with the balance used in other applications. Over the past 20 years, the American Forest and Paper Association (AF&PA) has made recovered paper an integral part of the paper and paperboard industry (AF&PA, 2010). More than 38% of all the raw material used to make new paper comes from recovered paper. In addition, the paper industry in the USA has coupled efforts to improve its recycling capabilities with initiatives aimed at improving access to recovered paper. These efforts have increased the amount of paper kept from the waste stream and put into new products and packaging.

Today, most recovered paper is used as a raw material in packaging grades such as carton board and paperboard, because the manufacture of these grades does not typically involve deinking and/or bleaching (Jaakko Poyry, 2009). Therefore, the processing is generally less costly and may also have fewer environmental impacts than when deinking and bleaching are required. Over 90% of recovered paper in the world is used in grades other than for printing and writing, such as newsprint, tissue, container boards and other packaging or board products. Approximately 6% of the global recovered paper supply is used in printing and writing grades, and this percentage is forecast to increase only slightly by 2025. Most of the forecasted increase is in container boards, carton boards and tissue paper (Jaakko Poyry, 2009).

The distance between the recovered fibre source (usually areas of large population density) and the mill site is a key factor to consider when using recovered paper as a raw material. It is typically more economical to have short transportation distances to make the life cycle of the paper more efficient, but also to minimize the carbon footprint of transporting raw materials. In graphics applications, customers often require paper with good printing qualities and paper that will run well, without breaking, on high-speed printing presses. These quality requirements often require the use of fresh wood fibre as a raw material, instead of deinked pulp from recovered paper. Sheet strength and printing quality can vary tremendously depending on the type and quality of deinked pulp used. Furthermore, poor runability on printing presses can lead to higher waste generation.

Clearly, the act of recycling paper is beneficial for the environment, but the responsible use of recovered paper as a raw material to make new products should take into account economic and environmental consequences. Sustainable use of recycled fibre means using it in the right locations and in the right paper grades. As a guideline, the United States Environmental Protection Agency (US EPA) has recommended significantly different levels of total recovered fibre in certain paper grades, ranging from 10% for printing and writing grades to 100% for newsprint and packaging (US EPA, 2007).

1.3 Benefits of Recycling

Recycling of waste paper has several benefits, both for humans and the earth (Bajpai, 2006; Putz, 2006; Sappi, 2011).

- The process of recycling protects the environment. Using recycled paper to make new paper reduces the number of trees that are cut down, conserving natural resources. Every tonne of recycled fibre saves an average of 17 trees plus related pulping energy. In some instances, recycling services are cheaper than trash-disposal services. Recycling paper saves landfill space and reduces the amount of pollution in the air from incineration. Businesses can promote a positive company and community image by starting and maintaining a paper-recycling programme. Parents can promote a clean environment and a healthy lifestyle to their children by teaching them about the benefits of recycling paper.
- By using waste paper to produce new paper, disposal problems are reduced. The savings are at least 30,000 L of water, 3000–4000 kW h of electricity and 95% of air pollution for every tonne of paper used for recycling. Also, 3 yd^3 of landfill space are saved. And in many cases, recovering paper for recycling can save communities money that they would otherwise have to spend for disposal.
- Compared with virgin paper, producing recycled paper involves between 28% and 70% less energy consumption. Also, less water is used. This is because most of the energy used in papermaking is the pulping needed to turn wood into paper.
- Recycled paper produces fewer polluting emissions to air and water. Recycled paper is not usually re-bleached and, when it is, oxygen rather than chlorine is usually used. This reduces the amount of dioxins that are released into the environment as a by-product of the chlorine bleaching processes.
- High-grade papers can be recycled several times, providing environmental savings every time.
- Producing recycled paper actually generates between 20% and 50% fewer carbon dioxide emissions than paper produced from virgin fibres.
- Because used paper is usually collected fairly near to recycling plants, manufacturing recycled paper reduces transport and carbon dioxide emissions.

- Recycling paper reduces the volume of waste while helping to boost the local economy through the collection and sorting of waste paper.
- Waste paper pulp requires less refining than virgin pulp and may be co-refined with hardwood pulp or combined hardwood/softwood pulps without significant damage
- The kinds of deinked pulp suitable for use in printing papers usually impart special properties to the finished papers compared with papers made from wood pulp, such as increased opacity, less curling tendency, less fuzziness, better formation, etc.
- Not all effects of recycling paper are positive ones. Recycling mills are known for producing sludge, which is the runoff that includes ink, adhesives and other unusable material removed from the usable fibre. But according to Conservatree, the materials in sludge would still end up in landfills or incinerator emissions if the paper was not recycled, and recycling mills have developed environmentally controlled methods of handling sludge. In some cases, paper recycling has real environmental and economic benefits and some cases it does not. Depending on the circumstances, paper recycling may use more resources than it saves, or cost too much to be of much benefit, depending on the circumstances. A lot depends upon the type of recovered paper being used and the type of recycled paper being produced. Because wood and recovered paper are excellent fibre sources and because advanced recycling technology allows papermakers to use recycled fibre in new ways, the possibilities for using recycled fibre in today's paper products are greater than ever. About 38% of the raw material used in US paper mills is recovered paper. In many cases, the quality of recycled paper products is very close to the quality of those made from new fibre. Paper manufacturers must choose the raw materials best suited to make their products. In some cases, new wood fibre is the better choice; in others, recycled fibre is preferable. It is up to the manufacturer to decide how to use the fewest possible resources to make quality products that meet consumers' needs.

1.4 Statistics

The importance of recovered paper as a raw material in the paper industry has increased markedly during the past decade. In 2011, paper recovery increased by 1.2 million tonnes, raising the US paper recovery rate to a record high 66.8% (www.paperrecycles.org), which is an increase from 63.5% in 2010 and 33.5% in 1990, the base year from which the industry's original 40% recovery goal was benchmarked (Table 1.1). The key figures show that the collected recycling paper increased from 26 million tonnes in 1991 to 57 million tonnes in 2011. Utilisation rose from approximately 26 million tonnes to 48 million tonnes. From 2000 to 2011 the exports to outside the Confederation of European Paper Industries (CEPI) increased from 3.5 million tonnes to 10.2 million tonnes, the imports from outside CEPI doubled from 753,000 tonnes in 2000 to 1.4 million tonnes in 2011. The utilisation rate – defined as percentage of paper for recycling utilisation compared with the total paper and board production – climbed from 39.1% in 1991 to 51% in 2011. In Europe at the same time, the recycling rate – the utilisation of paper for recycling and net trade for paper of recycling, compared with paper and board consumption – increased from 40% to 70%.

Data for the year 2011 indicates that 30% of the paper and paperboard recovered in the US went to produce containerboard and 11% went to produce boxboard,

Table 1.1 Paper and Paperboard Recovery Rate

Year	Recovery Rate (%)
1990	33.5
1991	36.7
1992	38.5
1993	38.7
1994	41.5
1995	44.0
1996	45.6
1997	44.2
1998	44.6
1999	44.5
2000	46.0
2001	48.3
2002	48.2
2003	50.3
2004	49.3
2005	51.5
2006	53.0
2007	56.0
2008	57.7
2009	63.6
2010	63.5
2011	66.8
2012	65.1

Source: Based on www.paperrecycles.org

Table 1.2 Uses of Recovered Paper

Packaging Type	Share of Total (%)
Newsprint	3.0
Tissue	8.0
Containerboard	31
Boxboard	12.0
Other	5.0
Net exports	41.0
Total	100

Source: Based on www.paperrecycles.org

which includes folding boxes and gypsum wallboard facings (Table 1.2). Exports of recovered paper to China and other nations absorbed 42% of the paper collected for recycling in the US in 2011. Although new supply of printing–writing and related papers declined by more than 5% in 2011, printing–writing paper recovery contracted just 1.2%. The recovery rate for printing–writing papers consequently

Table 1.3 Recovery of Printing–Writing Papers

Year	Recovery Rate (%)
1994	34.2
1995	36.4
1996	36.1
1997	34.8
1998	37.7
1999	38.9
2000	42.1
2001	39.6
2002	38.0
2003	40.5
2004	42.6
2005	44.1
2006	48.4
2007	53.5
2008	54.7
2009	61.0
2010	54.6
2011	56.8
2012	54.5

Source: Based on www.paperrecycles.org

increased to an estimated 56.8% in 2011, up by 2.2% relative to the 2010 level (Table 1.3).

After increasing by 11.2% in 2010, recovery of old corrugated containers/Kraft papers for recycling rose an additional 8.0% in 2011, to a record high 29.1 million short tonnes (Table 1.4).

Total recovery of news/mechanical papers declined 6.2% in 2011 as potential supply dropped 7.5% owing to reduced newspaper circulation and size (Table 1.5). With potential Old News Paper (ONP) supply down more than recovery, the recovery rate for news/mechanical papers edged up to 73% in 2011 from 72% in 2010.

Data for the year 2010 published by the US EPA indicates that paper and paperboard packaging accounts for more than 73% of all packaging materials recovered for recycling in the USA (Table 1.6).

The volume of paper recovered for recycling rose in 2011, while paper purchases declined. As a result, the volume of paper estimated to have been disposed of in landfills fell 9% in 2011, reaching the lowest level in decades (Table 1.7).

The European paper recycling rate reached an impressive 70.4% as announced by the European Recovered Paper Council (ERPC) in their annual monitoring report. The report shows that the total amount of paper collected and recycled in the paper sector remains stable at 58 million tonnes. This is the same as in the previous years, but with an increase of 18 million tonnes since 1998, the base year for the

Table 1.4 Recovery of Old Corrugated Containers

Year	Recovery Rate (%)
1993	54.5
1994	58.8
1995	65.0
1996	68.5
1997	67.7
1998	67.6
1999	65.7
2000	68.7
2001	70.2
2002	69.9
2003	72.8
2004	70.0
2005	73.5
2006	73.6
2007	75.0
2008	77.6
2009	82.0
2010	85.1
2011	91.2
2012	91.0

Source: Based on www.paperrecycles.org

first voluntary commitment the paper value chain set itself for increasing recycling in Europe. Since 2000 the recycling rate has increased by 18%, in part because of the excellent work of the ERPC (http://www.ipwonline.de/news.aspx?id=711).

A new reporting format includes more indicators in addition to the volumes and recycling rate. For example, the number of European countries exceeding a 70% recycling rate has increased to 13, whereas 12 EU countries still have under 60% recycling rates for paper, indicating further potential for increasing paper recycling in Europe. The number of cycles a paper fibre goes through in the loop reached, on average, 3.4 (compared with the global average of 2.4).

In addition to the quantitative progress, much qualitative work has been done to establish an ecodesign towards improved recyclability and in the area of waste prevention. The results include pioneering work to give recycling solid and scientific support, such as the adoption of scorecards to assess the recyclability of paper-based products.

The UK paper recycling rate has increased by about 5% on the preceding year, according to figures released by Confederation of Paper Industries (CPI) (www.recyclingportal.eu). The increase from 75.1% in 2010 to 78.7% in 2011 ensures that the UK exceeds the target set in the revised 'European Declaration on Paper Recycling (2011–2015)' to maintain current high levels in countries where it already reached levels of above 70%. The recycling rate could be ascribed more to a decrease in

Table 1.5 Recovery of ONP/Mechanical Papers[a]

Year	Recovery Rate (%)
1993	46.8
1994	49.8
1995	50.7
1996	49.1
1997	48.4
1998	49.5
1999	49.8
2000	52.3
2001	58.9
2002	60.1
2003	63.9
2004	62.3
2005	64.4
2006	70.0
2007	68.8
2008	69.2
2009	70.3
2010	72.0
2011	73.0
2012	70.4

Source: Based on www.paperrecycles.org
[a]Includes newsprint, uncoated mechanical papers and coated newspaper inserts.

Table 1.6 Packaging Recovery

Packaging Type	Share of Total (%)
Paper and paperboard	73.2
Glass	8.5
Metal	7.0
Plastic	5.0
All other	6.3
Total	100

Source: Based on www.paperrecycles.org

paper consumption than a significant increase in the amount of paper collected for recycling (CPI). According to the CPI Recovered Paper Sector Manager, 'Whilst the increase in the recycling rate is welcomed in principle, it is important to qualify the apparent performance improvement. UK collection of used recovered paper in 2011 was just over 8 million tonnes – an increase of 0.4% on 2010 – whereas consumption of paper and board products which entered the UK waste stream reduced by 4.2% compared to 2010'. With lower volumes available for collection, and only a modest

Table 1.7 Paper Landfilled

Year	Paper Landfilled (Million Tonnes)
1993	38.4
1994	37.9
1995	36.3
1996	34.4
1997	37.3
1998	38.4
1999	41.4
2000	40.0
2001	36.6
2002	37.7
2003	36.6
2004	39.9
2005	37.8
2006	36.9
2007	32.9
2008	28.1
2009	19.3
2010	20.3
2011	18.4

Source: Based on www.paperrecycles.org

increase in tonnage collected for recycling from existing sources, ensuring recycling performance is maintained in future will mean additional sources will be required. This will have a significant knock-on effect for cost and quality control for collectors, and may become increasingly difficult if demand from global paper mills falls, further suppressing collectors income. The formula used to calculate the 2011 UK recycling rate follows guidelines set out by CEPI. However, limited data exists for imports of paper packaging around goods imported to the UK for consumption, and is not deemed reliable enough to include in calculations.

Total global consumption of paper was 371 million tonnes in 2009. However, total paper consumption in North America declined 24% between 2006 and 2009. Consumption of paper per person in North America dropped from more than 652 lbs/year in 2005 to 504 lbs/year in 2009. North Americans still, however, consume almost 30 times more paper than the average person in Africa and 6 times more than the average person in Asia. In 2009, total paper consumption in China eclipsed total North American consumption for the first time. The paper recovery rate in the USA rose from 46% in 2000 to a record high 63.4% in 2009. In Canada the reported paper recovery rate in 2009 was 66%. Paper is the most commonly recycled product, and yet is still one of the largest single components of landfills in the USA, comprising over 16% of landfill deposits and equalling 26 million tonnes annually. This is down from 42 million tonnes in 2005, which represented 25% of the waste stream after recycling that year.

The recovered paper utilisation and collection rates do not necessarily reflect the true situation (fibre recycling). This is because in most cases the recovered paper that is collected, delivered to a paper mill and loaded into a pulper does includes not only fibre but also several other substances such as minerals, starch, additives, inks, coating materials, non-paper components and so on. In addition, the amount of process losses varies, depending on paper and board grade. For example, in packaging grades the average process loss varies from 8% to 12%, and in publication papers from 15% to 20%. In tissue production the losses can be between 35% and 40%. The amount of losses depends on the quality demands of the produced paper and board itself, as well as the quality and grade of the input recovered paper. Thus, official collection and utilisation rates do not exactly describe fibre recycling but give an average picture about paper and board rotation.

The UK is the largest net exporter of recovered paper, followed by Belgium and Germany. Finland recycles over 70% of the paper it consumes (Raivio, 2006). According to a new law, Finnish manufacturers are responsible for the waste management of their products. Finnish paper manufacturers are experienced recyclers. They aim for a 75% recycling rate. An increased effectiveness in Finnish paper collection is achieved by bringing recycling closer to the customer, by informing and educating, and by making paper collection an everyday environmental act. An EU waste disposal site directive and tight Finnish environmental laws guarantee the continuation of effective paper recycling in Finland.

China is the world's largest importer of recovered paper and the driver for the market in the USA. Massive expansion in paper and board production capacity in China is driving the use of recovered paper, and there is a trend towards greater use of mixed paper grades. About 60% of Chinese recovered paper imports come from the USA, with 20% from Japan and the remainder from Australia. Although China has put in place some initiatives to increase its domestic recovery rate, this will not be able to keep up with demand and its use of recovered paper will continue to grow. This demand is a major factor behind the recent announcement by the American Forest and Paper Association of an increase in the USA's paper recovery rate. However, future growth is dependent on a reliable supply of clean paper and advanced technology capable of handling the more contaminated mixed paper streams.

Owing to the large number and various sizes of organisations involved in the collection of recovered paper, there is no reliable basic collection data available in most countries. Normally, collection volumes are therefore calculated through recovered paper consumption and trade statistics. There are great differences in collection volumes and collection rates between countries. A few countries collect the main share of all recovered paper, globally. In 2006, the 13 most important countries collected 153 million tonnes or 78% of all recovered paper, globally. These countries include the USA, Japan, China and Germany, which together collected 55% of all recovered paper globally. In countries where the collection potential, namely the paper and board consumption per person and the level of environmental consciousness, is high, the collection rate is also high and well above the global average. These include industrialized countries such as the Republic of Korea, Japan, other developed Asian countries and most European countries. In the USA, the collection rate is slightly above the global average. In China and developing countries with a low paper consumption per person (and low environmental consciousness), the collection activity is well below the global average collection rate of 51%. Historically, the collection rate has only been calculated for total recovered paper. This is quite understandable as the collection rates have been modest and there has been no shortage of recovered paper. When developing recycling, easy sources such as trade, converting and industrial sources in densely populated regions are typically exploited first. However, effective collection requires use of household and office sources to achieve reasonable collection rates. Usually, companies planning new investments in production based on recycled paper are not interested in 'general' total recovered paper figures but in the localized availability of individual recovered paper grades such as white grades, old corrugated containers and old newspapers and magazines. Therefore the idea to calculate the recovered paper collection and collection rate for the main grades separately has been extensively discussed. The availability of this kind of information would make it possible to analyse and compare collection activity and availability of different grades in different countries or other geographical areas. This is especially important because of the widely predicted shortage of the best recovered paper grades in international markets. The recovered paper collection rate is the relation between paper collection volume and paper consumption. When calculating the collection rate for an individual recovered paper grade, the collection volume of a recovered paper grade should be compared with the consumption of corresponding paper and board grades. So far, collection rates have not been calculated for individual recovered paper grades because a generally accepted calculation methodology does not exist.

Recovered paper utilisation volumes and rates vary greatly by country. In 2006 the 13 most important countries used 155 million tonnes of recovered paper. This is about 79% of the global recovered paper utilisation. Already the most important four countries, namely China, the USA, Japan and Germany, utilise over 108 million tonnes, or 55% of global recovered paper. Not surprisingly, countries with vast forest resources and a strong wood pulp industry have based their raw material sourcing on virgin fibres. In these countries the recovered paper utilisation rate has been modest, i.e. below the global average. These countries include the USA, Canada, the Nordic countries, Russia and Brazil. The low utilisation rate in a country is not necessarily a result of a low recycling activity; it can reflect the fact that paper production considerably exceeds the paper consumption. For example, in the Nordic countries, paper production was about 28 million tonnes in 2006. During the same year paper consumption was 4.4 million tonnes. Despite a high collection activity (3 million tonnes and average collection rate, based on paper consumed in the region, of about 70%), the utilisation rate was only about 11%. Most paper produced was exported, consumed and recycled in other countries. Owing to the differing structure of internal markets, the utilisation rate of an individual country is not a valid metric describing recycling activity. In fact, because of the growth in international trade, the collection rate (rather than the utilisation rate) is a more reliable reflection of the recycling activity in a specific region or country. Countries with considerably high paper and board production but limited forest resources and pulp industry have effectively utilised recovered paper in their production. These countries include, for example

Japan, Germany, the UK, Spain, China, the Republic of Korea and Mexico. In general, the 'collection rate' provides the best benchmark for the importance given to 'recycling' by the society within a given region, state or country.

1.5 Challenges for Paper Recyclers

Some underlying trends are creating new technical challenges for paper recyclers and users (Friberg & Brelsford, 2002). World fibre demand will continue to increase, as will the use of recovered fibre. To meet this demand, suppliers of recovered paper must develop grades from lower-quality mixed paper streams. Furthermore, domestic recovery systems threaten to rely increasingly on 'single stream collection', where all recyclables (paper, glass and plastic) are commingled. Rising package/container exports from new Asian paper mills will increase the quantity of inferior-quality old corrugated containers in US recovered paper streams. To sustain and improve the industry's position, there is a need to improve profitability through marked innovations that reduce manufacturing costs and improve the quality of fibre delivered to the paper machine. The 2001 Technology Summit significantly changed the focus of and goals for the industry's recycling research programme. The new challenge is to 'make recycled fibre interchangeable with virgin fibre with respect to product quality and economics'. Now, more than ever, recycled fibre must directly compete with virgin fibre on all metrics: availability, strength potential, quality (uniformity and minimal contamination), runability, performance and cost. The two areas where targeted technology can be expected to facilitate the competitiveness of recycled fibre are improved quality and quantity of recovered paper delivered to paper mills, and improved mill processes including the development of next-generation fibre evaluation, defibrisation and decontamination technologies.

Issues causing problems in recycling paper are described below.

1.5.1 Deinking

One of the largest problems with recycling paper is the need to remove ink from recovered paper that has already been used for printing. Although this is not necessary for all types of paper, paper that does not have the ink removed will have a greyish tint. Grades of paper having high levels of whiteness and purity require the removal of printing inks. Additional chemicals must be used to remove the inks, which then must be washed away in large amounts of water. Sometimes the paper pulp is also bleached using hydrogen peroxide or, less ideally, chlorine, which, according to Waste Online, can sometimes combine with organic matter to produce toxic pollutants.

1.5.2 Adhesives

Pressure-sensitive adhesives, or self-adhesives, are the type of adhesive used on address labels and sticky notes, as well as peel-and-stick stamps. These adhesives do not dissolve in the water that is used to transform the paper into pulp so it can be recycled back into paper. The particles from the adhesive instead fragment into smaller particles, which deform under heat and pressure and can become lodged on papermaking equipment, creating weak spots in the final paper product or causing pieces of finished paper to stick together. According to California's Department of Resources Recycling and Recovery, pressure-sensitive adhesives cost the paper recycling industry approximately US$700 million a year.

1.5.3 Limited Life Cycle

Unlike some other materials that can be recycled an infinite number of times, paper can only can be recycled between five and seven times. The fibres in the paper get shorter each time they are recycled, eventually becoming too short to be made into new paper. Each time the paper is recycled, some new fibres must be added to replace the unusable fibres, so new materials must still go into the manufacture of recycled paper.

Friberg and Brelsford (2002) identified the main gaps in recycling technology that should be the industry's main research objectives, along with gaps in mill processes, the filling of which could improve and consolidate mill unit operations. The main gaps are significant losses from gross contamination and mixes of fibre types that recycling mills cannot handle. Once the inappropriate fibres are introduced into the pulper, subsequent mill process equipment cannot effectively remove or bleach them; lack of metrics for paper grades at the recycle plant or fibre types in the recycle mill; lack of methods to monitor and sort paper types; the paper industry itself perceives recycled fibre as inferior to virgin fibre; the supply structure leads to huge price volatility and long-term shortages and commercialization of technologies takes too long. These gaps can be filled by separating fibre types, removing gross contamination, increasing the amount of recovered paper and improving the working conditions.

The gaps in mill processes are that defibrisation and repulping methods have not radically changed in many decades, the lack of process knowledge to consolidate unit operations and the lack of novel physical applications with new technology transfer from other applications/industries, variation in recycled paper performance caused by variability in fibre characteristics, and a substandard recycled paper strength/cost ratio (Friberg & Brelsford, 2002). Target research areas to address these gaps focus on mill process improvement. New technology should displace multiple process modules at recycling plants with single units that provide higher yield, lower energy use, water reuse, simpler process flows that streamline existing systems, consolidate unit operations and reduce installed capital cost, better chemistry and fibre analysis.

The recycle session of the Agenda 2020 Technology Summit II stated that the long-term economic viability of US recycle mills depends on increased fibre recovery and improved fibre quality (Friberg et al., 2004). Groups from waste paper collection, paper manufacturing, recycled paper converting and government discussed ways to achieve this. It was stated that recycling technology is not yet sufficiently advanced to deal with current and likely future requirements in energy efficiency, operating cost reductions and improvements in fibre quality. Materials are not sorted properly, producing wastage in transport costs. Mills process inferior materials, leading to low quality yields, high energy consumption and high product cost. Improvements in recovery and sorting could secure an additional 5 million tonnes of good quality fibre. This would bring system savings of US$500 million per year. Five areas of research were suggested: development of automation in recovery, investment return assessment on sorting equipment, improved fibre recovery with wet-end equipment, better stickies measurement/removal equipment and five areas for improved government support. Alliance partnerships were recommended. Investment needed stands at US$20 million over 5 years, US$10 million of which would be centred on fundamental evaluation of mill process technologies and equipment.

Two European Commission COST initiatives related to recycling – E46 'Improvements in the understanding and use of deinking technology' and E48 'Limits of paper recycling' – have been undertaken (www.ingede.com/ingindxe/links-e.html). Objectives of E46 are development of a better understanding of the mechanisms involved, development of methods to improve the deinkability of inks and toners that are difficult to deink and assessment of developments in printing technology, ink and toner and their potential impact on deinking efficiency and the opportunity for the development of deinking-friendly inks or printing techniques. COST E48 is a European network of universities, research organisations and industrial partners engaged in paper recycling. It generates a trans-European cooperation aimed at developing possibilities for the future of paper recycling within Europe and the role recovered paper and board will play as a raw material source for the European paper industry.

References

American Forest & Paper Association (2010). <http://www.paperrecycles.org>. 12 February 2012.

Anon (2004a). Recycling rates hit new high in Europe. *Recycled Paper News*, *14*(11), 8.

Anon, (2004b). Waste paper collection increased by 53% in France in 10 years. *Pap'Argus* (152), 8.

Anon (2005a). *Recovered paper: Statistical highlights 2005* (pp. 15). American Forest and Paper Association.

Anon (2005b). *Special recycling: 2004 statistics* (p. 7). Confederation of European Paper Industries.

Bajpai, P. (2006). *Advances in Recycling and deinking* (180 pp.). U.K: Smithers Pira.

Department of Industrial Policy and Promotion (DIPP) (2011). *Discussion paper on collection and recycling of waste paper in India.*

Edinger, G. (2004). Waste paper. *Papier aus Osterreich*, *11*, 24.

Francois, D. (2004). Recycling of paper and board has the wind in its sails. *Caractere*, *604*, 6.

Friberg, T., & Brelsford, G. (2002). New directions in recycle. *Solutions*, *85*(8), 27.

Friberg, T., Gerhardt, T., & Kunzler, C. (2004). Ensuring the future of fiber recovery and utilization. *Solutions*, *87*(12), 33.

Jaakko Poyry. (2009). *World Fiber Outlook up to 2025*, 2009 ed., Vol. 1, Executive Report.
Jourdan, T. (2011). *Should we be pushing for quality over quantity on paper recycling?* <www.greenbiz.com/> .
Kenny, J. (2005). Recovered paper keeps on growing in Europe. *Solutions*, *88*(7), 28.
Moore, W. H. (2004). Chances are about 50–50 that you will recycle this paper: How can we improve those odds? *Corrugating International*, *28*(August), 22.
Putz, H. J. (2006). In H. Sixta (Ed.), *Recovered paper and recycled fibers, handbook of pulp (pp. 1147–1203)*. KGaA, Weinheim: Wiley-VCH Verlag GMbH & Co.
Raivio, H. (2006). 70% of paper gets a new life. *Painomaailma*, *1*, 22.
Robbins, A. (2003). Paper progresses towards recycling target. *Material Recycling Week*, *182*(17), 14.
Rooks, A. (2003). China casts large shadow on recycle market. *Solutions*, *86*(12), 39.
Sappi, E.Q. (2004). Insights. *Sustainable use of recycled fiber*, vol. 2.
Selke, S. (2004). Update on recycling of paper and paperboard packaging. *Food, Cosmetics and Drug Packaging*, *27*(11), 214.
Stawicki, B., & Barry, B. (2009). *The future of paper recycling in Europe: Opportunities and limitations*. <www.cost-e48.net/10_9_publications/thebook.pdf>.
US EPA. (2007). *Comprehensive procurement guidelines. Buy recycled series – paper products*. EPA530-F-07-039, October 2007, <www.epa.gov/osw>.
<www.paperrecycles.org>.
<www.recyclingportal.eu>.

2 Legislation for Use of Recycled Paper

2.1 Introduction

Attempts are being made by governments in industrialised countries to limit or to reduce the quantities of waste from homes, businesses and industrial operations. In most of the countries, the annual population growth increased less compared with the generation of municipal solid waste. So, avoiding waste and promotion of material recycling are the most important means of reducing the generation and growth of municipal and industrial waste. Legislation in various countries has attempted to promote material recycling and to decrease the waste generation that requires disposal in suitable facilities. Legislation can be classified into general regulations:

- Special regulations for certain paper or board grades by products
- Regulations relating to private households for reduction of municipal waste.

This area is subject to rapid change involving testing of many different political approaches. Some approaches oriented towards paper products include directives and ordinances, procurement policy guidelines, voluntary agreements, formulation of national recovery or utilisation rates of recovered paper and eco-labelling of paper. Other regulations that cover packaging materials or graphic paper products include packaging ordinance for packaging material, voluntary agreement of the graphic paper chain, stipulation of minimum contents of recycled fibres for example in newsprint and executive orders for purchasing graphic papers with a minimum content of recycled fibres by public funds (Göttsching & Pakarinen, 2000). These regulations set responsibilities for taking back used paper products, independently of the public disposal system, and recycling them. Instead, they may specify the content of secondary fibres used to manufacture a certain paper grade. Other efforts describe recycling objectives in general or stipulate the installation of collection systems aimed at household and municipal waste (Anonymous, 1996b).

In most of the developed nations, waste paper recycling is organised and operated by the municipal authorities, supported by suitable national policy normally based on the 'polluter pays' principle. Legislations are formulated in the form of directives, procurement policy guidelines as well as voluntary agreements. The collection mechanisms put in place are highly successful, as indicated by continuous improvement in the recovery rates of waste paper. These countries not only meet their domestic requirements but also export large quantities of waste paper. A few examples of such legislation are given below for illustration.

Recycling and Deinking of Recovered Paper. DOI: http://dx.doi.org/10.1016/B978-0-12-416998-2.00002-7

2.2 Legislation in the European Union

Countries of the European Union (EU) follow the European Packaging Directive No. 94/62 (EEC and 2004/12/EC2). In these countries, responsibility for collection and recycling of packaging waste lies with Packaging Recovery Organisation Europe, or PRO EUROPE, which is an umbrella organisation of 33 national producers. PRO EUROPE uses a 'Green Dot' as its registered trademark. A 'Green Dot' on packaging signifies that a financial contribution has been paid to a qualified national packaging recovery organisation. By contracting with the green dot system, the companies responsible for producing packaging entrust their take-back obligation to the scheme in return for an annual fee based on the type of packaging materials used and on the amount of packaging put in the market. The printing of the 'Green Dot' is an indication that the packaging producer financially supports the integrated system of collection and recycling of its packaging waste, which is mandatory in most EU countries.

In all the Member States of EU, economic operators within the packaging chain are responsible for packaging waste management and for providing data on the amount of recycled packaging put in the market. Most of the compliance systems need to be approved and are monitored by the Ministry for Environment or an independent body. The work of the compliance schemes is financed by fees collected from companies wishing to transfer the obligations imposed on them to the scheme. In general, the fee structure is based on weight or volume of the packaging material, per unit of packing or on a membership or fees based on turnover.

Most Member States have implemented Packaging Regulations (The Directive 94/62/EC on Packaging and Packaging Waste) in 1997. Only Greece has not yet transposed the EU Packaging Directive into national law. Depending on national waste management traditions, the regulation of packaging waste recovery is accompanied by voluntary agreements (Denmark, the Netherlands). Several Member States (Belgium, Denmark, France, Portugal and the UK) have transposed the EU Packaging Directive in regulating the recovery requirements and the environmental requirements in the design and manufacture of packaging ('essential requirements') in separate legal acts (DIPP, 2011).

In all Member States economic operators within the packaging chain (manufacturers, packers/fillers, distributors and importers) are responsible for packaging waste management, and for providing data on the amount of packaging put out in the market. Except for Denmark, the industry has built up organisations in all Member States to comply with the obligations imposed by national packaging regulations on behalf of the individual businesses affected. However, economic operators generally have the option of transferring their obligations to an external organisation (hereafter called compliance scheme) or fulfilling their obligations by themselves.

Most of the compliance systems need to be approved and are monitored by the Ministry for Environment or an independent body (e.g. packaging committee). The schemes co-ordinate the activities necessary for the recovery of packaging waste and have an essential interfacing role to play between the different actors within the packaging life cycle (industries, public legal entities, consumers, recycling and recovery operators). In Austria and the UK, a competition scrutiny system is explicitly applicable to these organisations to avoid monopolisation.

In eight Member States a 'green dot' system has been established. By contracting with the green dot system, the companies responsible for producing packaging entrust their take-back obligation to the scheme in return for an annual fee based on the types of packaging materials used, and on the amount of packaging put on the market. The printing of the 'green dot' is an indication that the 'packaging producer responsible' financially supports the integrated system of selective collection and recycling of its packaging waste. The green dot systems are predominantly in charge of the management of household/municipal packaging waste (DIPP, 2011).

The UK has adopted the concept of 'shared producer responsibility' for packaging waste. This refers only to the industries that produce or use packaging. Responsibility for recovery and recycling of packaging waste is divided among the commercial enterprises that form part of the 'packaging chain', raw material producers, packaging manufacturers, packer/fillers and sellers.

Except for Denmark and the UK, industry-based organisations are established in all Member States to take over the responsibility for management and recovery of municipal packaging waste. It is only in Belgium that the responsibility is only for municipal waste and industrial packaging waste, with two different organisations dealing with the two waste streams.

In Austria, Finland, Ireland, The Netherlands and Italy, the systems in place are responsible for both municipal and industrial packaging waste. In Germany, the activity of the nationwide Der GrünePunkt – Duales System Deutschland GmbH (DSD) system was restricted to sales packaging by the Federal Cartel Office. Systems for self-compliers have started operating in competition to the DSD since the amendment of the Packaging Ordinance in 1998.

Table 2.1 lists the main national packaging waste management organisations and summarises the responsibility of these systems according to municipal/industrial packaging waste (DIPP, 2011).

In principle, the private sector is responsible for the packaging they put out in the market. For definite packaging waste management activities, the responsibility is shared in most Member States between municipalities and industry. Although collection and sorting of municipal packaging waste is predominately undertaken by the public sector, the collection of industrial packaging waste and the recovery and recycling of both municipal and industrial packaging waste is by the private sector.

In Austria and Germany, obligated economic operators are explicitly required to organise the collection and sorting of domestic packaging waste and to comply with recycling targets for this waste stream. The packaging regulations in these countries set out criteria for the collection system, capacities and distances between collection points, extensions of the collection system, etc. The compliance schemes include contracts with municipalities (and private operators) for the services necessary in the context of separate collection and sorting of municipal packaging waste. Table 2.2 provides an overview of the share of responsibility (DIPP, 2011).

Table 2.1 Main National Packaging Waste Management Organisations and Summary of the Responsibility of These Systems According to Municipal/Industrial Packaging Waste

Country	Organisation	Responsible for Industrial Packaging	Responsible for Municipal Packaging	Green Dot
Belgium	Fost +	Yes	Yes	Yes
	Val-I-Pack	No	No	
Germany	DSD different	Yes	No	Yes
	organisations	(Yes)	Yes	
Denmark	Municipalities	Yes	Yes	No
Austria	Branch organisations	Yes	Yes	Yes
Finland	Pakkausalan Ympäristörekisteri PYR Oy	Yes	Yes	No
France	Eco-Emballages,	Yes	No	Yes
	Adelphe	Yes	No	
Ireland	Repak	Yes	Yes	Yes
Italy	Consorzio Nazionale Imballaggi (CONAI)	Yes	Yes	No
Luxembourg	Valorlux	Yes	No	Yes
The Netherlands	SVM-Pact	Yes	Yes	No
Portugal	Sociedade Ponto Verde, S.A. (SPV)	Yes	Yes	Yes
Spain	Ecoembalajes	Yes	No	Yes
	Ecovidrio	Yes	No	
Sweden	Reparegistret AB (REPA)	Yes	No	No
UK	Different organisation (e.g. Valpak)	No particular responsibility according to this classification		No

Source: Based on DIPP (2011)

There is separate collection of municipal and industrial packing waste in all Member States, but in varying degrees. In Austria, Denmark, Finland, Germany, the Netherlands and Sweden for example, a well-functioning reuse system, for example for glass, already existed, and glass and paper were collected separately for recycling.

The work of the compliance schemes is financed by fees collected from companies wishing to transfer the obligations imposed on them to the scheme. In general, the fee structure is based on weight or volume, type of packaging material, fee per unit of packaging or a membership registration fee based on turnover.

The activities of the compliance schemes are monitored by the ministries of Environment or other entities, for example the Interregional Packaging Commission in Belgium and the Agencies in the UK.

Several different systems have been implemented to collect recyclables from the general waste stream in developed countries. These systems lie along the spectrum of trade-off between public convenience and government ease and expense. The

Table 2.2 Share of Responsibility According to Activity

Country	Collection and Sorting (Municipal Packaging)	Recovery
Belgium	Municipalities	Fost Plus
Germany	DSD + other private organisation	Industry (guarantors)
Denmark	Municipalities	Industry
Austria	ARGEV + other private organisation	Branch organisation responsible for recycling (guarantors)
Finland	Municipalities	PYR
France	Municipalities	Eco-Emballages, Adelphe
Ireland	Municipalities	Repak
Italy	Municipalities	CONAI
Luxembourg	Municipalities	Valorlux
Portugal	Municipalities	Ponto Verde + entities of packaging and raw packaging material manufacturers
Spain	Municipalities	Ecoembalajes
Sweden	Material companies	Material companies
The Netherlands	Municipalities	Industry

Source: Based on DIPP (2011)

three main categories of collection are 'drop-off centres', 'buy-back centres' and 'curbside collection'. Source separation is the other extreme, where each material is cleaned and sorted before collection. This method requires the least post-collection sorting and produces the purest recyclables, but incurs additional operating costs for collection of each separate material. An extensive public education programme is also required, which must be successful if contamination among recyclables is to be avoided.

Germany has a three-pronged waste strategy: avoidance of waste; reuse or recycling of used paper products; and environmentally friendly disposal of waste. The EU has subsequently adopted this concept with its waste management directives in which the European Waste Guideline 91/156/EWG of 1991 and the EU Packaging Directive 94/62/EG of 1994 have special importance. The latter also places top priority on avoiding packaging waste and covers packaging made from paper and board, glass, metal, plastic and composite material. The other main principles are the reuse of packaging material, recycling of packaging material and other uses of packaging waste, and a consequent reduction in the amount of waste for final disposal. Table 2.3 shows Legislation on Recovery and Recycling of Recovered Paper in Germany.

The EU Member States must establish systems for take-back, collection and utilisation of waste as a secondary raw material. All parties involved in the production, conversion, import and distribution of packaging and packaged products must become more aware of the degree to which packaging becomes waste. They must

Table 2.3 Legislation on Recovery and Recycling of Recovered Paper in Germany

Waste Management Act (1986)
Prevention of waste
Recycling of waste
Environmentally friendly disposal of waste
Packaging Ordinance (1991)
Sets recovery targets for recycling (e.g. 70% recovery of paper and board sales packaging by 1999)
Proposed waste paper ordinance was replaced by:
Voluntary Agreement of the Graphic Paper Chain (1994)
Sets recovery targets for recycling
(sets 60% recovery of printed material and office paper by 2020)
Recycling and Waste management Act (1996)
Replaced Waste management Act of (1986)
And aims at comprehensive product responsibility
Municipal Waste Management Provision (1993)
By 2005 no solids must be landfilled that contain more than 5% organic matter

Source: Based on Putz (2006)

also take responsibility for this waste in accordance with the 'polluter pays' principle. Final consumers play a decisive part in the avoidance and use of packaging and packaging waste, and they must therefore be educated in this respect. The efficient sorting of waste at its source, for example in private households, has a major importance for a high recycling level.

Independent of the packaging material, a distinction exists among the following three categories of packaging: sales packaging, re-packaging and transport packaging. Sales packaging is offered to the final user or consumer at the sales outlet as a sales unit such as a bottle or can of beer, a box of washing powder or a cigarette box. Re-packaging contains a certain number of sales units given to the final user or consumer at the sales outlet, or serves only to stock the shelves. Removing this packaging does not affect the characteristics of the goods. Examples are a case, tray or pack of bottles or cans. Transport packaging facilitates the handling and transport of several sales units in such a way that direct contact with them and transport damage are avoided. Examples are corrugated boxes for carrying one or many purchased items. Containers for road, rail, ship or air transportation are not part of the category of transport packaging.

The objectives of EU Packaging Directive are that since 2001, national legislation must guarantee that at least 50% and at most 65% by weight of packaging waste must be used; within this utilisation objective, at least 25% and at most 45% by weight of the total packaging material contained in the packaging waste and at least 15% by weight of each individual packaging material must be processed by material recycling (Göttsching & Pakarinen, 2000).

All EU Member States had to translate the EU packaging directive into national law. The scope of measures implemented in the individual EU states ranged from issuing directives to reaching voluntary agreements with their national industries. All countries have set very high objectives for the collection of used packaging, averaging 70–80% of the total material (Kibat, 1996). In the year 2000, the material recycling objectives in the individual EU Member States exceeded the levels of the EU directive. All such regulations are based on the principle of producer responsibility. Producers, material suppliers, businesses and authorities bear the specific responsibility for disposal covering the entire life of a product, from its production until its final use. The producer is the most important player, because they set to a high degree the conditions for waste disposal. In the opinion of the EU Commission, the producer makes the decisions that determine how the product will be handled as waste. This covers the concept, use of specific materials, composition and marketing. With careful use of natural resources, renewable raw materials or harm-less substances, the producer can help to avoid waste and to design products that will be suitable for utilisation. The EU Commission has adopted a spirit of cooperation, emphasising the active participation of all parties involved to meet the objectives of the waste policy. The objectives will be reached only with the active cooperation of authorities, private and public organisations, environmental organisations and especially the general public as final consumer (Kibat, 1998). The latest EU concept for waste avoidance set prior options to the avoidance and the reduction of waste. This should be realised by extended duration times of products, an improved ability to reuse products and increased recycling. In 2000, and in accordance with this EU strategy, the Confederation of European Paper Industries (CEPI) and the European Recovered Paper Association (ERPA) published a voluntary European Declaration on paper recovery. The main target of this Declaration was to reach a recycling rate of the paper and board products consumed in Europe in 2005 of 56% (±1.5%). The increase in the recycling rate from 49% corresponded at that time (1999) to an additional recycling of recovered paper of 10 million tonnes or a 26.6% higher recovered paper utilisation compared with 1998. To guarantee the credibility and transparency, a European Recovered Paper Council was established. In the meantime the European Federation of Corrugated Board Manufacturers signed the declaration additionally. The European Federation of Waste Management and Environmental Services, the European Paper Merchants Association as well as the International Confederation of Printing and Allied Industries are also supporting the European Declaration. Apart the recycling rate to be obtained in 2005, the signatories of the declaration were also committed to achieve the following: further reduction of the generation of residues during all processes in the paper and board life cycle; further improvement of the efficient use of raw and auxiliary materials; optimisation of recovered paper collection systems by sharing expertise with those responsible for collecting recovered paper for recycling purposes; improvement of technical, operational and environmentally benign solutions by stimulating and supporting research and development; improvement of the awareness of paper recycling by informing the consumers about their role in closing the paper loop (Anonymous, 2000). The European paper industry is a good way to meeting the target, because the recycling rate achieved in 2003 was already 54% and the recovered paper use has increased since 1998 by 6.1 million tonnes (Anonymous, 2003).

There are, however, significant challenges still to be met not only in terms of collecting the volume of recovered paper needed, but also in remaining focused on the quality of paper collected. Stakeholders in the paper recycling chain are looking at initiatives that will promote separate collection streams for paper. Therefore, the industry has begun to prepare guidelines for the responsible sourcing of recovered paper, covering everything from collection, sorting and transportation to the storage and end-uses of recovered paper. As a result, the CEPI and ERPA have adopted a set of quality control guidelines to be used by suppliers and buyers of recovered paper (Anonymous, 2004a). Food contact papers are another challenge to be faced in this respect. The paper industry and its partners in the supply chain are also looking at new ways of improving the recyclability of paper and board products.

In addition to the European Packaging Directive, Germany has the Waste Management Act (1986), a Packaging Ordinance (1991) and a Voluntary Agreement of the Graphic Paper Chain (1994). Waste separation at the household level is a prominent feature of German waste management systems, which is regulated at the municipal level. Households dispose paper, cardboard, glass, biodegradable waste, light packaging (plastics, aluminium and tin) and the residual household waste, separately. In 1991, the German Government introduced the principle of producer responsibility for used packaging and placed a legal obligation on trade and industry to take back and recycle the packaging materials producers put into circulation. The consumers are required to follow sorting guidelines established by the municipalities.

Materials intended to come into contact with food are regulated under Food Contact Regulation (2004) and under Good Manufacturing Practice (GMP) Regulation (2006) neither of which has paper- and board-specific rules (Stawicki & Barry, 2009). The general requirement that materials for (direct or indirect) food contact 'must not transfer their constituents to the food in quantities that could endanger human health or change the composition or organoleptic characteristics of the food' applies to paper-based products, but no specific measure exists at European level to guide how paper and board is to comply with this requirement. Instead several national or regional standards are being used, which are sometimes unnecessarily restrictive or which simply ban use of recovered paper. To fill this gap at the European level, and to update existing standards with most recent scientific knowledge, the paper packaging chain is preparing an Industry Guideline for paper and board in food contact to set a single European standard acceptable for consumers and policy makers. The Industry Guideline is likely to set specific rules for managing recycling paper and board for food contact. Traces of non-intentionally added substances will occur in all packaging materials and their complete identification and total elimination is not possible. The paper industry will select its raw materials to ensure non-intentionally added substances in its products are present at extremely low levels. The Industry Guideline requires that an operator must perform a risk analysis of the manufacturing to identify all operations that have the potential to influence the suitability of the final product for food contact purposes; this risk analysis may, for instance, lead to exclusion of unsuitable sources or grades of recovered paper. A quality management system is required to manage the risk on an acceptable level. The continuing compliance with legislative and customer specifications has to be demonstrated and recorded.

The GMP Regulation, applicable as of August 2008, defines good manufacturing practice for all packaging to avoid food scares such as the ITX scandal in 2006. Food contact experts estimate paper and board packaging already complies with the regulation. However, the current paper industry GMP is under revision, to ensure consistency with the EU Regulation, and will be published. These voluntary industry measures, building on the legislative EU framework, ensure safe recycling operations for food contact purposes based on several key elements: selection of recovered paper of an appropriate quality; cooperation within the value chain to ensure design for recyclability of auxiliary materials in paper products (inks, adhesives, etc.); selection of appropriate processing technologies; operation of GMP, including Guidelines on responsible sourcing of recovered paper; testing of finished products to ensure that several known potential contaminants are absent from the recycled paper and board.

The PIRA Peer Review Report (March 2009) concluded on the Industry Guideline that 'it is very easy to argue that it provides a more useful set of Guidelines for the safe use of recycled fibres than is currently in place under any of the National Regulations', and aligned with the requirements of the EU legislations. 'A significantly more rigorous framework is offered for the approval of recycled fibres when compared to national legislation including fibre source, clean up processes and testing protocols, with the advantage over the Council of Europe Resolution that some of the associated complexity has been reduced'.

2.3 Legislation in Japan

Japan is the world's second-largest consumer of recovered paper. Considering the fact that Japan in the mid-1980s had already achieved high utilisation rates, a consideration of the political goals for recovered paper recycling is useful. The Waste Disposal Law enacted in 1970 was amended in 1991 when the Resource Recycling Facilitation Law was added. Article 1 describes the purpose of this law: considering that Japan relies on importing many important resources, but a large part of these resources are now being discarded without being used, the purpose of this law is to provide the basic mechanism for promoting the use of recyclable resources, and so promote the healthy development of the nation's economy. According to Ministerial Ordinance No. 53 of the Ministry of International Trade and Industry (MITI), a further increase in the utilisation rate of recovered paper of 55% was to be realised in domestic paper production by the end of the Fiscal Year 1994 (April 1994 to March 1995). Paper producers must also install processing systems for the utilisation of recovered paper and provide the necessary storage space for recovered paper; improve recovered paper processing technology in cooperation with machine manufacturers and chemical suppliers; develop a recovered paper utilisation plan for each fiscal year; and publicise statistical data on the percentage use of recovered paper. The MITI contribution to increase the utilisation rate of recovered paper consists of regulations that create financial and tax incentives on the recovered paper market. These include reduced corporate taxes, special rate of depreciation allowance, low credit rates for purchasing machines for pre-treating recovered paper such as bale presses and wiring machines, and a model for collection systems. Paper producers were offered reduced tax rates and special depreciation rates for machines for processing recovered paper and favourable loans for the installation of deinking systems. The MITI itself launched several advertising campaigns for the use of recycled fibre products. Two private associations also promoted use of recycled paper among consumers. The Paper Recycling Promotion Center introduced GreenMark Labels for magazines, toilet paper and books produced from recycled fibres. The Japan Environment Association developed the Eco-Mark project for products manufactured with recycled fibres. Unfortunately, because of the unclear definition of the recycled fibre contents in paper and board products, both programmes had only limited influence on the consumers' decisions. The objective of a 55% utilisation rate for recovered paper in 1994 was determined in 1989, when the rate was 50%. This rate was accepted without reservation by the industry as it represented a realistic goal. Nevertheless, a utilisation rate for recovered paper of only 53% occurred in 1994, despite the market having an over-supply of recovered paper. Some factors that contributed to this failure to reach the projected utilisation rate included reduced production share of those grades of packaging papers and board that contain large proportions of recycled fibres; increased demand for high-quality paper products; rising costs for processing recovered paper owing to environmental legislation; efforts by industry to increase profits; and inadequate legislation. This failure by the paper industry to achieve the goal of recovered paper use of 55% motivated the Ministry of Health and Welfare to accuse MITI of taking an excessively lenient approach towards paper producers. Economic studies about the use of recycled papers at private businesses showed that the consumers of recycled papers either demanded identical quality as with papers manufactured from virgin fibres, or much lower prices for them. The studies also revealed that a precondition for raising the level of use of recovered paper is improvement of public knowledge and understanding of recycled paper products. To overcome these problems, MITI established a Study Committee on Basic Issues of the Japanese Paper and Pulp Industry. This committee comprised representatives of newspaper, banking and trading companies, and the pulp and paper manufacturing, paper distribution, paper converting, printing and recovered paper industries. The industry calls for government support to establish a policy of improving the separation of recovered paper at the source; expand the collection of office papers by establishing a support system; expand consumer demand for recycled paper; negotiate an agreement between industry and consumers on standard qualities for printing and writing paper with a clear statement on the content of recycled fibres; promote increasing transparency on the domestic and export market for recycled paper products; subsidise improved recovered paper processing methods; re-evaluate the tax system to promote increased use of recycled products; establish a policy to reduce transport costs of recovered paper; develop alternative recovered paper uses other than recycling into paper and board products; increase recycling for energy recovery; this means that low-grade recovered paper can also have use for energy recovery; and increase international recycling with increased export of recovered paper (Carbonnier, 1996; Göttsching & Pakarinen, 2000). MITI supported the research and consulting activities of this committee, used it to establish a new political orientation and to help companies develop new strategies. In 1994, the Study Committee presented a new programme called Recycle 56. This aimed to achieve a recovered paper utilisation rate of 56% by 2000, which was exceeded by 1%. In fact, the programme was so effective that in 2002 a utilisation rate of 61% was achieved.

Japan's law for the promotion of sorted collection and recycling of container and packaging was enforced in April 1997 by the Ministry of Environment. As per provisions, the sorted waste is collected, stored and transported to the recycling companies by the municipalities. Manufacturers and business entities using containers and packages have to pay a recycling fee to the Japan Containers and Packaging Recycling Association, in accordance with the volume they manufacture or sell. Japan has managed the zero solid waste principle very effectively and minimised the use of scarce land space for landfills.

2.4 Legislation in the USA

There is no national legislation in the USA requiring the development of packaging recycling programmes or use of the Green Dot, as prevalent in Europe. Waste management regulations are the responsibility of each individual provincial and state government. The local waste management system design and operations are the responsibility of individual municipalities.

Several federal government Executive Orders are now in existence, but a general assumption is that the federal government will not introduce any radical changes during the next few years. Many individual states have issued laws and promulgated rules that promote recycling, reduce the amount of waste material or prohibit individual objects from being disposed of in landfills. So far, about 40 states have passed recycling laws or have implemented voluntary goals and directives. Most states have determined a recycling goal between 20% (Maryland) and 70% (Rhode Island) for used paper products. With its Act335, Wisconsin issued the most comprehensive and far-reaching recycling and waste material avoidance law as early as 1989. Besides numerous promotional and educational programmes, the law established Responsible Units, which oversee local recycling activities and are eligible for grants to maintain their administrative function (Anonymous, 1993,1998). In contrast to European Directives, which define specific recovery rates for paper and board packaging and partly also for printed material, a recycled fibre content for specific new paper grades or paper products is effective in the USA. This began in 1976 with the Resource Conservation and Recovery Act that led to thin introduction by the US Environmental Protection Agency (US EPA) of a set of Procurement Guidelines for government agencies in 1991. Dissatisfaction with these guidelines motivated the US

Recycling Advisory Council to promulgate a minimum recycled fibre content (including post-consumer recovered paper) for the paper procurements of the federal authorities. In 1998, President Clinton renewed the EPA acquisition guidelines by the Executive Order called Greening the Government by Waste Prevention, Recycling, and Federal Acquisition. According to this order, all writing, printing and office papers purchased after 1998 with public funds had to contain at least 30% post-consumer recovered paper, or not less than 50% recovered material including pre-consumer recovered paper. All price increases had to be offset by a reduction in the amount of waste material produced and by lower consumption. This constitutes a rare example of an officially decreed consumption limitation. In 1995, the EPA issued a Guidance Document on the Acquisition of Environmental Preferable Products and Services. This was the first comprehensive articulation of its Green Products policy. The document contains seven principles to ensure that environmentally friendly products receive greater consideration for orders placed by state procurement agents. This considered both quantity and quality. The first principle of this environmental preference did not consider whether one green product can substitute for another. Instead, it questioned whether a function must be improved, and realised this while minimising negative effects to the environment. The intention was that federal agencies should acquire more environmentally friendly products and services and initiate voluntary pilot projects. For paper and paper products, the EPA Office for Solid Waste has published a Paper Products Recovered Materials Advisory Notice (Anonymous, 1996a). This guidance note promotes paper recycling by centralised acquisition, and expands the markets for papers containing recycled fibres. Two important agencies for paper acquisition in the United States are the General Services Administration and the Government Printing Office. The EPA guidance notice has special importance in that it extends beyond governmental agencies and is also applicable on a wide scale for purchases in the private sector. Such purchases represent 95% or more of total paper requirements. In 1995 the US Postal Service faced the problem of stickies in paper mills by the more widely used pressure-sensitive postage stamps. Together with members of the pulp and paper industry, the adhesive industry, the converting industry and the USDA Forest Products Laboratory, a large research investigation was started to develop environmentally benign pressure-sensitive adhesives. The team specified Environmentally benign pressure sensitive adhesive (EBA) for stamps and label products, developed laboratory- and pilot-scale tests and set specifications for the allowable level of stickies from an EBA. Additionally, mill-scale trials with EBA were conducted to validate laboratory and pilot tests. Finally a list of identified qualified environmentally benign pressure-sensitive adhesives was published in 2001. Nevertheless, the use and acceptance of EBA paper labels is still in its infancy in the USA. That is the reason that the Initiative to Promote Environmentally Benign Adhesives was founded in 2003, as a consortium of adhesive manufacturers to promote the successful implementation of the US Executive Order 13148 Section 702, which rules the public order of pressure-sensitive adhesive label products (Oldack & Gustafson, 2005). EPA has defined conditions for the minimum recycled fibre content in different paper grades. A two-tier level exists consisting of post-consumer fibre content and total recycled fibre

Table 2.4 Recycling Advisory Council (RAC) and Environmental Protection Agency (EPA) Guidelines for Government Paper Products

Products	EPA		RAC 1992	
	Total Recycled Fibre Content (%)	**Post-Consumer Fibre Content (%)**	**Total Recycled Fibre Content (%)**	**Post-Consumer Fibre Content (%)**
Printing and writing	10 ± 50	10 ± 20	40 ± 50	10 ± 15
Newsprint	40 ± 100	40 ± 85	40	40
Tissue	20 ± 100	20 ± 60	60 ± 80	50 ± 70
Construction paper	n/a	n/a	80	65
Packaging paper	5 ± 40	5 ± 20	40	20
Uncoated paper: corrugated	30 ± 50	30 ± 50	40	35
Uncoated paper: other	n/a	n/a	100	60
Coated paperboard	100	45–100	90	45

Source: Based on Anonymous (1996b); Putz (2006)

content. According to the American Forest & Paper Association, almost all pre-consumer recovered paper is collected and recycled. Therefore, the EPA believes that the two-tier approach leads to greater use of post-consumer recovered paper. So far, the EPA has not defined environmentally preferable paper in an even stricter sense. For example, it has not taken into account major environmental influences on paper production such as the type of forests from which the virgin fibre furnish is derived. Table 2.4 gives the minimum content of recycled fibres for different paper grades. It includes a comparison of the recycled fibre content proposed by the Recycling Advisory Council for government acquisitions since 1992 with the EPA guidelines. At first glance, the EPA conditions appear to be more strict. However, the statistical values must be interpreted very carefully as the definitions differ somewhat (Anonymous, 1996b). Besides general laws, ordinances and guidelines, many federal laws exist in the USA to avoid waste material. Typically, these laws cover the fields of separation of discarded paper at the source, procurement preference and recycled fibre content in newsprint. They achieve the aim of the EPA to reduce the amount of waste material for disposal in landfills or in incineration plants. These laws include mandated separation of newspapers by households; federal, state and municipal procurement policies favouring recycled materials. Laws require consumers of newsprint to use recycled fibre containing newsprint for a certain portion of their total requirements. The goals to be achieved as recycled fibre content in newsprint in the state regulations are mainly between 30% and 50% and are mandatory or voluntary. The various states have no general definition, either for the penalties or for the basis of the recycling goals. Such regulations therefore force mills in the USA or Canada to invest in paper recycling plants to produce papers with the stipulated proportion of recycled fibres. Customers of paper are limited in their choice of supplier. A lack of methods also exists to monitor the secondary fibre content in paper and board grades produced with certain proportions of recovered paper. From the European viewpoint, legislation stipulating contents of recycled fibres in certain paper grades is not acceptable by the paper industry and the consumers of paper products. With this background, guidelines for a national recovered paper collection rate, recycling rate or both make more sense because they give the paper industry considerably greater freedom in its efforts to achieve the targets, particularly concerning technical and economical matters. Depending on the availability of recovered paper grades and the technical possibilities of recovered paper processing, corporate decisions could ensure the recycled fibre content of future paper production on a national level. This also allows newsprint mills that operate near major cities to make predominant use of recovered paper as fibre material, whereas mills located in rural, thinly populated areas can continue to produce newsprint from virgin fibres. Both paper grades can be marketed everywhere and compete directly with one another. For this reason, the American Forest & Paper Association (AF&PA) specified a recovered paper collection rate of 50% of paper consumption by 2000 as a goal for the American paper industry, but this failed by 4% (Anonymous, 2004b). In 2002, a recovery rate of 48% was realised and the new target set by AF&PA was 55%, which was achieved in 2012.

References

Anonymous (1993). *Waste reduction and recycling initiative*. PUBL-IE-041, REV 2/93, Wisconsin Department of Natural Resources, Madison.

Anonymous (1996a). *Information from internet web-site of EPA*. Available online: <www.epa.gov>.

Anonymous (1996b). *Towards a sustainable paper cycle*. London: International Institute for Environment and Development (IIED).

Anonymous (1998). *Information from internet web-site of American Forest & Paper Association*. Available online: <http://www.afandpa.org>.

Anonymous (2000). *European Declaration on paper recovery*. CEPI/Brussels.

Anonymous (2003). *European Declaration on paper recovery*. CEPI/Brussels.

Anonymous (2004a). *Responsible management of recovered paper: Guidelines on responsible sourcing and quality control*. CEPI/Brussels.

Anonymous (2004b). *Information from internet web-site of American Forest & Paper Association*. Available online: <http://www.afandpa.org>.

Carbonnier, S. (1996). *Paper recycling and the waste paper business in Japan. Sub-Study No. 16 for IIED study: Towards a sustainable paper cycle*. Tokyo.

Department of industrial policy and promotion (DIPP) (2011). *Discussion paper on collection and recycling of waste paper in India*.

Göttsching, L., & Pakarinen, H. (2000). *Recycled fiber and deinking*. Helsinki: Fapet Oy.

Kibat, K. D. (1996). Altpapier und Produkt-verantwortung in Europa. *Wochenbl. f. Papierfabr.*, *124*(14/15), 656.

Kibat, K. D. (1998). Altpapiereinsatz im Spek-trum der europäischen Abfallpolitik. *Wochenbl. f. Papierfabr.*, *126*(16), 742.

Oldack, R. C., & Gustafson, F. J. (2005). Initiative to promote environmentally benign adhesives (IPEBA): Solving a sticky issue. *Progress in Paper Recycling*, *14*(2), 6–8.

Putz, H. J. (2006).. In H. Sixta (Ed.), *Recovered paper and recycled fibers, handbook of pulp (pp. 1147–1203)*. KGaA, Weinheim: Wiley-VCH Verlag GMbH & Co.

Stawicki, B., & Barry, B. (2009). *The future of paper recycling in Europe: Opportunities and limitations*. <www.cost-e48.net/10_9_publications/thebook.pdf>.

3 Collection Systems and Sorting of Recovered Paper

3.1 Collection

The collection of recovered paper is performed by (i) collection in private households and small commercial enterprises, (ii) collection from industrial and business operations sites where unwrapping is performed, offices, authorities and administration and (iii) returns of recovered paper from converting facilities such as printing houses and corrugated board industry and over-issues (Putz, 2006). Depending on the origin of the collected recovered paper, a clear difference exists between pre- and post-consumer recovered paper (Table 3.1) (Mulligan, 1993).

Collection of recovered paper depends on the collection region, its characteristics and customers for the collected recovered paper. Recovered paper and board collection may be classified as industrial collection from businesses and private collection from households and individuals (Bilitewski, Berger, & Reichenbach, 2001). The basic collection system has been reported by Stawicka (2005).

Several different systems have been implemented to collect recyclates from the general waste stream (Table 3.2). These systems lie along the spectrum of trade-off between public convenience and government ease and expense.

Kerbside collection is a service provided to households, typically in urban and suburban areas, of removing household waste. It is usually done by personnel using purpose-built vehicles to pick up household waste in containers acceptable to or prescribed by the municipality. For paper and board kerbside collection, the waste has to be properly prepared and packed. Some countries require the use of special bags or the municipality might provide a special container or the public are asked to secure papers (with string) for collection. Kerbside collection encompasses many subtly different systems, which differ mostly on where in the process the recyclates are sorted and cleaned. The main categories are mixed waste collection, commingled recyclables and source separation. A waste collection vehicle generally picks up the waste. At one end of the spectrum is mixed waste collection, in which all recyclates are collected mixed in with the rest of the waste, and the desired material is then sorted out and cleaned at a central sorting facility. This results in a large amount of recyclable waste, especially paper, being too soiled to reprocess. However, it has advantages as well: the city need not pay for a separate collection of recyclates and no public education is needed. Any changes to which materials are recyclable is easy to accommodate as all sorting happens in a central location. In a commingled or single-stream system, all recyclables for collection are mixed but kept separate from other waste. This greatly reduces the need for post-collection cleaning but does require public

Recycling and Deinking of Recovered Paper. DOI: http://dx.doi.org/10.1016/B978-0-12-416998-2.00003-9

Table 3.1 Pre- and Post-Consumer Recovered Paper

Pre-Consumer Recovered Paper
Material from manufacturing, other waste from the papermaking process and finished paper from obsolete inventories. Pre-consumer recovered paper collection, in the form of over-issues and converting residues such as shavings and cuttings, requires pick-up systems, traditionally organised by the dealers in recovered paper, who use containers at individual points of generation. Most pre-consumer collections are handled in the pick-up mode
Post-Consumer Recovered Paper
Comprises paper, board and fibrous material recovered from retail stores, material collected from households, offices, schools, clubs, etc. This kind of raw material can be collected at each individual point of generation (pick-up collection system), from kerbsides by collecting associations, and material can be also carried to a centralised collection point by consumers

Table 3.2 Methods for Collection of Waste Paper

Kerbside
Blue bins
Public containers
Recycling yards
Collection shops
Recycling centres
Drop-off recycling parks

education on what materials are recyclable. Source separation is the other extreme, where each material is cleaned and sorted before collection. This method requires the least post-collection sorting and produces the purest recyclates, but incurs additional operating cost for collection of each separate material. An extensive public education programme is also required, which must be successful if recyclate contamination is to be avoided. Source separation used to be the preferred method because of the high sorting costs incurred by commingled collection. Advances in sorting technology, however, have lowered this overhead substantially – many areas that had developed source separation programmes have since switched to commingled collection. Unfortunately, strongly fluctuating prices for recovered paper have almost completely discouraged kerbside collections in Germany. In Switzerland, however, such collections of recovered paper continue to have major importance for the national paper industry. The difference may be because the Swiss paper industry has continuously supported and financed the kerbside collection, even when it was more economic for the paper manufacturing companies to buy the recovered paper from the market. *Blue bins* are convenient, easy-to-use wheeled bin collection for recycling paper and board from households. The blue bin is provided to households to allow separation and storage of paper and cardboard for recycling. The contents of the bin are collected by special services and taken to a materials recovery facility where they are sorted. There are different sizes of blue bins. Bins are collected from the kerbside every few weeks, depending on the service, according to scheduled collection dates.

The system of *public containers* is based on placing containers of different colours for separation of different wastes within the community. Containers must be accessible to all and the spacing of containers should be set to be within easy reach of all citizens. The *recycling yards* system requires the waste producer to carry the recycling materials to a specific location, where containers for different categories of recyclable materials are available. *Collection shops* are places where paper and cardboard can be sold. Shops pay for grades of recovered paper: mixed, white, old corrugated cardboard (OCC) and old newspapers (ONP) and old magazines (OMG). To maximise the price paid for the recovered paper, people are encouraged to sort it before selling. Usually this kind of collection technique does not need any further sorting. *Recycling centres* are designed for all types of waste, usually operated by municipalities. These are manned sites, where people can bring their recyclable wastes. Recycling centres frequently contain material reclamation facilities. *Drop-off recycling parks* are special places, where people can deposit separated paper, board or other wastes. Drop-off recycling parks resemble other recycling facilities with recycling bins, but there are containers underground for storing larger amounts of waste. Drop-off centres require the waste producer to carry the recyclates to a central location, either an installed or mobile collection station or the reprocessing plant itself. They are the easiest type of collection to establish, but suffer from low and unpredictable throughput. For drop-off systems, containers larger than $5\,m^3$are used. These are placed at central, publicly accessible locations and are emptied at regular intervals. They can be either single-compartment containers intended only for collecting paper, or multi-compartment containers for collecting different materials such as paper, glass and metal that remain separate within the container. Besides the greater space that is necessary for multi-compartment containers, the decision about whether to install single- or multi-compartment containers depends on the availability of suitable vehicles and the occurrence of the materials concerned at the installation sites. Containers can simply be emptied into a suitable refuse collection vehicle for single-compartment containers or exchanged, as in the case of drop-off containers. For multi-compartment containers, the refuse collection vehicle must suit the multi-compartment system of the container to avoid mixing the individual materials. With multi-compartment containers, exchanging or emptying the entire container when collecting the materials is necessary, despite the fact that individual compartments may not be totally full. This means the collection frequency must consider the material whose compartment fills first. Standard vehicles are useful for emptying single-compartment containers by the exchange method. Overflowing containers must be avoided; otherwise, the tendency to throw materials in still-empty containers for other materials increases, or the sites can turn into rubbish dumps. Single-compartment containers offer the basic advantage that the collection intervals of single containers can be linked to the occurrence of material. Containers that fill more quickly can be emptied more frequently than those that fill more slowly. Larger containers can also be selected. In principle, separate collection of used graphic paper products and packaging material is possible with at least two different containers or a multi-compartment container.

The collection of pre-consumer recovered paper in the form of over-issues and converting residues such as shavings and cuttings uses pick-up systems that traditionally are organised by the recovered paper dealers using containers at individual points of generation. Pick-up systems collect recovered paper from industrial and business operations. Containers are exchanged or emptied at regular intervals, either with or without the benefit of stationary compressing compactors. Collection vehicles that compress refuse are seldom used. The traditional recovered-paper trading industry is facing increasing competition from waste management companies, who tend to offer their services to industrial and business operations and to the retail trade, not only for paper but also for the collection of other materials such as glass, plastics or wood. Thus, they provide a greater convenience and flexibility. Consequently, most pre-consumer collections are handled in the pick-up mode.

Recovered paper from operations where unwrapping occurs belongs to post-consumer recovered paper. In these facilities, the same collection systems are installed as for collection of pre-consumer recovered paper, namely in the pick-up mode. In contrast, many different collection systems are available for recovered paper collection from private households. The choices made are based on population structure (population density and rural or urban regions), housing structure (inner city, high-rise developments, single-family housing) and the customary use for the collected recovered paper. A distinction exists here between pick-up and drop-off systems, both of which have been in long-term practice in Germany.

In highly developed countries with established high collection rates, the ratio of material generated between these two sources, as a percentage total collection, is approximately equal (Stawicki & Barry, 2009). The situation changes in countries with low collection rates, where collection systems mainly focus on industrial sources. Industrial and commercial paper and board collection is much more easily performed and often referred to as the 'low hanging fruit'. Countries with high collection rates have a capacity to collect more than 90% of the used paper from industrial collection sources. In countries where the collection system is in the early stages of development, the industrial paper collection is the primary source of recovered paper. Additional effort is made to collect paper from paper, printing and converting firms, which provides a high-quality recovered raw material. It is therefore important to keep collecting this material separately. In contrast, household and public collection is more difficult and a multitude of different collection methods has been developed. There are several collection systems already in existence. One of the most common techniques of collecting paper used in most countries is the blue bins and similar container systems. Less economically developed countries are now introducing similar systems. Unfortunately, in the early stages the efficiency of these systems is not high. It is affected by an insufficient number of containers and the inconvenience of accessibility in parallel with low consumer awareness. Well-established systems, such as in Germany, where paper collection through blue bins was introduced much earlier and has become a part of everyday life, give the best results. Unfortunately, blue bins are not suitable for board, as its stiffness and volume hinders collection. Therefore, in Spain a 'door-to-door' system is being introduced for collecting board. Kerbside collection is practised in many countries and gives satisfying results. In comparison with recycling yards or recycling centres, where people have to transport their wastes by car, this method is more accessible for people.

Table 3.3 Current Waste Paper Collection Mechanisms in India

Source Items	Collected	Collected by	Quantity Collected (in Million Tonnes/ Annum)
Collection from households	Old newspaper and magazines Notebooks and textbooks	Weekend hawkers	1.50 0.5
Annual scrap contracts of printers, publishers and converters	Paper trimmings, print rejects, overprint/misprint sheets and other waste	Contractors	0.25
Scrap contracts with industries, offices, libraries	Old corrugated cartons, examination answer sheets, library records, old office and library records, etc.	Contractors	0.5
Total			2.75

Source: Based on DIPP (2011)

In Bulgaria, the main portion of recovered paper originates from collection shops. People manually sort paper themselves to secure a better price. The advantage of this kind of collection system is that no further sorting is needed.

In India the collection of waste paper is mainly performed by the informal sector, i.e. by rag pickers and door-to-door collectors/vendors. As much as 95% of the collection of waste paper in the country is performed by the informal sector. The value chain comprises the direct collectors from various source points and small shops – where primary sorting of the waste into different categories takes place – and zonal segregation centres owned by wholesalers where the waste material gets collected from small shops and is baled for dispatch to the end users. The current mechanism adopted for collection of waste paper in India is given in Table 3.3.

Clearly, the existing institutional mechanisms are weak and lead to considerable leakages. The life-cycle analysis of different grades of paper, given in Table 3.4, indicates the potential for sizeable enhancement of recoveries, particularly for copier and creamwove paper from offices and newspaper and packaging from households.

It has been said that the paper arising from industrial and commercial sources is the easiest, cleanest and most economical to collect. Nevertheless, the predicted demand for recovered paper and board is anticipated to grow substantially, outstripping the quantity of material available by this route; this is why it is so important that other sources and collection methods are being developed. The type of recovered paper can be classified as follows:

- Composition

Recovered paper is classified into main groups, which in turn consist of several grades

Table 3.4 Recovery Potential for Waste Paper

Grades of Paper	Potential Source of Generation	Generation/ Consumption (%)	Type of Waste	Collection Rate (%)
Writing/printing				
Copier paper	Offices	50	Post-consumer	20
	Business establishment	40		
	Others	10		
Creamwove	Printing house	20	Pre-consumer	100
	Paper traders	5		
	Households	20	Post-consumer	20
	Schools/colleges	10		
	Offices	25		
	Business establishment	10		
	Others	10		
Packaging paper	Converting house	15	Pre-consumer	100
	Households	20	Post-consumer	50
	Offices	5		
	Business establishment	50		
	Others	10		
Newspaper	Publishing house	20	Pre-consumer	100
	Distributors	5		
	Households	40	Post-consumer	30
	Offices	10		
	Business establishment	15		
	Others	10		

Source: Based on Indian Recycled Paper Mills Association (IRPMA) (2011), DIPP (2011)

- Pre- and post-consumer recovered paper

Depending on where the recovered paper was extracted from the supply chain, it is defined as pre-consumer recovered paper or post-consumer recovered paper.

To maintain a satisfactory level of collection and purity of collected material, it is not only necessary that suitable locations for the containers (clean, accessible by automobile and close to home) should be ensured, but also that a balanced division of activities (promotion of separate collection, sorting at source, etc.) is encouraged (Gottsching & Pakarinen, 2000; Stawicka, 2005)

In the Netherlands, the most effective system for collection of recovered paper in terms of the quality of collected secondary raw material is public collection with the use of 35 m^3 containers. This is because in this system consumers bring the collected material to identified collection points in relatively small amounts (1–5 kg), pre-sorted at households. Consumers seem to be quite well informed on what kind of paper products are suitable for recycling, and what needs to be done to improve the collection. A similar situation, although less efficient, takes place with kerbside collection. On the contrary, drop-off recycling parks are the least efficient. Here consumers bring collected recovered paper in larger amounts (5–20 kg), less often and less regularly, and the material is not well controlled and more contaminated. The Diftar system is considered more efficient and a more honest way of charging consumers for generation of waste. However, it has had a negative influence on recovered paper as secondary raw material collected from households. Higher levels of contamination were observed with the Diftar system. It should be noted that recovered paper, along with glass, is one of the waste streams with 'free disposal'; it would seem that consumers mix other 'chargeable' waste materials (often detrimental to paper) with the paper to avoid the extra payments. This does not happen in municipalities without the system. This explains the high contamination levels shown by results from collection systems using Diftar. Obligatory payments for wastes (other than paper) can cause the growth of contamination in reusable material containers. Higher levels of contamination are also a side effect of the constant growth in recovered paper. While considering priorities, if mills can cope with some contamination in recovered paper, the quantity of collection takes precedence over the quality. Nevertheless, improvements in reducing contamination are expected and required. Recent developments in newspaper, magazine and brochure delivery towards moisture protection have resulted in the application of polyethylene films around publications. Easy handling for the consumer had the disadvantage of being an obstacle for recovered paper collectors and the paper industry. Material packed in this way (sealed papers) is difficult to process, although it very often contains a valuable fibrous fraction inside. Sealed paper has proved to be the main contamination, up to 2% by weight, of the recovered paper. The lowest contamination with sealed paper (0.6%) was found in public collection systems with 35 m^3 containers in municipalities without Diftar. The sealed paper content in materials collected with kerbside collection and at drop-off recycling parks is higher, up to 2.8% by weight, than that collected by public containers.

It can be concluded that the low quality of today's collected and recovered paper is a result of introducing paper-based products that are difficult to recycle, as well as the addition of other non-paper contaminants to the paper by consumers. It is much easier to prevent contamination with these non-paper components, which are detrimental for processing in the paper industry and difficult or impossible for recycling paper-based products. The increased collection rates have had a negative influence on the quality. Sorting can help, but only to a certain level; however, separate collection is an absolute necessity. Collection is still developing, in some countries very rapidly. This means that collection of recovered paper is an international issue. In this respect, public awareness campaigns should be done. Influencing the awareness of consumers and rewarding for separate collection of high-quality secondary raw material are crucial factors in alleviating the social limits of paper recycling. Even with the most efficient collection systems for recovered paper, the collected secondary raw material will still contain certain amounts of contamination. This can be removed by further sorting processes; however, because of certain technical and economical factors, contamination is not removed and it becomes a fraction of recovered paper that is finally generated as solid waste within stock preparation lines of recycling paper. Table 3.5 gives the basic collection methods for paper recovery.

Table 3.5 Basic Collection Methods for Paper Recovery

Single Stream Collection

Allows participants to put all recyclable materials into one collection container. In the case of paper, all grades are mixed together. These materials are then collected and separated, usually at a central point such as a materials recovery facility. Paper collected in single-stream systems may be further separated into various paper grades. For single-stream recycling to work, the processing facility must sort the recyclable materials properly and thoroughly to meet market specifications. Two basic collection methods for paper recovery are common.

Advantages

- Makes it easy to add new materials to the collection system
- Increases the amount of recyclables collected
- Reduces the number of collection trucks needed

Sorted Stream Collection

Requires participants to place each recyclable material in the appropriate collection bin when they first discard the item. Recovered paper can be collected separately by grade (e.g. white office paper, newspapers, magazines and corrugated cardboard boxes) or, more commonly, collected as mixed paper separated from other recyclable materials

Advantages

- Lower levels of contamination at the source
- Higher quality and more valuable recovered material
- Lower costs to process the recovered paper

No direct correlation is found between the collection system and the collection rate. According to an investigation done within the framework of COST Action E48, neither the type nor the share of the different collection systems used seemed to have a significant influence on collection rates. Similar collection rates can be seen in different countries with different collection systems, whether they be municipal organisations, municipal plus private organisations, or municipal plus private plus other organisations. A well-functioning, robust collection system is essential. The quality of recovered paper, however, is greatly affected by the collection system. Collection systems generally evolve with collection rates. In countries with low collection rates, the major source of recovered paper is commercial and industrial; consequently, the collection systems are adapted to this type of recovered paper. In this case, only large containers at the site of the arising are used. However, to achieve high collection rates, office and household recovered papers also need to be collected. Consequently, new collection systems, such as 'blue' containers, become installed in larger cities. When the collection rates start to increase, more containers are installed, and new and more complex collection systems are introduced, such as the 'door-to-door' collection of commercial board, 'pick-up' collection of household recovered paper, etc.

Within COST Action E48, the environmental awareness of citizens has been identified as one of the most crucial drivers for achieving higher collection. In countries with low collection rates, the most important stage in increasing collection of recovered paper is the development of environmental awareness through education and campaigns (Stawicki & Barry, 2009). Collection systems in these countries, although less developed, are similar to those used in other countries with much higher collection rates; however, citizens do not engage with the recycling ethos and the collection systems are of almost no use. This is true in some East European countries in which a lot of effort is made to provide, for example, 'blue containers' in large cities, which are not used; consequently, the collection rates are notably lower. Environmental awareness particularly influences the collection from domestic households. In countries with established collection rates that are very high, in which a very positive environmental attitude is widely developed, improvements in the collection systems can have some influence on increasing the collection rate.

The key to any recycling programme is participation, which is greatly influenced by motivation. People can be motivated by extrinsic or intrinsic rewards. Usually, extrinsic rewards consist of a payment for collected materials. This solution can be very effective, but extrinsically motivated behaviour does not continue on its own when the inducement is withdrawn. On the other side, intrinsic rewards fulfil a person's need to have an impact on their world, producing satisfaction to individuals (Stawicki & Barry, 2009). For these reasons, intrinsic behaviour tends to last longer than extrinsically motivated behaviour. An extensive study in Sweden (Berglund, 2003) showed that the moral motives significantly lower the cost associated with households' recycling efforts. Furthermore, moral motives can, in some cases, be the cause of inefficient policy outcomes when introducing economic incentives to promote recycling efforts.

Studies show that the more knowledgeable people are about recycling, the more likely they are to do it and to feel satisfied with their actions. Thus, regardless of the motivation, environmental education, awareness campaigns and positive examples are effective means for developing environmentally responsible behaviour in people (Garces, Lafuente, Pedraja, & Rivera, 2002; Kollmuss & Agyeman, 2002). In Europe, the main sources of paper collection are 50% from printing, converting, trade and industry, 40% from households and 10% from offices. Future potential for increasing collection rates is mainly related to households, as industrial sources are already used to a significant extent (Ringman & Leberle, 2008). Considering household collection consists of numerous small sources, which creates pressure on the costs and quality of recovered paper, environmental education and raising awareness are very important factors for increasing the collection rate of recovered paper in this area. Environmental education can be accomplished by different means: audio-visual programmes for schools, civic and community groups; educational curricula and class projects; community events, fund-raisers; and contests. All of these have been tried with varying degrees of success. Environmental awareness can also be promoted by different channels: press conferences and kits, especially when an awareness campaign is starting, public service announcements, mass media – broadcast, newspaper, magazine, press releases, printed materials – direct mail, newsletters, inserts in utility bills, door hangers, posters and bumper stickers.

In the survey on general parameters influencing the future competitiveness of paper recycling in Europe, the partners of the COST Action E48 were asked to rank the importance of different ways of improving paper collection efficiency in their country (Grossman, Bobu, Stawicki, & Miranda, 2007). General results, expressed as an average of 19 countries, show that a society's environmental awareness is the dominant prerequisite for efficient recovery of used paper and board products. This survey was based on detailed questionnaires, which included all aspects that could influence the competitiveness of paper recycling. Because the answers to various questions were quantified by related values, it was possible to see the evidence of the relationships among factors that impact the competitiveness of paper recycling. Thus, good correlations were found between environmental awareness, the quality of recovered paper and collection rate.

Household collection can be substantially improved by increasing the environmental awareness of consumers. Statistical evaluation of the 'CEPI study on the collection systems in Europe' leads to a similar conclusion. The CEPI study includes an investigation into collection systems in 30 countries, based on current statistical data and questionnaire answers. The Task Force of COST E48 Action combined the answers with statistical data to find driving forces and influences on the collection rates of used paper and board. One of the conclusions of the statistical analysis confirmed that environmental consciousness is the main driving force for increasing collection rate (Blanco, Carpetti, Faul, Miranda, & Strunz, 2006). Thus, as more people and organisations become aware that they can make a positive difference by the separate collection of used paper products, more secondary raw material will be available for the paper industry, contributing to sustainable development.

For the different systems, the specific amount of paper collected per inhabitant depends on many different marginal conditions, such as collection frequency, population density and the sizes of the containers used (Stawicki & Barry, 2009). The efficiency of different collection systems is given in Table 3.6 (Bilitewski, Hardtle, & Marek, 1994) (values are per person and year).

Densely populated regions should use a pick-up container system, although the space required for an additional paper container at home requires consideration. In less populated areas, drop-off container systems are useful. Finding the most suitable locations for the containers (clean and attractive, accessible by automobile and close to home) is especially important to ensure that the collectable volume and the standard of purity of the recovered paper satisfy the requirements. Bundled paper collections in densely populated areas are always suitable if containers cannot be installed in sufficient numbers. A few regions of Germany collect used papers from private households as one component of the so-called multi-component recyclable container collection. Here, one container is used in private households for collecting several recyclable materials such as printed products, packaging materials including liquid packaging, plastic, metal and textiles. Glass is usually collected separately because of the danger of breakage and the subsequent occurrence of glass splinters in the paper and board fraction, even after manual sorting. Although paper recovery by this multi-component container collection is possible, it requires especially intensive and consequently expensive manual sorting. Generally, contamination of the moisture-absorbing recovered paper with liquid and foodstuff residues from liquid packages, plastic bottles and tin cans can also occur. The paper industry therefore has major

Table 3.6 Efficiency of Different Collection Systems

Drop-off container system: 32 kg
Kerbside collection: 43 kg
Recycling centre: 45 kg
Pick-up container system: 53 kg

reservations about the use of recovered paper from this collection system. In accordance with European standard EN 643, recovered paper from this collection system requires express identification as such. Recovered paper from multi-component collection systems has not been used for materials and articles of paper and board for contact with foodstuffs that must comply with international or national hygiene regulations. It is not permissible to mix recovered paper from multi-component collection systems with other recovered paper grades, without special identification (Gottsching & Pakarinen, 2000).

Despite all the efforts to recover paper as a raw material for the paper industry, domestic waste also includes paper and board, though in increasingly smaller quantities. In the USA, according to the Environmental Protection Agency (EPA) definition, municipal solid waste consists of durable goods, containers and packaging, food scraps, yard trimmings, and miscellaneous wastes from residential, commercial and industrial resources (Stawicki & Barry, 2009). In 1996 in the USA, municipal solid waste contained about 38% paper and paperboard, of which about half was recycled. Finally, about 19% paper content remained in municipal solid waste and went to landfill or to incineration plants (Andersen, 1997). In Germany, according to the most recent National Household Recovery Analysis (in 1985), the share of paper, paperboard and composite paper contained in household waste was only about 18%. This result was achieved much sooner than in the USA because of the much higher level of separate paper recovery in Germany, and the different categories of waste counted as domestic waste. Despite the fact that there has been no National Household Recovery Analysis since 1985, all recent results from individual German cities reveal even lower figures (Bilitewski, Hardtle, et al., 1994). According to Intecusan, the actual average proportion of paper products in household waste in Germany is estimated to be between 9% and 10%. In principle, the paper industry takes a negative stance about the use of recovered paper that is sorted from general domestic waste collections. This is also reflected in EN 643 by the specification that paper recovered from general refuse is not suitable for use in the paper industry

3.2 Sorting, Handling and Storage of Recovered Paper

The primary challenge in recycling paper is to obtain raw material with the highest purity. Sorting paper into compatible grades is a necessary step before recycling. In recycling, waste papers are segregated into various grades because they are subjected to different recycling processes. Highly sorted paper streams will facilitate high-quality end product, and save processing chemicals and energy. A grade refers to the quality of a paper or pulp. The basis of grade is weight, colour, usage, raw material, surface treatment, finish or a combination of these factors (Goyal, 2007). In most materials recovery facility plants, the recovered waste papers are sorted as (i) computer print-out (CPO), (ii) white ledger (office paper), (iii) coloured ledger (office paper), (iv) cardboard (OCC), (v) newspaper (ONP), (vi) magazines (OMG), (vii) coated sheet (glossy) and (viii) mixed office paper (Rahman et al., 2011). Waste paper sorting systems are classified into manual and automated systems. Automated paper sorting systems offer significant advantages over manual ones in terms of fatigue, throughput, speed and accuracy.

White, unprinted and woodfree, bleached pulp based on recovered paper is considered one of the highest-value grades. Recovered paper and board derived from printing and converting operations is usually clean, and easy to collect regularly in reasonable quantities. Post-consumer recovered paper is more difficult to manage. It may be sorted at source, such as newspaper and white printed paper and board at households or brown corrugated cases at supermarkets. Paper and board that are not sorted at source and are mixed have a lower value and are more likely to contain contraries, like broken glass, plastics and other kinds of solid contaminant. It has been found that as more used paper and board products are collected and recycled, they tend to contain higher proportions of contaminants. Mixed collection can be cheaper but simultaneously reduces the value of the material for recycling (Moore, 2002). Despite many developments, sorting recovered paper subsequent to collection continues to be a predominantly manual, labour-intensive activity at an increasing level as recovered paper becomes more heterogeneous and contaminated with non-paper components. If recovered paper is a component of a commingled collection of different materials for example, its sorting will be especially troublesome and will require sophisticated separation technologies. In the first instance, sorting systems can be divided into manual, mechanical or automatic. Manual sorting can be performed by small sorting companies with a relatively low productivity in the range of 5000–20,000 tonnes annually. Larger sorting centres use a combination of all three systems. It is said that complete elimination of manual sorting is impossible and unprofitable. The number of people required to sort depends on the sorting equipment applied and type of recovered paper. Mechanical separation technologies were developed specifically for fibre recycling and have recently become more widely used in the belief that they are more cost effective for sorting papers into their marketable components. Nevertheless, to achieve both a high level of productivity and satisfactory quality of recovered paper, it is essential to apply a suitable sorting system. Nowadays many separation lines are available on the market. The main factors that determine their characteristics are efficiency and the quality of recovered paper that can be processed. State-of-the-art sorting processes for recovered paper usually comprise four steps. The first consists of charging the sorting plant and ensuring a constant flow. In the second step, mechanical sorting is performed. The third step is performed by a sensor-based sorting unit, where unwanted objects are identified and separated. The fourth and last step is manual sorting, performed to remove residual unwanted materials. Additionally, loosening and singling stages may be implemented along the process to provide better conditions for single-sorting operations. Manual sorting is still the most common method for removal of unwanted materials. It is often performed without any further preparation except flow control. Mechanical solutions are well established and are used to support the manual sorting. In mechanical sorting the separation is based on physical properties like size, stiffness or weight. Several screen systems, ballistic separators (rotating screens), gap techniques (acceleration of materials over or into a gap between conveyor belts) and special applications such as the paper spike or deinking screens are currently operated (Bilitewski, Heilmann, & Apitz, 1994; Gottsching and Pakarinen, 2000; Stawicka, 2005). The preparation of the unsorted paper is decisive for the efficiency of the single-separation steps of paper recovery. For an efficient sorting, the recovered paper has to be loose and free of plastic bags, in particular. Typically, personnel are required to check the input to sorting plants and manually open any visible plastic bags or similar items. In standard mechanical screening, large objects are often mistaken: posters are often removed where they should be kept in and corrugated board is accepted when it should be removed. When it comes to sensor-based sorting, only the surface of an object is assessed. Paper products are often non-heterogeneous themselves, for example the cover of a magazine is made from a different type of paper and is converted in another way than the pages. Magazines and commercials packed in plastic foils may mistakenly be identified as plastic only. In such cases, sensors only detect the properties of the thin layer, but the main raw material content is not assessed. Moderate shredding can improve this situation (Bilitewski, Heilmann, et al., 1994; Gottsching and Pakarinen, 2000; Stawicka, 2005). To achieve greater harmonisation in the paper industry, to improve the implementation of the EN 643 standards and to facilitate commercial relationships between paper mills and recovered paper merchants in Europe, special attention should be paid to developing and improving collection methods, sorting systems and sorting techniques (CEPI, 2002).

Current methods of manual sorting are tedious, slow and expensive. The throughput from these is also very low. A typical manual conveyor operates near 75 ft/min (about 0.40 m/s). In addition, recently conducted research has concluded that labourers working at manual sorting facilities are exposed to microorganisms, organic dust and fungi, which can cause severe infections (Sigsgaard, Malmros, & Nersting, 1994). Hence, the sorting process needs to be automated to provide efficient recycling and a safer sorting process. Sorting at higher speeds also improves the economic viability of an automated process. Target speeds in the industry are about 1200 ft/min (6 m/s), with a throughput of 15 tonnes/h. High-speed automation of the sorting process will improve the cost efficiency and increase the amount of paper recycled significantly. Faibish, Bacakoglu, and Goldenberg (1997) proposed an automated paper recycling system where ultrasonic sound is used to separate different grades of papers. However, because of contact manipulation and sensing, the system is too slow (80 ms per sub-frame) for industrial applications. Ramasubramanian, Venditti, Ammineni, and Mallapragada (2005) developed a lignin sensor, which is working well for separating newsprint samples from others. However, the lignin sensors are influenced by colour and sensor distance from the sample. Hottenstein, Kenny, Friberg, and Jackson (2000) proposed a sensor-based sorting approach in which a brightness sensor (reflected light intensity at 457 nm) is used to sort papers primarily into three categories: white papers containing optical brighteners, white papers without optical brighteners and the rest. Venditti, Ramasubramanian, and Kalyan (2007) developed a stiffness sensor that is used to sort recovered paper into paperboard and others. However, it cannot distinguish between a stack of newsprint and a single paperboard. Sandberg (1932) proposed a sorting device to separate paper objects from contaminants. The attempts of Bialski, Gentile, and Sepall (1980) and Grubbs, Kenny, and Gaddis (2001) have not been successful, mainly because of the absence of reliable sensing systems to distinguish between grades. Khalfan and Greenspan (2006) introduced an optical paper sorting method in 2002. It used diffuse reflectance to identify a sheet of paper as either white or non-white. Their proposed paper sorting system segregated papers into white and ground wood paper based on the amount of lignin content. Eixelberger, Friedl, and Gschweitl (2003) proposed an optical paper sorting method to separate waste paper into two classes based on the radiation reflected from the surface of the papers. Bruner et al. (2003) proposed one optical paper sorting method to separate waste papers into bright white paper and others based on amount the fluorescence present on the surface of paper objects. Doak, Roe, and Kenny (2007) proposed optical paper sorting methods to separate different grades of paper based upon at least one characteristic of colour, glossiness and the presence of printed matter. Heinz (1998) proposed using visible light, ultraviolet light, X-rays and/or infrared light to illuminate the paper for sorting. They used mechanical pickers, thus indicating that the system would operate at relatively low speeds. Owing to inadequate throughput and some major drawbacks of mechanical paper sorting systems, the popularity of optical paper sorting systems has increased. However, the implementation of all the prevailing optical paper sorting systems is very complex and expensive. All the systems segregate only two types of paper at a time. Moreover, no image-processing or intelligent techniques are used to extract the features or characteristics from the paper objects. Rahman, Hannan, Scavino, Hussain, and Basri (2008, 2009) have developed a learning-based vision sensing system that is able to separate the different grades of paper using statistical reasoning. The remarkable achievement of this proposed method is the identification and versatility for all grades of paper using only one sensor, which is the best among the prevailing techniques of automated paper sorting systems. Very recently, Ramasubramanian, Venditti, and Gillella (2012) identified the key parameters that must be measured, discussed the design of a sensor system and described the integration of the output from the sensors to interpret the type of sample using a fuzzy inference algorithm. Results show that the sensor system proposed is capable of identifying the samples at 90% accuracy compared with manual identification. The sensor system can be integrated into a conveyor and actuation system for automated sorting.

The handling of sorted recovered paper depends on the following factors: transport conditions of the recovered paper operation; storage conditions at this operation and at the mills; and additional agreements between the dealer of secondary raw material and the mill. Handling may include shredding, which usually is done on office papers and confidential listings. The straightforward approach is to load the sorted recovered paper as loose material onto large trucks and transport it directly to the paper mill. Owing to the large space requirements resulting from the low settled bulk of such paper, it is stored in large quantities only in the paper mill, where appropriate unloading equipment and storage bunkers are installed in accordance with safety regulations. These storage facilities must always be covered by roofs. The trade and supply of loose recovered paper involves almost exclusively recovered paper for deinking. Settled apparent density is approximately 330 kg/m^3. In Germany, most of the recovered paper for deinking is supplied as loose material; the other recovered paper grades are delivered primarily in bales. To produce bales, the sorted recovered paper enters the hopper of a stationary baling press where it is compressed hydraulically at pressures of 8 MPa. This produces compact bales having a typical size of 1.2 m × 1.0 m × 0.8 m, weighing 500–600 kg. These are then tied with wire within the baling press. With larger baling presses and higher pressures, bales of larger size and weight can also be produced. The individual bales can be simply transported and stacked by forklift trucks equipped with bale clamps. Bale density ranges from 250 to 900 kg/m^3, depending on the recovered paper grade and pressing conditions. Additional transport can use almost all modes such as truck, train and ship. For this reason, the trade, transport and storage of recovered paper in bales is a global business (Gottsching & Pakarinen, 2000).

Processing of waste paper in a paper mill requires flexible plants with large capacities suitable for receiving delivered waste paper and simultaneously processing bales and loose waste paper. The required plant capacity range is approximately 100–2000 air-dried metric tonnes/day. Multi-line plants may even exceed this figure. The operation should be fully automatic, requiring no personnel and a minimum of maintenance. The latest design of equipment, combined with the latest control techniques and tailor-made layout concepts, requires the close cooperation of paper producer, consultant and supplier. The target of the design of new waste paper plants is to realise a similar degree of automation as is common for paper machines. Typical equipment for fully automatic waste paper preparation plants includes conveyor techniques for waste paper, automatic bale de-wiring, continuous bale breaking, sorting systems for waste paper, storage systems for large capacities, feeding of drum pulpers, feeding of pulpers, and auxiliaries. With the introduction of this key equipment, it is possible to automate fully the waste paper handling of a paper mill. It is obvious that existing and new plants will require larger capacities and more reliable equipment to provide the required high-quality furnish at low production cost. This needs the close cooperation of the paper producer, consultant and supplier from the very beginning of a project to develop the plant in accordance with local circumstances and the latest available technology.

Recovered paper, especially in bales, is stored in plain areas, under a roof or in the open air, whereas loose recovered paper is stored in special bunkers or enclosed facilities with roofs. Mixed recovered paper that is traded at a low price is often stored outdoors. Inside storage is used more with higher quality and therefore more expensive grades. Recovered paper stored outdoors is exposed to weather such as rain, snow, frost or sunshine.

References

Andersen, S. L. (1997). In S. Abubakr (Ed.), *Recycling* (pp. 3–6). Atlanta, GA: Tappi Press.

Berglund, C. (2003). *Households perceptions of recycling effects: The role of personal motives*, Doctoral Thesis, Lulea University of Technology.

Bialski, A., Gentile, C., & Sepall, O. (1980). Paper sorting apparatus, US Patent No. 4, 236, 676.

Bilitewski, B., Berger, A., & Reichenbach, J. (2001). Collecting systems for paper and cardboard packaging in Germany 2000.

Bilitewski, B., Hardtle, G., & Marek, K. (1994). *Waste management*. Berlin: Springer-Verlag. 609 pp.

Bilitewski, B., Heilmann, A., & Apitz, B. (1994). *Wissenschaftliche untersuchungund begleitung von modellversuchenzur getrennten erfassung graphischerpapiere. Phase I: Erfassung und Bewer-tung vorhandener Sammelsysteme*. Dresden, Germany: INTECUS.

Blanco, A., Carpetti, G., Faul, A., Miranda, R., & Strunz, A. (2006). *CEPI study on collection systems in Europe presented at COST E48 Workshop*, Milan, Italy, 10–11 May 2006.

Bruner, R. S., Morgan, D. R., Kenny, G. R., Gaddis, P. G., Lee, D., & Roggow, J. M. (2003). System and method sensing white paper, US Patent No. 6,570,653.

Confederation of European Paper Industries (CEPI). (2002). EN643 – European List of Standard Grades of Recovered Paper and Board.

Department of Industrial Policy and Promotion (DIPP). (2011). Discussion paper on collection and recycling of waste paper in India.

Doak A. G., Roe M. G., & Kenny G. R. (2007). Multi-grade object sorting system and method, US Patent No. 7,173,709.

Eixelberger R., Friedl, P., & Gschweitl K. (2003). Method and apparatus for sorting waste paper of different grades and conditions, US Patent No. 6,506,991.

Faibish, S., Bacakoglu, H., & Goldenberg, A. A. (1997). An eye-hand system for automated paper recycling. *Proceedings of the IEEE international conference on robotics and automation* (pp. 9–14), Albuquerque, New Mexico.

Garces, A., Lafuente, M., Pedraja, M., & Rivera, P. (2002). Urban waste recycling behavior: Antecedents of participation in a selective collection program. *Environmental Management*, *30*(3) (on-line).

Gottsching, L., & Pakarinen, (2000). Papermaking science and technology, Book 7: *Recycled Fiber and Deinking*. Fapet OY, Finland.

Goyal, H. (2007). Paper grades. <http://www.paperonweb.com/ppmanf.htm>. Accessed June 2012.

Grossman, H., Bobu, E., Stawicki, B., & Miranda, R. (2007). Factors influencing the competitiveness of paper and board recycling in Europe. *International symposium: Present and future of paper recycling – technology and science*, Bilbao, Spain, 24–25 May 2007.

Grubbs, M., Kenny, G. R., & Gaddis, P. G. (2001). Paper sorting system, US Patent No. 6,250,472.

Heinz, G. K. (1998). Method for sorting waste paper, European Patent, EP0873797.

Hottenstein, F. A., Kenny G. R., Friberg T., & Jackson M. (2000). High-speed automated optical sorting of recovered paper. *Proceedings of the TAPPI recycling symposium*, Vol. 1 (pp. 149–158), Atlanta, GA.

Khalfan Z., & Greenspan S. (2006). Optical paper sorting method device and apparatus, US Patent No. 7,081,594.

Kollmuss, A., & Agyeman, J. (2002). Mind the gap: Why do people act environmental and what are the barriers to pro-environmental behavior? *Environment Education Research*, *8*(3), 239–260.

Moore, G. (2002). Making the most of the urban forest, pulp and paper international, May 2002.

Mulligan, D. B. (1993). Sourcing and grading of secondary paper. In R. J. Spangenberg (Ed.), *Secondary Fiber Recycling*. Atlanta, GA: Tappi Press. (Chapter 8).

Putz, H. J. (2006). In H. Sixta (Ed.), *Recovered paper and recycled fibers, handbook of pulp* (pp. 1147–1203). KGaA, Weinheim: Wiley-VCH Verlag GMbH & Co.

Rahman, M. O., Hannan, M. A., Scavino, E., Hussain, A., & Basri, H. (2008). Development of an intelligent computer vision system for automatic recyclable waste paper sorting Engineering Postgraduate Conference (EPC) 2008, Universiti Kebangsaan Malaysia, 21–21 October.

Rahman, M. O., Hannan, M. A., Scavino, E., Hussain, A., & Basri, H. (2009). An efficient paper grade identification method for automatic recyclable waste paper sorting. *European Journal of Scientific Research, 25*(1), 96–103.

Rahman, M. O., Hussain, A., Scavino, E., Basri, H., & Hannan, M. A. (2011). Intelligent computer vision system for segregating recyclable waste papers. *Expert Systems with Applications, 38*(8), 10398–10407.

Ramasubramanian, M. K., Venditti, R. A., Ammineni, C. M., & Mallapragada, M. (2005). Optical sensor for noncontact measurement of lignin content in high-speed moving paper surfaces. *IEEE Sensors Journal, 5*(5), 1132–1139.

Ramasubramanian, M. K., Venditti, R. A., & Gillella, P. K. (2012). Sensor systems for high-speed intelligent sorting of waste paper in recycling progress in paper recycling. *Tappi Journal, 11*(2), 33–39.

Ringman, J. & Leberle, U. (2008). Recycling facts. <http://www.cepi.org> January, 2012.

Sandberg, N. H. (1932). Sorting device for waste paper, US Patent No. 1,847,265.

Sigsgaard, T., Malmros, P., & Nersting, L. (1994). Respiratory disorders and atopy in Danish refuse workers. *American Journal of Respiratory and Critical Care Medicine, 149*(6), 1407.

Stawicka A (2005). Non-paper fraction in recovered paper. Present and future methods of reduction, Master thesis, Poznan. Poznan University of Technology.

Stawicki, B., & Barry, B. (2009). The future of paper recycling in Europe: opportunities and limitations. <www.cost-e48.net/10_9_publications/thebook.pdf>. Accessed January 2012.

Venditti, R. A., Ramasubramanian, M. K., & Kalyan, C. K. (2007). A noncontact sensor for the identification of paper and board samples on a high speed sorting conveyor. *Appita Journal: Journal of the Technical Association of the Australian and New Zealand Pulp and Paper Industry, 60*(5), 366–371.

4 Process Steps in Recycled Fibre Processing*

4.1 Introduction

Paper recycling has been practised for many decades in numerous countries around the world. The paper industry is among those that almost perfectly meet the expectations of society for the sustainability of their raw material base, the environmental compatibility of their processes and the recyclability of their products, and it has done so for a long time. The recycling process is becoming increasingly efficient, with rapid developments being made in deinking processes for the reuse of secondary fibres. Research continues and improved methods of deinking are being introduced (Borchardt, 2003). Currently more than half of all paper is produced from recovered papers. Most recovered papers are used to produce brown grade papers and boards. However, in the past two decades there has been a substantial increase in the use of recovered papers to produce, through deinking, white grades such as newsprint, tissue and market pulp. Deinking is a sophisticated recycling process (Borchardt, 1997). High-grade papers can be produced by using this technique. The components that cause a reduction in brightness, i.e. the inks, are removed, and the additives used during printing and converting are removed. From the recycling point of view these additives are contaminants. They include various types of adhesive (binding materials, labels and tapes, etc.) staples, plastic films, inks, varnishes and all the components of the pulp that cannot be used to produce paper. In some cases, fillers must also be removed. In Tables 4.1 and 4.2, the characteristic particle size and specific gravity of some typical contaminants are presented (Holik, 2000).

The important steps involved in recycling are – production of recycled pulp from recovered papers and manufacturing a paper by using this pulp alone or mixed with other pulps which can be virgin or recycled. The second step is basically not very different from the production of paper from virgin pulp but this is not the case for the first one. The techniques used for the production of deinked pulp (DIP) are completely different than those used for the production of pulp from wood. (Bajpai, 2006; Ben & Dorris, 1999; Bennington & Wang, 1999; Crow & Secor, 1987; Dash & Patel, 1997; Fabry & Carre, 2003; Holik, 2000; Johnson & Thompson, 1995; Moss, 1997; Seifert & Gilkey, 1997).

Recycling technology is the combination of the various treatments performed to produce a pulp from recovered papers and to clean it for its use on a paper machine to produce paper. The deinking technology includes all the main steps of the recycling technology, but special treatments are used to remove the ink. Deinking is

*Some Excerpts taken from Bajpai Pratima, "Advances in Recycling and Deinking", 2006 with kind permission from PIRA International UK.

Recycling and Deinking of Recovered Paper. DOI: http://dx.doi.org/10.1016/B978-0-12-416998-2.00004-0

Table 4.1 Specific Gravity of Contaminants in Recovered Paper Processing

Type of Contaminant	Specific Gravity (g/cm^3)
Metal	2.7–9.0
Sand	1.8–2.2
Fillers and coating particles	1.8–2.6
Ink particles	1.2–1.6
Stickies	0.9–1.1
Wax	0.9–1.0
Styrofoam	0.3–0.5
Plastics	0.9–1.1

Source: Based on Holik (2000)

Table 4.2 Particle Size of Contaminants in Recovered Paper Processing

Type of Contaminant	Particle Size (μm)				
	<1	<10	<100	<1000	>1000
Metal					✓
Sand		✓	✓	✓	✓
Fillers and coating particles	✓	✓	✓		
Ink particles	✓	✓	✓	✓	✓
Stickies	✓	✓	✓	✓	✓
Wax	✓	✓			
Styrofoam				✓	✓
Plastics			✓	✓	✓

Source: Based on Holik (2000)

performed in two steps. The first step involves ink detachment from the surface of the disintegrated fibres, which is performed during pulping (slushing). In the second step, the detached ink particles are removed from the pulp slurry by washing or flotation. Ink detachment is performed by the combination of mechanical forces (agitation during slushing) and chemical action due to chemical added in the pulper, which contributes to the disintegration of recovered paper. The chemical environment can be controlled by alkaline agents or can be neutral. Alkaline deinking is being widely used and is considered more efficient for ink detachment than neutral deinking. However, it generates significant chemical oxygen demand, which is caused by the dissolution of carbohydrates and organic additives present in the fibrous material.

The equipment and chemicals required for selection of deinking processes are determined by the type of ink present in the waste paper. The size of ink particles to be removed becomes the primary criterion. Figure 4.1 shows the different deinking processes to be adopted based on the particle size of the ink. Washing is most effective for removing small particles (<10 μm), flotation for the medium-sized particles (10–100 μm) range. For removal of large ink particles (>100 μm), screening and centrifugal cleaners are used (McCool, 1992, 1993; McCool & Silveri, 1987).

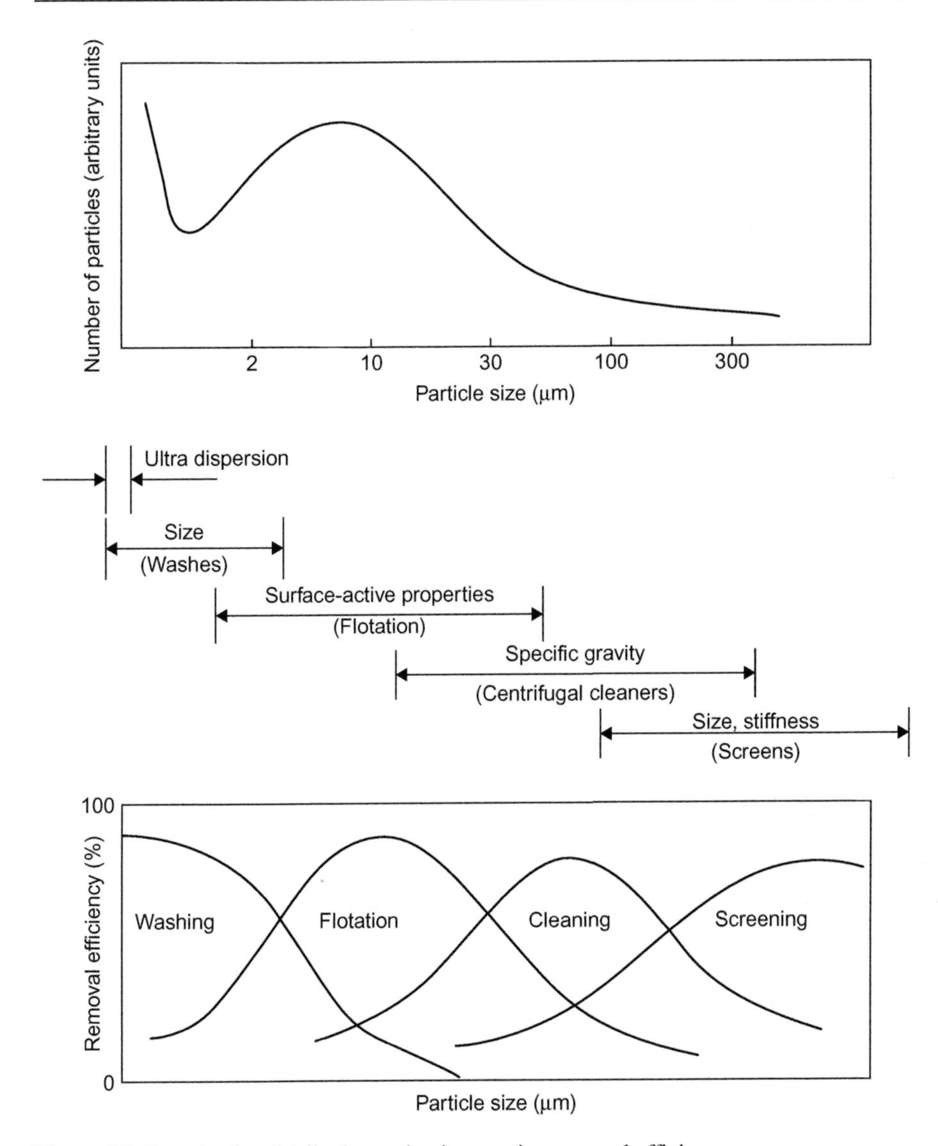

Figure 4.1 Particle size distribution and unit operation removal efficiency.
Source: Reproduced from Ackermann et al. (2000) with permission from Fapet Oy, Finland.

Mixed waste papers present technical and economic challenges to the paper recycler. Of the wide variety of fibres and contaminants present in the paper stock, toners and other non-contact polymeric inks from laser-printing processes are the most difficult to deal with. These inks are synthetic polymers that do not disperse readily during conventional pulping processes, nor are they readily removed during flotation or washing. So, recycled papers contaminated with non-contact inks have a lower value. To process mixed office waste (MOW), deinking systems have become more complex in recent years, with advances in the type and increases in the number of stages of process equipment used. Most systems use several sets of flotation cells, and at least one dispersion step mainly to address the non-impact printing inks. Several systems even have three sets of flotation and two sets of dispersion and/or kneading equipment. Flotation cell design and dispersion technology have advanced in the 1990s; screening technology offers fine screen slots ranging down to 0.15 mm or even 0.10 mm in width. New types of more efficient reverse cleaner have been introduced, as well as improved washing technology for better removal of dispersed ink and ash. Water clarification is also an important part of the deinking process in these systems, balancing yield and overall water use against contaminant.

4.2 Process Steps and Equipment

4.2.1 Pulping

Pulping is the first step. The purpose of pulping is to defibre the paper, detach ink from fibres and produce dispersed ink particles of the size and geometry that can be efficiently removed in subsequent ink removal steps. Deflaking of typical papers used for recycled newsprint production is rapid even at neutral pH (Broullette, Daneult, & Dorris, 2001). In laboratory flotation deinking of a 70:30 old newspaper (ONP):old magazine (OMG) blend, the most significant variables in producing the highest DIP brightness and lowest effective residual ink concentration (ERIC) values were the highest pulping consistency and pH, the lowest pulping temperature and the highest flotation temperature and longest flotation time (Borchardt, 2001). Generally consistent results have been reported by other researchers (Fabry, Carré, & Crémon, 2001).

Excessive ink dispersion during pulping should be avoided through careful choice of pulping conditions and process chemicals (Borchardt & Schroeder, 2002). Smaller ink particles have a greater tendency to redeposit on fibres than larger ones. The very small ink particles produced when pulping paper printed with water-based (flexographic) inks are extremely difficult to remove (Pirttinen & Stenius, 2000) and can deposit inside, the fibre lumen. The same amount of ink present in DIP as very small particles has a greater adverse effect on brightness than when present as larger ink particles (Walmsley, Yu, & Silveri, 1993). Ink particles less than 2.5 μm in diameter have the most deleterious effect on brightness (Pirttinen & Stenius, 2000). Flotation, which is now used more widely than washing as the primary ink removal unit operation, is more effective in removing somewhat larger ink particles (greater than about 20 μm in diameter). Water recycling can return dispersed small ink particles to the process, resulting in increased ink redeposition and lower pulp brightness. Both the rate of ink detachment from fibre and the rate of ink redeposition during pulping are first-order processes (Bennington & Wang, 2001). The lower the shear factor during pulping, the slower are both ink fragmentation and ink redeposition and the more ink is removed in flotation (Fabry, Roux, & Carré, 2001). Operating at lower shear rates and at higher consistency, the primary purpose of drum pulping is to limit fragmentation of stickies, plastics and other contaminants. Today drum pulpers are used in most new deinking plants and modernisation projects (Patrick, 2001). Reducing the pulping pH to near-neutral values can reduce newspaper ink dispersion (Pirttinen & Stenius, 2000; Pélach, Puig, Vilaseca, & Mutjé, 2001). Image analysis indicates that increasing the sodium hydroxide concentration (and thus pH) increases the formation of very small ink particles. Interestingly, although increased pulp consistency at alkaline pH increases brightness, better brightness is obtained at neutral pH at lower pulp consistency (Ackermann, Putz, & Góttsching, 1999). Sodium sulphite has been suggested for use in neutral deinking to replace sodium hydroxide, sodium silicate and EDTA in the pulper (Lapierre et al., 2002). The primary variables determining ink removal efficiency were pulper conditions: pulp consistency, temperature and the interacting effects of pulper temperature and consistency and of pulper temperature and water hardness.

Paper transportation and storage temperatures can reach 50°C during summer months (Castro, Daneault, & Dorris, 1999) and accelerate ink autoxidation, increasing the associated adverse effects on deinking. Extensive studies in Canadian, US and European newspaper deinking mills have this 'summer effect' on ink detachment, fragmentation and removal (Carré, Magnin, Galland, & Vernac, 2000). Lower DIP brightness appears to be primarily caused by higher ageing temperatures causing stronger ink adherence to the cellulose fibres because of deeper penetration of the ink into the fibres, oxidation leading to some chemical bonding of the ink vehicle or binder resin to the fibres, and heat-promoted evaporation of the ink vehicle. To deal with these adverse effects, shorter pulping times at higher pulping temperatures using increased charges of alkaline chemicals and deinking agent in the pulper are recommended (Carré et al., 2000). Deinking becomes more difficult with ONP age as cross-linking of binder resins gradually makes inks more difficult to detach from fibres. (Le Ny, Haveri, & Parkarinen, 2001; Kaul, 1999). The autoxidation chemical reaction appears to be responsible for ink chemical changes that reduce deinking efficiency (Castro et al., 1999). Addition of antioxidants to offset ink formulations can retard autoxidation. Conventional vegetable-oil-based inks are generally more susceptible to autoxidation processes than mineral oil-based inks and exhibit a greater tendency to form very small ink particles. (Carré et al., 2000; Dorris & Pagé, 2000). However, some vegetable-oil-based inks designed to have a low autoxidation rate exhibit deinkability comparable to that of mineral oil-based inks (Haynes, 2000).

Various steps involved are as follows: feeding the system with a pre-determined rate; break down of the raw material into individual fibres; removal of solid contaminants such as foils, stickies and printing ink from the fibres; removal of solid contaminants at an early stage that might otherwise break down into excessively small particles; and mixing of process chemicals into the suspension such as deinking chemicals or bleaching agents. The first step is often limited to coarse slushing and heavy particle separation to save energy. Fine disintegration with a disk screen or deflaker usually follows.

For recovered paper with very high wet strength, a disperger is usually necessary to break down the flakes into individual fibres. Using thermal, chemical or both types of treatment during the pulping process is often more cost-effective (Holik, 2000). Rapid and complete wetting of the incoming recovered paper is extremely important for efficient slushing. Not much information is available about slushing mechanisms. It has been proposed that acceleration, viscosity and clinging such as on the rotor generate deflaking forces. Sudden motion and resultant inertia apply acceleration forces to the flakes. To apply viscosity forces, the suspension of solids in water must have a velocity difference with the flakes. Viscosity forces are responsible for the fibre-to-fibre friction that causes flake disintegration due to shear stress. High shear forces exist at the interfaces between slow moving stock in the pulper and the individual pulp jets that emanate from the rotor at high velocity. The flakes wrapping themselves over the rotor, screen plate holes or impact bars cause clinging forces. To obtain a defibring effect, these forces must act in pairs and in opposite directions.

Various kinds of device are used for pulping. These are low consistency pulpers, medium consistency pulpers and drum pulpers. The choice of pulper depends on various parameters including efficiency and energy consumption in defibring kinetics to minimise residual paper flakes and minimising the breakup of contaminants to improve their removal efficiency. Today, most secondary fibre raw material for deinking plants is repulped, in batch pulpers or continuous drum pulpers, at 12–18% consistency. Usually in a deinking line, high-consistency pulpers are preferred. This is because of higher concentrations of a given charge of chemicals, steam savings, and because of gentle operation, which tends to preserve contaminants in the largest possible form. Higher consistency also tends to generate fewer fines than low consistency operation. Two devices are commonly proposed: an high consistency (HC) batch pulper and a drum pulper. On the one hand, the batch pulping is said to improve the ink detachment during this stage by applying high energy levels. On the other hand, the drum pulping keeps plastic, stickies and contaminant intact because the forces involved in such a type of pulper are gentle (successive drops). The main advantages and drawbacks of each pulper type are reported in Table 4.3. However, discussions and interest in the deinking performance differences between drum pulping and HC batch pulping have continued since the first introduction of the drum pulper in the deinking line over 30 years ago. Klar and Burkhardt (2000) reported the case of a tissue mill where a drum pulper replaced a conventional HC pulper. The final pulp characteristics (shives, dirt count and stickies) were very similar. The main differences observed were the following: the HC pulper induced lower speck contamination at the end of pulping stage; the drum pulper induced larger particles of stickies at the end of the pulping stage (this would induce a more efficient removal of them). In a drum pulper, the dwelling time is constant whereas it can be adjusted in the HC pulper. Merza and Haynes (2001) suggested that the pulping time flexibility of batch pulper improved deinking performance compared with the drum pulper. A pulping time from 30 to 60 s and a dual deinking agent system gave the best deinking pulp quality and minimised the 'summer effect' according to those authors (it must be noticed that the conventional HC pulping time in a North American deinking mill is about 20 min). Indeed, the work performed on the pulping stage recommended a strong reduction in pulping time in the pulper (Ben & Dorris, 1999; Bennington, Smith, & Sui, 1998; Fabry, 1999; Fabry, Carré, et al., 2001; Fabry, Roux, & Carré, 2000; Haynes, 1999; Merza & Haynes, 2001): such a strategy allows reduction of the

Table 4.3 Advantages and Drawbacks of the Conventional High Consistency Pulper

Advantages	
Drum Pulping	**HC Batch Pulping**
50% less energy consumption	Increase in fibre-to-fibre interaction
Heavy contaminant removal	Variation in pulping time
Gentle action	Controlled ink fragmentation
High consistency	Small space requirements
Continuous operations	High or low consistency
Self-contained detrasher	
Drawbacks	
Drum Pulping	**HC Batch Pulping**
Extended pulping time for defibring	High energy cost
Increase in ink reattachment	Contaminant removal
Higher initial costs	Detrasher limitations
Space requirements	Maintenance costs

Source: Based on Merza and Haynes (2001)

ink fragmentation and redeposition during pulping, which improves ink removal by flotation (Fabry, 1999; Fabry, Carré, et al., 2001).

High-consistency batch pulpers are compact machines, which typically hold 10 tonnes of paper per batch. The maximum available batch size is about 20 tonnes. After adding the secondary fibre raw material, water and deinking chemicals, the pulping cycle lasts for 15–30 min to achieve complete defibration of the furnish. Ink particles are loosened from the fibre or coating, and are now suspended in the slurry. At the end of the batch cycle, pulp is extracted through a perforated plate in the pulper bottom or through a large side-opening in the pulper wall. In the latter case, a discharge module, with a perforated extraction plate usually about 6 mm in diameter, would affect the first separation of coarse contaminants. High-consistency drum pulpers operate on a continuous basis. They provide uniform retention time for defibring as the charged paper gradually moves along the length of the drum. Near its end zone, the drum is perforated to allow dilution and extraction of the paper slurry, while gross contaminants are discharged at the axial end. This type of pulper is often chosen because contaminants seem to be discharged from it in larger form. It is debatable whether this feature is truly significant, because the other pulping systems are also gentle and reject large particles.

The development and rapid implementation of drum pulpers into the recycling process has dramatically improved both efficiency and quality in recent years (Borchardt, 2003; Patrick, 2001). Before drum pulpers, recycling plants primarily used low-consistency/high-shear vat pulpers with low-positioned, high-speed rotors that tended to disintegrate anything they came in contact with, including plastics,

Figure 4.2 Metso OptiSlush drum pulper.
Source: Reproduced with permission from Metso.

stickies, etc. This technology, still used today, was followed by high-consistency vat pulping that is much more gentle and creates more fibre-to-fibre rubbing action. Drum pulpers, compared with low-consistency vat pulpers that beat contaminants to as small a size as possible so they can be subsequently washed out, are a continuous process and operate at high consistency. Drum pulpers are even more gentle than their predecessor high-consistency vat units, having 'shelves' or partitions inside that lift and drop the pulp many hundreds of times as it passes through the continuous horizontal tube. It is the impact of this falling action and the accompanying fibre-to-fibre rubbing that gently accomplishes pulping in a drum unit. Drum pulping is so gentle that things such as glued magazine backs remain intact and are easily removed in subsequent screening stages. Also, some trials have shown that ink particles remain larger with drum pulping, making them easier to remove in flotation deinking cells. Drum pulpers use a very similar chemistry as vat pulpers. Although drum pulpers have been around for a decade or more, they have gained acceptance and popularity only in the past few years. Today, drum pulpers are used in almost all new deinking plants and modernisation/upgrade projects.

Figures 4.2–4.5 show a drum pulper and high-consistency pulper from Metso and Andritz.

4.2.2 Deflaking

Deflaking machines are normally used, which typically defibre flakes by fibre-to-fibre rubbing or hydraulic shear rather than by close bat-to-bar clearance. Deflaking breaks down any fibre lumps or bits of undefibred paper into individual fibres. The flakes to be broken down contain sized, coated and wet strength grades. Disk screens also have a certain deflaking effect but to a lesser extent. Both deflakers and disk screens can handle stock consistencies of 3–6%.

For recovered paper that is more difficult to break down, discontinuing slushing at a high flake content and then using a deflaker is more cost-effective. With decreased flake content, the energy demand for further deflaking in the pulper increases rapidly. Significant energy savings are possible, by interrupting slushing at a suitable point and continuing with a deflaker. The logical outcome of this is a processing

Figure 4.3 Andritz FibreFlow drum pulper.
Source: Reproduced with permission from Andritz.

Figure 4.4 Metso OptiSlush HC pulper.
Source: Reproduced with permission from Metso.

sequence comprising coarse deflaking in a pulper followed by fine deflaking with a deflaker. Another method to save energy with a clean suspension is by screening with subsequent feed forward of the heavily flaked screen reject through a deflaking stage. Stocks that are extremely difficult to break down are usually treated with chemicals at higher temperatures to reduce fibre bonding for easier deflaking. A

Figure 4.5 Andritz FibreSolve FSU high-consistency pulper.
Source: Reproduced with permission from Andritz.

disperger is also useful for efficient flake reduction. Deflaker efficiency depends greatly on contamination of the recycled fibre pulp. Because deflakers have fine-grade fillings for more intensive breakdown of flakes, coarse trash easily blocks them. For this reason, they require protection by efficient upstream cleaning and screening. To save technical outlay and costs, disk screens are often useful instead. Deflakers break down flakes principally by forces induced by wrapping, viscosity, acceleration or a combination of these forces in a similar way to slushing in pulpers. The forces are much higher in deflakers than in pulpers. The probability is also much higher for flakes to break down because they are forced through shear and impact zones in the deflaker fillings. Nevertheless, the deflaking forces are limited. For grades with too much wet strength, higher dispersing forces are necessary. Reducing

Figure 4.6 Andritz deflaker.
Source: Reproduced with permission from Andritz.

flake strength with high-temperature treatment and chemicals is often more cost-effective. The chemicals used for this are alkaline or acidic, depending on the type of wet strength agent.

Screens, particularly disk screens, also have a certain deflaking effect. The greatest deflaking effect is observed in screens with additional impact bars on the disk near the rotor blade ends. Disk screens have primary use for coarse screening of suspensions with a high flake and debris content. The advantage of disk screens over deflakers is that they are insensitive to debris that they separate simultaneously; their disadvantage is a limited deflaking effect. This is adequate for mixed recovered paper processing in the production of packaging paper and board. It reduces reject rates in the secondary stages of coarse screening.

Figure 4.6 shows a deflaker from Andritz.

4.2.3 Refuse Removal

This is an important step to remove contaminants from deinking systems. Most plastic and other coarse contaminants are removed at this step. Systems using batch pulpers all have some sort of removal device, which is generally a horizontal vessel with extraction holes from 6 to 10 mm in the extraction plate. Approximately 2% of the incoming material is rejected at this point of the system. Additional reject handling equipment, such as rotating drum screens and refuse presses, is associated with the removal equipment to dispose of the coarse rejects effectively.

4.2.4 High-Density Cleaning

High-density cleaning is used to remove staples and other heavy materials from the waste paper stream. A vertical cyclone cleaner is used to remove this material, which generally is less than 1% of the total incoming feed material. Separation efficiency is dependent on the feed consistency, pressure drop and other factors. In the past several years, the slot size of screens has decreased from the typical 0.008 in., with 0.004 in. becoming common in new and modernised mills. Forward centrifugal cleaners are particularly useful in removing large toner ink particles. Reverse centrifugal cleaners are being less used in deinking mills for two reasons. First, the density of stickies and other contaminants has gradually increased to near that of water. Second, screens are removing contaminants more effectively. Given the limited benefits and the energy cost of running reverse cleaners, some mills have shut them down. The primary exception to this is old corrugated container (OCC) mills, which have a high concentration of lightweight contaminants such as waxes.

4.2.4.1 Coarse and Fine Screening

The task of screening in recycled fibre lines is to remove the impurities as early as possible in process with maximum line capacity and yield. Screening of DIPs should be separated into coarse and fine screening steps. The former is quite well established through standard secondary fibre technology, whereas fine screening saw great technological progress in the early 1990s. Several coarse screening steps are applied after transforming the raw material into a slurry. The first step may be in the pulper itself, where acceptable pulp flows through coarse extraction holes and large refuse particles are retained. A similar operation occurs in pulper extraction or discharge modules, which are often placed after the high-consistency pulpers. They operate batch-wise and lend themselves to cycled operation, including a washing step to minimise the fibre content of the rejects stream. The accepted slurry from these steps is usually collected in a chest from where a continuous process begins. This uses coarse screens, which feature hole-perforated screening elements, with hole diameters ranging from 2.0 to 1.0 mm. Today, such screens operate easily and reliably at consistencies between 2% and 4%. Coarse screens sometimes have slotted cylinders, with slot width ranging from 0.2 to 0.3 mm.

In deinking, today's challenge is to remove the smallest possible particles. Many are pliables or stickies, which are rounded or spherical in shape; sometimes they are rolled into small balls. Screens to remove such particles feature cylinders with very small slots, 0.15–0.10 mm in width. The industry might use even smaller slots; however, these would cause too much rejection of fibres and are still impractical. Although the trend is towards such very small slots, the available published data is limited. Generally, however, the benefit of small slots to reduce the number and size of residual sticky contaminants in the final pulp is well accepted. All fine slotted screens with cylinders of 0.15–0.10 mm slots have strongly profiled cylinder inlet surfaces. The rotor operates on the inlet side of the screen plate to create turbulence and to prevent plugging and fractionation. Although some screens operate with radially inward flow, most use an outward flow. Cylinders can be made from plates into which the inlet profile accepts reliefs, and the actual fine slots are milled, in successive steps. Alternatively, slots are created by the wedge wire principle in which precision-drawn triangular wires are welded or clamped together, with spaces – the slots – between them.

Screening and cleaning have improved radically. The centrifugal cleaner units are typically smaller in diameter than their counterparts of a few years ago, and are designed to remove larger-sized contaminants generated by the drum pulper units as well as to operate at higher consistencies. Later in the process when 'things get thinned out', centrifugal cleaners are smaller to remove sand and other small contaminants. Today's screens represent another major evolutionary development, alongside drum pulpers. About 10 years ago, the state of the art for screening systems was 0.008-in.-wide slots. Since then, that technology advanced to 0.006 in. slots and then to 0.004 in. slots. This represents a significant step forward during the past 10 years. Reducing slot widths by half allows many more contaminants to be taken out of the recycled pulp stream. Also, finer screening slots act to improve overall efficiency and 'work' of the system. With these improved screens, there is no need to run a rotor so fast that it tears contaminants into little pieces. Use of 0.004 in. screens in sorted office paper (SOP) deinking plants has dramatically improved plant operating efficiency.

For centrifugal cleaners, a fairly recent trend has been the steady disappearance of reverse flow cleaners in recycling plants. Reverse cleaners were used to take out lightweight contaminants. This trend has occurred mainly because today's plastics and stickies have evolved to near neutral in relation to the specific gravity of water – with some possibly being on the heavier side. Some ONP and SOP plants that have reverse cleaners are now shutting them down because they are not worth the energy cost to run them. However, a few OCC plants still use reverse cleaners extensively because their fibre stream continues to have a concentration of lightweight contaminants such as waxes. Improvements in screening have been instrumental in allowing plants to cut back reverse-flow cleaner operations. These screening improvements are in fact among the major recycling process developments in recent years. They are the result, mainly, of vendors doing a better job of responding to the industry's needs, and getting better over time.

Figures 4.7–4.9 show a Metso HC Cleaner, an Andritz ModuScreen F for coarse screening, an Andritz ModuScreen C&F screen for coarse and fine screening operations and a Metso OptiScreen ProFS Fine screening system.

Figure 4.7 Metso HC cleaner.
Source: Reproduced with permission from Metso.

Figure 4.8 (A) Andritz ModuScreen F for coarse screening. (B) ModuScreen C&F screen for coarse and fine screening operations.
Source: Reproduced with permission from Andritz.

4.2.5 Flotation

Flotation is an important step in removing Tappi (Technical Association of the Pulp and Paper Industry) (over 230 μm), sub-TAPPI visible dirt (over 50 μm) and sub-visible particles (under 50 μm). Flotation consists of removing hydrophobic particles (particularly ink particles), by encouraging the ink particles to coalesce with air bubbles and floating them up to the surface for removal. Flotation aids are used to improve the attachment of ink particles to air bubbles.

Figure 4.9 Metso OptiScreen ProFS Fine screening.
Source: Reproduced with permission from Metso.

Flotation cells designs have evolved to provide such features as air injection nozzles that improve the size distribution of air bubbles. The improved brightness provides increased operating flexibility by reducing the amount of OMG that must be added to ONP. The advantage of reducing the furnish OMG content is the increased fibre yield that results. When flotation deinking with fatty acids, the presence of calcium ions increased both water-based ink and toner ink removal efficiency by increasing ink attachment to air bubbles (Acosta, Scamehorn, & Christian, 2003). Mill studies confirmed that the single most important parameter affecting ink removal efficiency and filler yield was water hardness (Moe & Røring, 2001). Also, high surfactant dosage to the pulper and a high pH (a high alkaline dosage combined with a low silicate dosage) improved flotation ink removal. In laboratory studies, in addition to flotation time, the primary variables determining ink removal efficiency were pulper conditions: pulp consistency, temperature, and the interacting effects of pulper temperature and consistency and of pulper temperature and water hardness (Borchardt, 2001). Visualisation studies of inks on bubble surfaces can improve understanding of the effects of ink chemistry, flotation cell design, process chemicals and process conditions on ink attachment to and detachment from air bubbles. Visualisation studies using toner inks (Thompson, 1997) and models of offset and flexographic inks (Davies & Duke, 2002) have been reported. These studies used quite large, stationary air bubbles. Unpublished results indicate that it is possible to visualise ink particles on air bubbles during flotation in laboratory cells and ink dispersion during pulping using a focused beam reflectance measurement instrument. With a longer probe, it should be possible to visualise these phenomena in commercial mill pulpers and flotation cells. Atomic force microscopy is another advanced technique used to study interfacial forces and other interactions in deinking systems (Pletka et al., 1999; Thompson, 1997).

In flotation laboratory tests, increasing air flow resulted in a modest increase in ink removal efficiency (Pelach et al., 2001). Injecting silicone oil with air into flotation cells can result in higher flotation brightness gains (Gomez, Acuña, Finch, & Pelton, 2001). This may be due to either increased gas hold-up resulting in smaller or slower rising air bubbles or to increased ink attachment to oil-coated bubble surfaces. Ink autoxidation associated with the summer effect has adverse effects on the removal efficiency of flotation ink (Dorris & Pagé, 2000). The oxidised compounds found in inks after autoxidation are more polar and increase gas hold-up in flotation cells. This inhibits air bubble coalescence; the smaller bubbles increase froth stability resulting in a greater loss of fibres, fines and water in flotation. Brightness drops about 1.4 points for each 10% increase of old telephone directories (OTDs) in ONP:OMG (Cao, Heise, Tschirner, & Ramaswamy, 2001). OTD contains about 20% higher ink concentration than ONP and requires higher water hardness for good flotation deinking efficiency. Using a series of alcohol ethoxylate surfactants, the surfactant with the highest critical micelle concentration produced the highest flotation ink removal efficiency at surfactant concentrations above the critical micelle concentration (Johansson, Pugh, & Alexandrova, 2002).

In the 1990s, flotation gained wide acceptance for all grades of deinking. Traditionally in North America, most of the systems installed in the 1960s and 1970s and into the 1980s had been washing systems. In the late 1980s and 1990s, however, most new deinking systems featured flotation as the primary ink removal device instead of washing, although several new washing systems were installed as well. In addition, several older systems added flotation to an existing washing system. Several commercially important cell designs have been used (Darrin Rhodes, 1998; Eguchi, 2005; Holik, 2000; Seifert & Gilkey, 1997; Serres, Colin, & Grossmann, 1994) including Beloit PDMs, Ahlstrom flotation cell, Black Clawson Flotators, Fibreprep Lamort cells, Voith sulzer EcoCell, Voith E-Cell, Must Cell (multistage, Metso Paper/Fibre), IIM Flotator, Mac Cell, Prome-Mac, Foamer (Krofta), Kvaerner flotation column, Shinhama Hi- Flo Cell, Sunds Swemac, Sulzer Escher Wyss, MT-Flotator (Ishikawajima Heavy Industries). Figures 4.10 and 4.11 show flotation cells from Metso and Andritz.

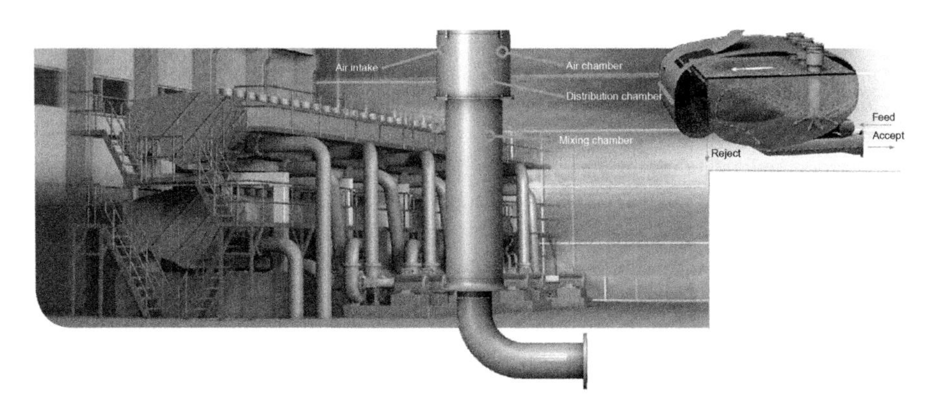

Figure 4.10 Metso OptiCell flotation cell.
Source: Reproduced with permission from Metso.

Most higher-grade deinking systems have two loops, with flotation in both of them; several cases even have three flotation steps. In general, flotation has been shown to be less effective on larger TAPPI-visible ink particles than on smaller ones. The factors that affect flotation are as follows:

Water hardness

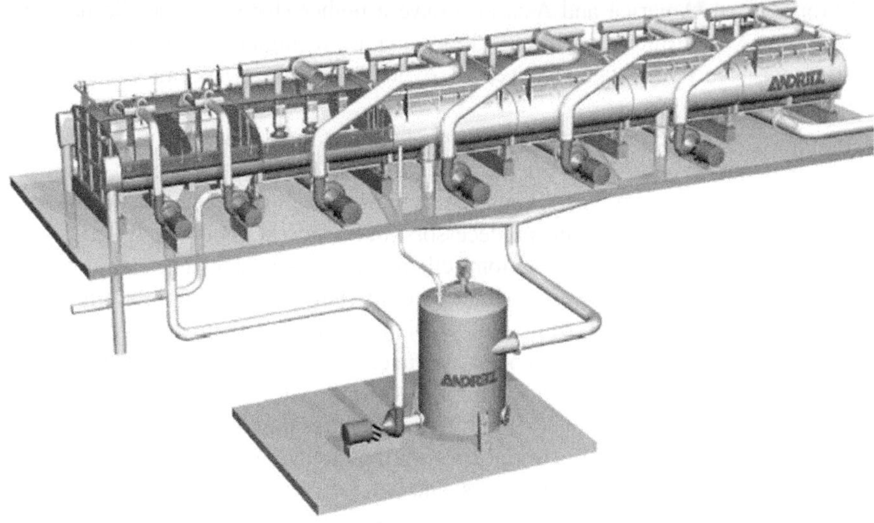

Figure 4.11 Andritz SelectaFlot™ Flotation cell.
Source: Reproduced with permission from Andritz.

Maximum flotation is achieved when all hardness agents are flocculated and a slight excess of free active soap is present at the surface. For effective ink removal, the calcium ion level should be 150 ppm (as $CaCO_3$).

pH

To get adequate foam stability, the pH should be maintained above 8.5.

Consistency

The flotation cell consistency should be kept below 1–2% because yield suffers when consistency increases.

Particle size

Although flotation is less sensitive to particle size than washing, it still plays a vital role in deinking. For the particles below 5 μm, Brownian movement counteracts the adhesion of particles to the air bubbles.

Dual time

A dual time of 8–10 min is usually sufficient for flotation deinking depending upon the flotation cell design.

Temperature

A temperature around 40–45°C is recommended to get a brighter pulp.

The first deinking system working according to the flotation principle was installed in the late 1950s in the USA. The first European operation started production at a Dutch paper mill in 1959. This unit had a capacity of 10 tonnes/day for the production of hygienic papers. The largest share of the global deinking capacities is in Europe; North America and Asia also have a higher share whereas the deinking capacities in Latin America, Africa and Oceania are negligible. Globally, the use of DIP in newsprint dominates, followed by hygienic papers and printing and writing papers.

Today, every modern deinking plant uses both flotation and washing deinking to get the best out of both. Many of the older, less effective flotation cells have been shut down in the past few years, replaced by newer, high-efficiency models that on the surface might look like their predecessors, but inside are completely redesigned units. For example, one new flotation cell incorporates a redesigned air injection nozzle that significantly improves bubble size distribution. This important improvement is not readily visible from the 'outside'. Although there have been some improvements in washing technologies, advances in flotation deinking have acted to reduce overall dependence on washing in general, with resultant gains in yield. In fact, some mills that still use side hill screens are mainly using them for fibre recovery. Yield is critically important to deinking/recycling plants today, driven by the current high cost of recovered papers, and will be even more crucial in the future. Evolved flotation technologies along with today's improved surfactants and other recycling chemicals have allowed greater ink removal so that plants do not have to do as much washing. In addition to using the best flotation technology available, mills in the recent past were having to use highly efficient washers to remove ink that still remained in the pulp. The unfortunate thing about high-efficiency washers is that they remove most of the ink but take a lot of fibre with it. Anytime the pulp is washed extensively, the fibre is lost. With improved flotation technologies available today, mills can switch to less efficient washers and reap a big gain in yield. In fact, many ONP recycling plants are now using vacuum washers, which would not have been possible without the newer, high-efficiency deinking cells. These washers, being used just for final side washing, retain many more fillers and fines, and thus yield is greater. The fact is that with today's flotation technologies, washing does not have to be as good as it once was. Most SOP pulp producers and even some ONP producers use multiple deinking/washing lines, generally in tandem. The SOP process can be especially complex, and quite often involves two or even three flotation deinking/washing lines in tandem. Because of the excessive washing in these kinds of processes, North American SOP yield is typically in the 60–70% range, compared with ONP plants in the 80% range and OCC even higher. In fact, some OCC plants are at or approaching system closure and enjoy extremely high yield.

4.2.6 Washing

Because relatively high yield losses are associated with washing, reducing yield loss would appear to be a valuable area of research. With the increased ink removal effectiveness of flotation, vacuum washers, which are less efficient in removing ink but provide significantly lower yield losses, are increasingly being used. Advances in washing technology include the widespread use of high-speed belt washers, the introduction of pressure screen washers and advances in drum washers. Newsprint systems traditionally used counter-current washing in North America, compared with a preference for flotation in Europe. Since the 1980s, the trend in the USA and Canada has also swung towards the use of flotation for newsprint systems. Some 'washing only' systems have been installed, and flexo inks have left a role for washing in most systems. Many washing systems clarify only a part of the wash water (filtrate) stream to maintain a relatively high yield in the system, essentially recirculating some ink in the whitewater until it become more concentrated before going to the clarifier. High-speed washers also have varied wire design, wire speed, feed consistency and retention aids as ways of maintaining both yield and quality.

Tissue systems have seen the widespread use of high-speed washers for removal of ash, ink and fines, to increase brightness, strength, paper machine retention and softness of the sheet. The washing step is usually done in two stages to produce pulp with a final ash in the 1.0–1.5% range. The main disadvantage is the low yield of these systems (as low as 60%), depending on the feed furnish being used.

The role of washing in MOW systems has been investigated. Washing has been a controversial step in MOW systems for fine paper and market pulp systems. Several systems have high-speed washers, such as those made by Black Clawson or Voith Sulzer, whereas others rely on more conventional washers, such as the vacuum disc or drum. The latter have the lowest loss and virtually no washing effect. Although gravity drums might be considered a compromise between retention and washing efficiency, long fibre loss is significant. In high-intensity washers, microstickies can be washed out, together with dispersed ink and ash. Certain strategies have been used with high-speed wire washers to increase the fibre retention across the washing step, such as increasing the feed consistency, slowing wire speed, bypassing the clarifier and using tighter wires.

4.2.7 Thickening

Screw presses or belt presses are commonly used for thickening. Presses are mainly used to thicken before a dispersion or kneading step, but are also used to create a water barrier at the end of the system to separate the deinking system from the paper machine water. Belt presses have the advantages of perhaps slightly higher retention than screw presses, less severe plugging problems and perhaps less major maintenance. Screw presses have the advantages of tolerating a wider range of feed consistency and no wire to change, but they require major routine maintenance because the screw flights can wear. Both belt and screw presses have minimal washing effects as their retention is quite good owing to high feed consistencies and slow speeds.

4.2.8 Kneading and Dispersion

This is an important step in the deinking process. It is done with kneaders and dispersers. Kneading and/or dispersion are critical steps in processing MOW to reduce TAPPI and visible ink levels. Reductions in TAPPI dirt of up to 90% have been seen in kneading steps. Also, size reduction of sticky particles has been seen in kneaders and refiners. Both units have power inputs generally in the range 60–80 kW h per tonne (3.0–4.0 hp day per US tonne), and operate with 25–35% consistency. Many systems also use their disperser or kneader as a high-consistency peroxide mixer, getting a better bleaching response at 30% consistency in the disperser than at 10–15% in the pulper.

Kneaders and dispersers break contaminants into smaller particles so they can be more easily removed by other steps in the deinking process. They also render the contaminant invisible. Dispersers can be thought of as refiner devices with small clearances generally rotating at high speeds. Examples are units from Beloit, Krima and Voith Sulzer (the EW type). Kneaders are horizontal, single- or double-shaft units, much slower than the refiners with longer retention times and much larger clearance. Examples are units from Ahlstrom, Black Clawson, Maule, Shinhama and Voith Sulzer (the Voith type).

Kneaders are used for low-speed dispersion and dispersers are used for high-speed dispersion. Kneading occurs in the middle of the cycle before flotation and washing, whereas dispersion occurs near the end before pulp storage. Kneading does not affect freeness, whereas dispersion causes a moderate freeness decrease of between 50 and 100 mL Canadian standard freeness. Kneading separates the ink and contaminants from the fibre, resulting in higher brightness levels, whereas dispersion homogenises and disperses the ink, resulting in low brightness levels. Kneading involves high capital expenditure but low operational costs for high-quality end

Figure 4.12 Metso OptiFiner disperger.
Source: Reproduced with permission from Metso.

results, whereas dispersion involves low capital expenditure and higher operational costs (Cochaux, Carre, Vernac, & Galland, 1997; Galland, Carre, & Vernac, 1998; Niggl & Kriebel, 1997). In the presence of bleaching chemicals, both high-speed dispersion and low-speed kneading improve ink detachment and post flotation efficiency. Use of sodium silicate and caustic soda can prevent ink redeposition on the fibres during kneading.

The disc disperger is used for the dispersion of conventional printing inks, and for deinking newsprint, cardboard, tissue paper, coated paper and linerboard (Niggl & Kriebel, 1997). The kneader is used for laser print dispersion, and for the deinking of coated paper, cardboard and tissue paper. Both run at a temperature of 90°C. The disc disperger can be run at higher temperatures to reduce bacteria and dirt specks, without harming the fibres. The kneader is more cost efficient as it can be run at lower temperatures.

McKinney and Roberts (1997) have reported that dispersion and kneading have a substantial impact on the physical properties of fibres, partly because of their impact on curl. These are common unit operations in recycling processes. Pilot-plant trials were performed to assess the effect of kneading and dispersion on the ease of removal of inks and stickies. A range of physical tests, including one for fibre curl, were also performed on pulp samples. It was found that under the conditions used in the trials, dispersion reduced curl, whereas kneading increased it.

Figures 4.12 and 4.13 show a disperger from Metso and Andritz.

4.2.9 Fine Cleaning

Fine cleaning is done with fine forward cleaners and fine reverse cleaners. Fine forward cleaners are widely used for deinking applications. They operate most efficiently when the operating consistency is low (0.6–0.7%), and when the particles are large and of high specific gravity. Pulp treatment before the cleaners is

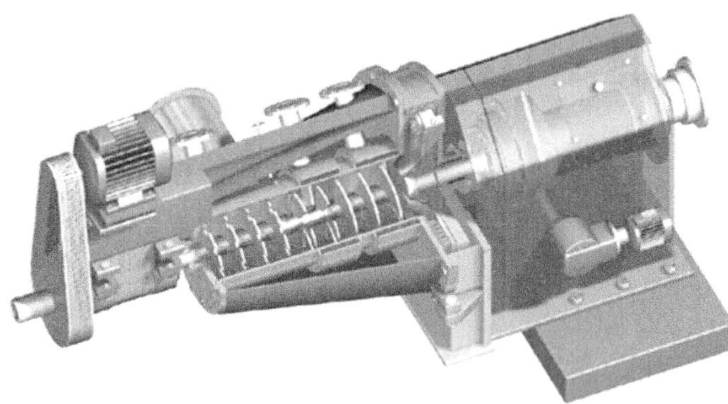

Figure 4.13 Andritz CompaDis™ disperger.
Source: Reproduced with permission from Andritz.

thus important, so as not to reduce particle size. In fact, some technologies try to agglomerate particles chemically, especially printing ink, to make these more easily removable, to the extent that the operating consistency might be raised as high as 1.2%. Unfortunately, many particles are of a specific gravity near that of water, i.e. 1.00–1.10, with only minimal centrifugal effect and correspondingly low removal efficiency in a cleaner. Many stickies are just in this range, thus providing a still unsolved problem. Chemical agglomeration with heavier particles has been tried for stickies. Some success may eventually come from progress in this direction.

True reverse cleaners have a tangential inlet and a central outlet for light rejects at the base of the cone, and an accepts outlet at the apex. Typically, the diameter at the base is 75–100 mm. The volumetric reject rate of such cleaners is very high and light contaminants are carried away in a very dilute high-volume stream. To concentrate them further, subsequent reverse cleaning stages must be arranged. The recommended

Table 4.4 Typical Indicators of DIP for Newsprint, SC and LWC Papers and Tissue

(a) Newsprint	
Raw materials	100% ONP, 50–70% ONP/50–30% OMG
Percentage DIP material in end-product	20–100% DIP
Brightness increase	45–52 → 52–68 ISO
Ash content	8–25% → 5–17%
Yield	80–86%
Energy consumption	300–350 kW h/tonne
(b) SC and LWC Papers	
Raw materials	50–60% ONP, 50–40% OMG
Percentage DIP material in end-product	20–80% DIP
Brightness increase	50–55 → 68–72 ISO
Ash content	20–30% → 5–25%
Yield	70–75%
Energy consumption	350–400 kW h/tonne
(c) Tissue	
Raw materials	100% ONP, 50–70% ONP/50–30% OMG, 100% MOW
Percentage DIP material in end-product	100% DIP
Brightness increase	42–70 → 52–84 ISO
Ash content	8–35% → 2–5%
Yield	55–75%
Energy consumption	350–400 kW h/tonne

Source: Based on Toland (2003)

operating pressure drop for this ranges from 50 to 100 psi (3.5–7 bar). Reverse cleaning systems were used in the 1970s and 1980s. Because of their overall high-energy requirements, they were eventually displaced by through-flow cleaners. Until recently, enticed by the lower energy requirements, the industry has learned to live with the less than satisfactory removal efficiencies of the through-flow and centrifuge cleaners.

4.2.10 Deinking Processes

Deinking lines comprise a combination of the various techniques described previously. The number of stages in the process depends on the quality requirement of the DIP to be produced and on the grade of the furnish. To produce the target pulp, the important parameter is the choice between high capital cost and high operating cost. Older mills include only one deinking loop and try to use sorted recovered papers, whereas newly designed deinking lines for the production of high-quality DIP include two or sometimes three deinking loops and use the lowest possible grades of recovered papers. Table 4.4 shows typical indicators of DIP for newsprint, supercalendered (SC) and lightweight coated (LWC) papers and tissue.

Most modern plants have several deinking loops and use post-deinking after bleaching to produce higher-quality pulp for papers requiring high brightness, low stickies and low speck contents. The methods applied vary according to end product and restrictions on factors such as cost and water use. There are also some differences in deinking philosophies and process design between European, North American and Pacific Rim countries (Ferguson, 1994). In Europe, the public is more tolerant and the end use is generally kept in mind. As a result, a brightness and a higher dirt count can be accepted, compared with Japan where the character-based writing is more sensitive to random dirt specks or to North America where DIP must offer the same quality as virgin pulp. In Europe, high-consistency batch pulpers or drum pulpers are used, kneaders or dispersers are used at the end of an old process, whereas they are used before a post-deinking loop in more recent processes. The light restriction on water use is reflected in the use of disc filters as thickeners rather than true washers. In Japan, the characteristic technology includes soaking towers and kneaders. Soaking tower technology was originally installed to reduce noise during at night and to take advantage of off-peak electricity costs. Extended soaking times, in alkaline environments, improve the separation of fibres but also the separation of fibres from inks and plastics. Kneaders are placed either near the beginning or the middle of the process to facilitate ink detachment and to move the ink-particle size into the range that can best be removed by flotation and washing. In North America, the first recycling processes were based on washing technology. When flotation was introduced, the technology and deinking processes were European. Nowadays a combination of European and Japanese technologies can be found, and multiple stages of flotation, washing and dispersion/kneading have been introduced to handle the lower-quality recovered papers. A new reverse cleaner system has been introduced, which promises much greater efficiencies for difficult particles. Removal efficiencies of greater than 95% are claimed for lightweight debris of up to 0.995 specific gravity. The high volumetric reject flow from the cleaner, which carries the lightweight contaminants in very dilute form, is treated in a new through-flow cleaner. It is optimised to operate at the unusually low consistency of this stream. A low volume rejects stream takes the contaminants out of the system. The latest DIP lines combine kneading, soaking and flotation and typically have two process loops and three water loops (Darrin Rhodes, 1998). Some mills have installed dispersion in the second loop with and without washing, and other mills are using the wetlap machine to wash out ink and stickies that have been dispersed. Newer systems will have two different types of bleaching, with one in the middle of the process and one at the end. In Japan, trials have also been conducted into ink separation by dry pulverisation, the use of cleaners for toner ink removal as large particles after coagulation in a suspension, and resin recovery of minute particles (Tamotsu, 1997). Defibrising and the efficient separation of ink from fibre are required to make high-quality mechanical pulp from waste newsprint. Pulpers, kneaders and dispersers are being combined as defibrisers. Separate treatments are selected for easily dispersed inks and coagulated solid inks.

Metso Paper, with its complete range of secondary fibre processing technology, can supply the most optimised process for all applications (Hatia, Heimonen, & Saarteinen, 2004). The OptiSlush HP high-consistency batch and continuous pulper system separates impurities from the main pulp flow by pumping the pulp through a screen plate with 8–12 mm holes. The drum pulpers are designed for pulping various recycled fibre grades. The OptiScreen family covers coarse screening, medium-consistency fine screening and multi-stage low-consistency fine screening. OptiBright multi-stage flotation cell technology is based on aeration technology combined with new ideas for the cell structure, offering tools for controlling all of the flotation sub-processes. OptiThick (ash washing with two-sided dewatering technology) offers higher capacity per unit width. The conical OptiFiner dispersion technology offers several benefits, including more uniform pulp flow through the plates and a larger treatment area. The OptiBright bleaching system provides optimal conditions with equipment including a heater, a medium- or high-consistency chemical mixer, a medium-consistency pump and a medium- or high-consistency tower. White water clarification and sludge removal with the OptiDaf microflotation concept complements Metso Paper's range of products.

A model predictive control (MPC) and optimisation, incorporated in an automated process control (APC) system, can be used to meet the targets of deinking plants, namely brightness and ERIC and sometimes ash content, higher throughput and yield and lower chemical cost (Austin, 2004). A multivariable model of the process with constraints on each input (MV) and output (CV) is the first stage of MPC. The MPC controller determines the control action for each MV. The APC controllers implemented on the Aylesford deinking plant comprise 34 MVs, 3 feedforward variables (FF) and 24 CVs. Compared with a conventional distributed control system, the APC system results in reduced brightness variation, better brightness control, improved yield, reduced chemical consumption and controlled or maximised throughput. APC is in its early days in the pulp and paper industry, but benefits appear very attractive.

As the demand for recovered fibre continues to grow around the world, equipment suppliers are developing the systems capable of making the same or better quality paper and board from a lower-quality furnish. Reduction of stickies and fibre losses, and lower specific energy consumption, are other key demands. At the same time, newsprint and board lines continue to grow in size, creating new challenges in the deinking plant. Some of the latest technologies are being installed around the world (Anonymous, 2006). In China at APP Ningbo mill, Zhejiang Province, eight fibre lines (five for recovered fibre and three for virgin pulp) feed the 700,000 tonne/year recycled cartonboard unit, the world's largest paper machine. Metso Paper supplied both the deinking plant and the paper machine (PM). The mill can process 2630 tonne/day of secondary fibre as well as some 615 tonne/day of virgin fibre (mostly softwood) to feed the mammoth board machine. The recycled fibre is a mix of DIP, ONP, MOW, SOW and OCC. Technology in use includes OptiSlush drum or continuous pulpers, OptiScreen coarse and fine screens, OptiBright flotation cells and OptiFiner and low consistency refiners. DIP brightness is 78–80 ISO. Recently, Shandong Huatai Paper began commercial production on a newsprint and LWC paper machine at its Dongying mill in Shandong Province. Voith Paper supplied both the 400,000 tonne/year PM 11 and a 1000 tonne/day deinking line to feed the machine. The scope of supply covered the deinking line from the dump chest through to the finished stock storage tower, including complete engineering, the approach flow based on the Advanced Wet End Process Concept, preparation of chemicals and automation. Another 400,000-tonne/yr newsprint mill in Shandong province was started at the end of 2006 at Shandong Chenming Paper Holdings' number 4 mill in Shouguang city. The deinking line, largest in the world was supplied by Metso Paper with a capacity of 1,500 tonnes/day. The furnish comprised 70% old newspapers and 30% old magazines.

In Denmark, Dalum Papir has upgraded the 15-year-old disperger system in its deinking plant in Naestved with a high-efficiency disperger from Voith Paper. The aim was to improve further the optical quality of the DIP, which is produced mainly from higher-quality recovered paper grades, including office papers. The upgrade is the second performed by Voith at the plant, following the addition of MultiFoil rotors and C-bar screen baskets. Dalum Papir makes high-grade coated and uncoated graphic papers at its mill in Odense from the DIP produced at Naestved. In Germany, UPM was set to complete an upgrade of the DIP line at its newsprint and SC paper mill in Schongau, Upper Bavaria. Voith Paper was awarded the contract to install four new EcoCell primary flotation cells and revamp the screening system with three MultiScreen machines, each equipped with C-bar screen basket and MultiFoil rotor for expanding IC and LC slot screening operations. UPM expects the scheme to boost the plant's DIP capacity, as well as to reduce stickies and enhance brightness after pre-flotation.

In India, Malu Paper Mills in Nagpur has installed a 150 tonne/day deinking line from Andritz to feed a 52,500 tonne/year newsprint and printing/writing paper machine. Andritz's scope of supply includes the first FibreFlow drum pulper for India; a two-stage coarse screening system with Andritz ModuScreen CR screens followed by an Andritz three-stage medium-consistency cleaner system with Vortex Control reject nozzles; the latest Andritz SelectaFlot flotation technology for pre- and post-flotation, including the Andritz FoamEx foam breaker; fine screening equipment to remove contaminants and stickies with minimum energy consumption and extremely low yield loss; and a CompaDis dispersing system. Emami Paper has ordered a 300 tonne/day two-loop deinking system from Voith Paper including EcoMizer cleaning and C-bar fine screening for its Balasore mill in Orissa. Voith has also delivered a 250 tonne/day deinking system to Rama Newsprint in Barbodhan, Gujarat state, and a plant capable of processing 200 tonne/day of recovered office paper for the production of printing/writing papers to Century Pulp and Paper, Lalkua, Uttaranchal state. Again these will be two-loop systems. In the case of Century Pulp and Paper, a VarioSplit washer will be installed in the first loop for de-ashing to help secure maximum stock quality. In addition, the mill will operate a kneading disperger in the second deinking loop to reduce dirt specks further and positively influence specific volume of the end product. The engineering strategy of all three plants will be based on Voith's EcoProcess philosophy, which is designed to provide a significant reduction in energy requirements.

In Mexico, Fabrica de Papel San Francisco S.A. de C.V (FAPSA) installed a 250 tonne/day deinking line which includes a ScreenONE screening system, cleaning and washing equipment and a MAC flotation cell. It processes MOW to produce tissue paper. In Nigeria, Star Paper Mill started up a new 60–65 tonne/day Recard crescent former tissue machine at its plant in Aba. The 1600 m/min PM is fed with 100% recovered fibre through a Comer deinking line. The machine's output will mostly be used to make toilet paper.

References

Ackermann, C., Putz, H. J., & Góttsching, L. (1999). Alkaline pulping: Effect of pulping and chemical conditions affect the results. *Pulp and Paper Canada, 100*(4), T109–T113.

Ackerman, Ch Göttsching, L, & Pakarinen, H (2000). Papermaking potential of recycled fiber. In L. Göttsching & H. Pakarinen (Eds.), *Recycled Fiber and Deiking*. Finland: Papermaking Science and Technology. Chapter 10, pp (book 7).

Acosta, D. A., Scamehorn, J. E., & Christian, S. D. (2003). Flotation deinking of sorted office papers using sodium octanoate and sodium dodecanoate as surfactants and calcium as the activator. *Journal of Pulp and Paper Science, 29*(2), 35–41.

Anonymous, (2006). Developments in deinking. *Pulp & Paper International, 48*(7), p. 22.

Austin, P. (2004). Advanced process control and optimisation of a modern deinking plant: Paper recycling technology. *Eighth Pira International conference* (13 p.). Prague, Czech Republic, Paper 10, 17–18 February.

Bajpai, P. (2006). *Advances in Recycling and deinking* (180 pp.). U.K: Smithers Pira.

Ben, Y., & Dorris, G. M. (1999). Irreversible ink redeposition during repulping, II: ONP/OMG furnishes. *Fifth research forum on recycling PAPTAC* (pp. 7–14). Ottawa, 28–30 September.

Bennington, C. P. J., Smith, J. D., & Sui, O. D. (1998). The effect of mechanical action on waste paper defibering and ink removal in repulping conditions. *Journal of Pulp and Paper Science, 24*(11), 341–348.

Bennington, C. P. J., & Wang, M. H. (2001). A kinetic model of ink detachment in the repulper. *Journal of Pulp and Paper Science, 27*(10), 347–352.

Borchardt, J. K. (1997). An introduction to deinking chemistry. In M. R. Doshi & J. M. Dyer (Eds.), *Paper recycling challenge* (Vol. 2). Appleton, WI: Doshi & Associates.

Borchardt, J. K. (2001). Use of flotation in removing ink from recovered paper. *Presented at the 87th annual meeting of the Pulp and Paper Technical Association of Canada* (pp. 71–79). 30 January, Preprint Book A.

Borchardt, J. K. (2003). Recent developments in paper deinking technology. *Pulp and Paper Canada, 104*(5), 32.

Borchardt, J. K., & Schroeder, T. J. (2002). Preventing ink over-dispersion improves old newspaper deinking results. *Preprints of the 88th annual meeting of the Pulp and Paper Technical Association of Canada* (B139–B145). Montreal, Canada, 30 January.

Broullette, F., Daneult, G., & Dorris, G. (2001). Effect of initial repulping pH on the deflaking rate of recovered papers. *Preprints sixth research forum on recycling*. Magog, QC, 1–4 October.

Cao, B., Heise, O., Tschirner, U., & Ramaswamy, S. (2001). Recycling old telephone directories. *TAPPI Journal, 84*(1), 98.

Carré, B., Magnin, L., Galland, G., & Vernac, Y. (2000). Deinking difficulties related to ink formulation, printing process and type of paper. *TAPPI Journal, 83*(6), 60.

Castro, C., Daneault, C., & Dorris, G. M. (1999). Use of antioxidants to delay the aging of oil-based inks. *Pulp and Paper Canada, 100*(7), 54–59.

Cochaux, A., Carre, B., Vernac, Y., & Galland, G. (1997). What is the difference between dispersion and kneading? *Progress in Paper Recycling*, *6*(4), 89.

Crow, D. R., & Secor, R. F. (1987). The ten steps of deinking. *TAPPI Journal*, *70*(7), 101.

Darrin Rhodes, J. D. (1998). Evolution of high grade deinking in Japan. *Paper Technology*, *39*(9), 37.

Dash, B., & Patel, M. (1997). Recent advances in deinking technology. *IPPTA Journal*, *9*(1), 61.

Davies, A. P. H., & Duke, S. R. (2002). Visualizations of offset and flexographic inks at bubble surfaces. *TAPPI Journal*, *1*(3), 41.

Dorris, G. M., & Pagé, N. (2000). Natural and accelerated aging of old newspapers printed with black mineral oil inks and colored vegetable oil inks. Part 1: Deinkability. *Progress in Paper Recycling*, *9*(4), 14–26.

Eguchi, M. (2005). High performance flotator for smaller ink particles removal: MT-Flotator. *Japan TAPPI annual meeting*. Nigata.

Fabry B. (1999, December) Etude de la rhéologie des suspensions fibreuses concentrées dans le but d'améliorer le recyclage des papiers, *Ph. D. thesis*, *INPG*.

Fabry B., & Carre B. (2003). Pulping and ink detachment. *Advanced training course on deinking technology*. Grenoble, March.

Fabry, B., Carré, B., & Crémon, P. (2001). Pulping optimization: Effect of pulping parameters on defibering, ink detachment and ink removal. *Sixth research forum on recycling PAPTAC* (pp. 37–44). Magog, QC, 1–4 October.

Fabry, B., Roux, J. C., & Carré, B. (2000). Pulping: A key factor for optimising deinking. *53ème Congrès ATIP*. Bordeaux, 17–19 October, session 8.

Fabry, B., Roux, J. C., & Carré, B. (2001). Characterization of friction during pulping: An interesting tool to achieve good deinking. *Journal of Pulp and Paper Science*, *27*(8), 284.

Ferguson, L. D. (1994). Comparison of North American, European and Pacific rim deinking technologies. *TAPPI pulping conference*. San Diego, CA.

Galland, G., Carre, B., & Vernac, Y.(1998). About the benefit of using chemicals in the hot dispersion stage. *EUCEPA symposium 1998*, Florence.

Gomez, C. O., Acuña, C., Finch, J. A., & Pelton, R. (2001). Aerosol-enhanced flotation deinking of recycled paper. *Pulp and Paper Canada*, *102*(10), 28–30.

Hatia, M., Heimonen, J., & Saarteinen, J. (2004). Latest developments in recycled fiber processing. *14th ACOTEPAC conference and exhibit*. Medellin.

Haynes, R. D. (1999). Measuring the summer effect in North American Newsprint deinking mills. *Fifth research forum on recycling PAPTAC* (pp. 25–36). Ottawa, 28–30 September.

Haynes, R. D. (2000). The impact of the summer effect on ink detachment and removal. *TAPPI Journal*, *83*(3), 56–65.

Holik, H. (2000). Unit operations and equipment in recycled fiber processing. In L. D. Gottsching & H. Pakarinen (Eds.), *Papermaking science and technology* (7, pp. 88). Fapet Oy. Helsinki, Finland.

Johansson, B., Pugh, R. J., & Alexandrova, L. (2002). Flotation de-inking studies using model hydrophobic particles and non-ionic dispersants. *Colloids and Surfaces A: Physicochemical and Engineering Aspects*, *170*(2–3), 217–230.

Johnson, D. A., & Thompson, E. V. (1995). Fiber and toner detachment during repulping of mixed office waste containing photocopied and laser-printed paper. *TAPPI Journal*, *78*(2), 41.

Kaul, K. P. (1999). Aquasol: A new process to deink old newsprint. *TAPPI Journal*, *82*(8), 115–129.

Klar, A. K., & Burkhardt H. (2000). Pulping technology and recovered paper qualities. *Ninth PTS-CTP deinking symposium*. Munich, 9–11 May.

Lapierre, L., Dorris, G., Pitre, D., Bouchard, J., Hill, G., & Pembr, C. (2002). Use of sodium sulphite for deinking ONP:OMG at neutral pH. *Pulp and Paper Canada*, *103*(1), T8–11.

Le Ny, C., Haveri, M., & Parkarinen, H. (2001). Impact of RCF raw materials quality on deinking performance. Pilot and mill scale study. *Preprints 11, sixth research forum on recycling*. Magog, QC, 1–4 October.

McCool, M. A. (1992). Deinking and separation technology. *Pira International conference*. Leatherhead.

McCool, M. A. (1993). Flotation deinking. In R. J. Spangenberg (Ed.), *Secondary fiber recycling*. TAPPI Press.

McCool, M. A., & Silveri, L. (1987). Removal of specks and non-dispersed ink from a deinking furnish. *TAPPI Journal*, *70*(11), 75.

McKinney, R. W. J., & Roberts, M. (1997).The effects of kneading and dispersion on fibre curl. *TAPPI recycling symposium*. Chicago, IL.

Merza, J., & Haynes, R. D. (2001). Batch pulping versus drum pulping: The impact on deinking performance. *Sixth research forum on recycling PAPTAC* (pp. 45–51). Magog, 1–4 October.

Moe S. T., & Røring A. (2001). Theory and practice of flotation deinking. *Preprints sixth research forum on recycling*. Magog, QC, 1–4 October.

Moss, C. S. (1997). Theory and reality for contaminant removal curves. *TAPPI Journal*, *80*(4), 69.

Niggl, V., & Kriebel, A. (1997). Dispersion – The process stage for improving optical properties. *Papier*, *51*(10), 520.

Patrick, K. (2001). Advances in paper recycling technologies. *Paper Age*, *117*(7), 16.

Pélach, M. A., Puig, J., Vilaseca, F., & Mutjé, P. (2001). Influence of chemicals on deinkability of wood-free fully coated paper. *Journal of Pulp and Paper Science*, *27*(10), 353–358.

Pirttinen, E., & Stenius, P. (2000). The effect of chemical conditions on newsprint ink detachment and fragmentation. *TAPPI Journal*, *83*(11), 1–14.

Pletka, J., Gosiewska, A., Chee, K. Y., Mcguire, J. P., Drelich, J., & Groleau, L. (1999). Use of the atomic force microscope for examination of interfacial forces in recovered paper deinking systems. *Progress in Paper Recycling*, *8*, 55–67.

Seifert, P., & Gilkey, M. (1997). *Deinking: A literature review*. Pira International. Leatherhead Surrey, U.K. (p. 139).

Serres, A., Colin, P., & Grossmann, H. (1994). Recent advances in deinking optimization. *PTS deinking symposium*, Munich.

Tamotsu, Y. (1997). Latest technology in DIP process of waste paper. *Asia Pulp and Paper Technology Markets*, *34*(4), 14. (Spring 1997).

Thompson, E. V. (1997). Review of flotation research by the cooperative recycled fibers program, department of chemical engineering, University of Maine. In M. R. Doshi & J. M. Dyer (Eds.), *Paper Recycling Challenge* (Vol. 2). Appleton, WI: Doshi & Associates.

Toland, J. (2003). Developments in deinking: Rounding up some of the latest trends in the recovered paper sector. *Pulp and Paper International*, *45*(4), 25.

Walmsley, M., Yu, C. J., & Silveri, L. (1993). Effect of ink specks on brightness of recycled paper. *Proceedings of the TAPPI recycling symposium* (pp. 417–441). Atlanta: TAPPI Press.

5 System and Process Design for Different Paper and Board Grades*

5.1 Introduction

The complete process to be applied to recovered paper to meet the quality requirements of recycled pulp is determined by the grade of the recovered paper and the demands of the final paper product (European Commission, 2001; Schwarz, 2000; Stawicki & Barry, 2009). Almost half of the total recovered papers are recycled for the production of fluting and liner, mainly from old corrugated carton (OCC). Besides these brown recovered paper grades, white recovered paper grades, i.e. mainly old newspapers (ONP) and old magazines (OMG) and some high-quality grades, are reused in newsprint, tissue and other graphic paper grades and in whip top layers of packaging papers. So, these are deinked. Mixed recovered paper grades are used with OCC and for grey solid board layers and thus are not deinked.

Various product characteristics require cleanliness and brightness properties from the recycled fibre (RCF) pulps and the process concepts vary accordingly. For example, deinking is not required in many board grades. On the contrary, a very efficient multistage process is required for high-speed paper machines, thin paper or for grades where brightness is important. The degree of sophistication of the whole process depends on the furnish used and the paper grade to be manufactured. Therefore, it is not reasonable to describe 'one typical' recovered paper processing system.

The recycling of corrugated board starts with clean post-consumer carton boxes and converting rejects. The recycling process involves re-pulping, deflaking and refining of the kraft-fibre-rich raw material. The steps in the contaminant removal process have gained importance as the recycling rate has increased and with the development of hot-melt glues and plastics in packaging. The hot-dispersing process has been developed to enable the recycling of bitumised and waxed cartons. The fractionation process has also been developed for the recycling of packaging papers to limit the refining and hot-dispersing action to the long-fibre-rich fraction, and to reduce the damage to the already shorter-fibre fraction, which also contains contaminants and residual flakes that disperse.

Today, with a recycling rate of more than 90% in the case material sector, there is much less need to refine multi-recycled fibres. The need to preserve fibres and to save energy has also led to shut down many hot-dispersing treatments. This can be attributed to progress in fine screening as it is more effective, in a recycling perspective, to remove the contaminants, like hot-melt glues, instead of keeping them dispersed in the paper.

*Some excerpts taken from Bajpai, 2006 with kind permission from Pira International, UK.

Recycling and Deinking of Recovered Paper. DOI: http://dx.doi.org/10.1016/B978-0-12-416998-2.00005-2

Deinking lines are more complex than OCC recycling lines, because, in addition to the re-pulping and contaminant removal process steps, the pulp has to be deinked, i.e. inks have to be detached from the fibres and removed, and optionally bleached to reach the increasing brightness requirements. Removing inks that are difficult to deink, like flexo prints and some digital prints, and 'stickies' from pressure-sensitive adhesives, are currently the main challenges in the field of deinking (European Commission, 2001).

Some examples of typical recycling and deinking lines for different applications are presented below.

5.2 RCF Stock Preparation for Newsprint and Improved Paper Grades

Newsprint is a coarse grade of paper used mostly for printing newspapers. The raw material used is a typical deinked pulp (DIP) consisting of a 50:50 mixture of newspapers and magazines. The system is characterised by a two-stage flotation and bleaching process combined with an intermediate dispersion (European Commission, 2001).

For upgraded newsprint qualities with higher demands on brightness, a reductive stage with hydrosulphite may follow thickening. For the production of graphic papers from recovered paper, water loop design and water clarification are especially important. Other pulp components used for these paper grades generally include thermomechanical pulp (TMP). For basis weights below 45 g/m^2, the strength potential of the RCF and TMP mixture is usually not adequate. Chemical fibres must also be added. For lightweight coated (LWC) and supercalendered (SC) grades, certain quantities of chemical pulp may require addition for the same reason. Using TMP as a furnish component does not influence the process stages of the recovered paper line. Water clarification must be adapted to the increased content of colloidal substances caused by TMP. Chemical pulp as the third component has no effect on RCF processing. Figure 5.1 shows a total system for newsprint and improved paper grades stock preparation line (Schwarz, 2000).

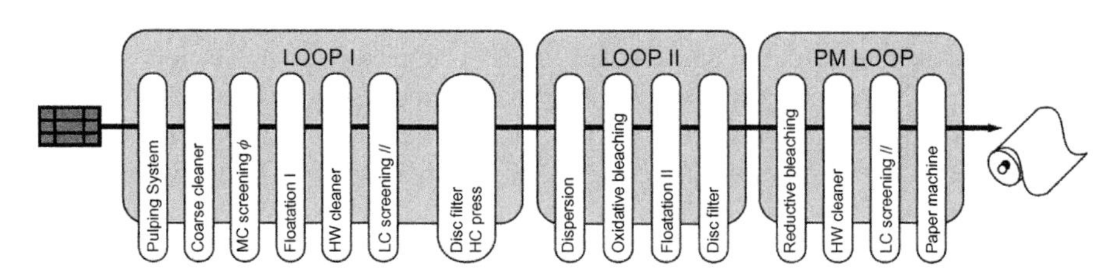

Figure 5.1 RCF stock preparation for newsprint and improved paper grades.
Source: Reproduced from Schwarz (2000) with permission from Fapet Oy, Finland.

When using the standard range of mixtures of ONP and OMG to manufacture new newspaper, one factor is relatively certain: the characteristics of the deinked furnish fibre are similar to or perhaps even better than those of the virgin furnish. The coarseness of the ONP fibres is, by necessity, no worse, because the virgin pulp has gone through additional refining, on-machine screening, drying and calendering – steps that would make initially coarse fibres smaller. Deinked OMG fibres are finer to start with, whether originating from chemical or mechanical pulp. In fact, the long fibres of OMG often allow reduction of a small percentage of chemical pulp when such is required, for product strength, paper machining or press runnability (Anonymous, 1992; Chorley, 1990). Newsprint with DIP content has somewhat greater density as the fibres tend to conform more easily in the sheet plane. Smoothness is obtained more easily, and less calendering is often required. If the deinking process leaves ash in the pulp, the same may help final product opacity (Krauthauf, 1992). Accordingly, printability has also not been a problem (Schmid, 1994). Occasionally, the brightness of DIP may be low, particularly when using old raw material which has yellowed. When ink is thoroughly removed, bleaching with peroxide easily brightens the pulp to newsprint standards. Cleanliness of DIP is important and can be satisfied with good process practice. Otherwise, breaks in the paper machine and in the newsprint pressroom occur. Troublesome contaminants, such as stickies from labels and magazines, can be removed sufficiently well by fine slotted screens. The main focus in European recycled newsprint papers is to increase filler contents, which, while helping offset printing quality factors, can be detrimental to other printing quality requirements (Rankin, 2005). Technological developments in the production of DIP have allowed the German paper industry to increase its use of recycled paper, and this now accounts for all German newsprint (Jensen, 2005).

Mixtures of ONP and OMG can also be used to produce increasingly better printing papers. However, the demands on the pulp increase, in line with the enhanced printability requirements. Simple ink and contaminant removal is no longer sufficient to make the DIP suitable for such grades. Many fibres are too coarse, being similar to virgin newsprint pulp fibres. Thus, the pulp must be treated further, or raw materials must be adjusted. Treatment may include extra fine screening, fractionation and refining (Pimley, 1991; Sauzedde, 1993). Contaminant removal also becomes more critical as the sheet finish increases. On-machine soft calendering has perhaps the least additional cleanliness demands (Fetterly, 1991; Matzke, 1995; Watanabe, 1991). For coating with a roll coater, holes in the paper must be avoided. For coating with blade coaters, and/or super calendering, residual contaminants must be extremely small, so as to completely hide in the paper which is often less than 0.1 mm thick. Obviously, even screens with 0.1 mm slots do not provide 100% removal of potential problem contaminants. Higher-quality newsprint treated with soft nip on-machine calenders can tolerate significant percentages of DIP, and is produced commercially (Drier & Schuelke, 1992; Frostman, Henning, Harding, & Laube, 1996; Stockler & Grossmann, 1994). Those DIP systems, which aim for higher-quality paper than newsprint, feature dispersion followed by post-flotation, as a minimum additional treatment. SCA's mill in Aylesford, UK, produces newsprint from 100% recycled paper. Its deinking system is the fastest in the world (Anonymous, 1997). On

31 May 2005 PanAsia Paper opened a new US$315 million greenfield newsprint mill in Zhaoxian, Hebei Province, China. The operating company is Hebei PanAsia Longteng Paper Co Ltd (HPLC). This mill is using a Metso OptiConcept 330,000 tonnes/year machine using 100% DIP and producing both 45 and 48.8 gsm newsprint (Rodden, 2005). The Metso paper machine is equipped with an OptiFlo headbox and OptiFormer LB with high dewatering capacity and good paper profiles, formation and evenness. HPLC installed three power boilers that burn B-C oil; each produces 70 tonnes/h of steam used to dry paper. The mill's electrical needs are 40–50 MW, all purchased. The mill has installed an activated sludge system for effluents treatment. Fresh water consumption is 11 m^3/tonne. Stora Enso's Langerbrugge mill in Belgium has a PM4 paper machine producing 400,000 tonnes/year, which started production in 2003. It produces newsprint from 100% recovered paper processed on a new DIP line (Gray, 2003). The line includes a flotation plant supplied by Voith, comprising two EcoCell pre- and post-flotation lines operating in parallel, as well as two EcoDirect disperger systems. The pre-flotation lines each have a 735 tonnes/day design capacity, with a 666 tonnes/day capacity for each of the post-flotation lines. The two dispergers are equipped with equalising and dilution screws, each one designed to process 679 tonnes/day DIP.

5.3 RCF Stock Preparation for Market Pulp Systems

Market pulp is pulp produced for sale on the open market to paper mills, as opposed to that produced in an integrated or affiliated mill for its own consumption. When using recovered paper for the production of market DIP, it must be processed in such a way that not only coarse contaminants, but also printing inks, stickies, fines and fillers have to be removed. This considerable reduction in ash and fines means about 30–100% more recovered paper is needed compared with the finished stock, which means a relatively high amount of waste has to be handled and treated. There are examples (e.g. Niederbipp mill, Switzerland) where all rejects, including sludge, are incinerated on site, generating steam for mill consumer points, resulting in a reduction of residues to ash which is used as an aggregate in the building industry (cement). The major difference of these grades compared with newsprint is the requirement of de-ashing (removal of the fines and fillers) due to the requirements on PM runnability and softness and absorbency of the end product. As an example, depending on the grades of recovered paper used as raw material, the ash content may vary between 15% and 38% (in the case of coated wood-free papers).

The direct impact of the ash content on the amount of solid waste generated during recovered paper processing should be considered when comparing figures on the amount of solid waste. Potential customers for market pulp include major printing and writing producers in the USA and Canada. Market pulp systems have shown the largest growth. System sizes have generally been large, to take advantage of economies of scale. The European market has also been targeted for deinked mixed office waste (MOW), as the level of recycling MOW currently is not as strong in Europe as in the USA. Many mills aim to replace bleached hardwood kraft pulp with their

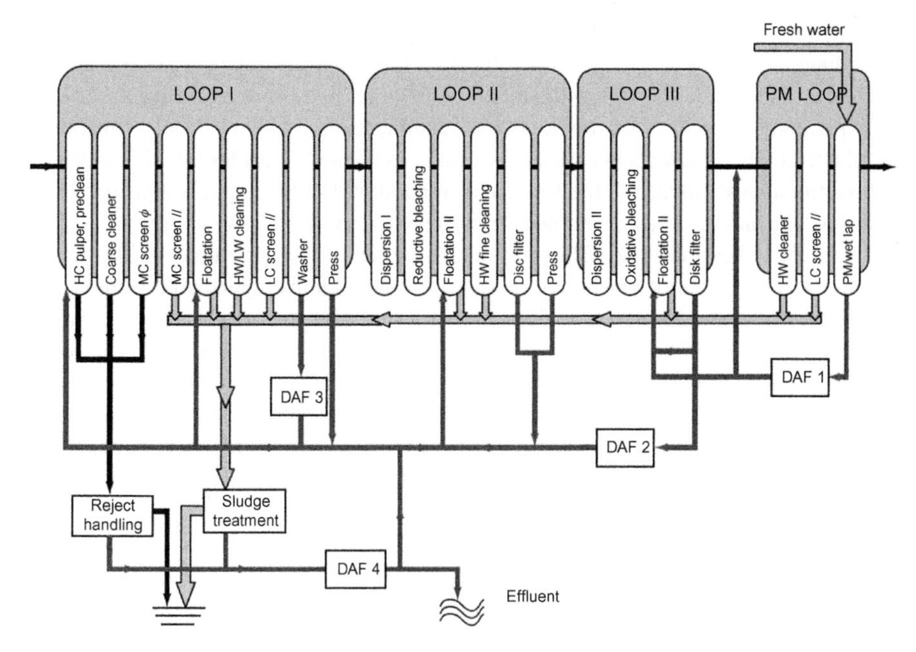

Figure 5.2 RCF stock preparation for market pulp.
Source: Reproduced from Schwarz (2000) with permission from Fapet Oy, Finland.

recycled fibre qualities. Several surveys have been undertaken to evaluate deink pulp properties from different deinking systems (Anonymous, 1993). Because they use a similar starting product to the fine paper systems, and must produce a clean pulp to be used in printing and writing grades, many market pulp systems resemble the fine paper MOW deinking systems in their degree of complexity and sophistication. Several equipment sequences for systems using MOW to produce market pulp are being used. Again, they all vary in some way from each other.

Production of market DIP from RCF is the most demanding process (Schwarz, 2000). A main requirement is the exclusive use of wood-free grades as furnish using sorted office waste. The mechanical fibre content should be below 3% as much as possible and must never exceed 5%. To meet the highest quality demands, the stock preparation line includes an additional (Figure 5.2) (Schwarz, 2000). Loop I is similar to that in the process example for writing and printing paper grades. Loops II and III include two additional full-flow flotation stages. Two-stage dispersion occurs in the second and third loops, which are separated from each other by a thickening stage to 30%. Only this repeated sequence of dispersing and flotation fulfils the high cleanliness requirement at the end of the system. An additional advantage of the three-loop system for recycled fibre processing is that it allows a reductive–oxidative bleaching sequence (dithionite–peroxide). The first reductive bleaching stage in loop II is less susceptible to heavy metals than oxidative bleaching. It is also insensitive to catalase that is destroyed by heat treatment in the dispersion stage. Peroxide bleaching at the end of the system therefore occurs under optimum conditions with clean stock and loop water from which any heavy metals have already been removed. Residual peroxide does not require decomposition. It helps with stock conservation by restricting bacterial growth. When planning RCF lines of this kind for manufacturing a chemical pulp substitute, careful accounting is always necessary for cost-effectiveness. Besides the high machinery investment costs incurred, the low yield of about 65% also requires consideration. Careful design and the latest technology contribute to long-term profitability.

The Great Lakes Pulp and Fibre mill and American Fiber Resources mill are two recycling mills capable of producing market DIP that is indistinguishable from virgin pulp (Ferguson, Shaw, DeBerry, & Henriksson, 2001). The process and chemistry used at the mills is designed to handle 100% MOW papers, and both mills have a pulp dryer at the end of the process to produce dry lap pulp. The high-quality DIP has physical and optical properties equal to and in some cases better than a hardwood pulp. The twice-dried fibres of the dry lap form of the DIP pulp drain faster than virgin hardwood pulp, meaning paper machine operators can increase machine speed and/or have a lower drier temperature. The DIP pulp requires the lowest level of refining energy to reach headbox freeness, and co-refining the DIP with hardwood or with hardwood and softwood combined does not significantly damage the fibres. The burst index is higher than bleached chemithermomechanical pulp (BCTMP) and the bulk is similar to BCTMP and higher than the hardwood kraft pulps tested. DIP has a higher brightness than BCTMP and a competitive brightness to the hardwood kraft pulps, as well as having the lowest dissolved and colloidal materials content of any of the pulps tested.

Stora Enso Sachsen Mill in Eilenberg, Saxony, Germany, has produced 70,000 tonnes/year of flash-dried market DIP from old magazines and newspapers since 1994 (Pfitzner, 2003). A modern two-loop deinking plant recycles the recovered paper using equipment from Voith, Andritz-Ahlstrom and Kadant Lamort. Recovered paper is pulped in a drum pulper, and the pulp is screened and passed to high-consistency cleaners. Pre-flotation is followed by cleaning to remove heavyweight contaminants. The second recycling loop incorporates post-flotation stage as well as three primary and two common secondary flotation stages. Dewatering stages use disc filters. A fluffer disperses fibre bundles and separates fibres after the double-wire press. A two-stage flash drier produces a high dry content of 80–85% while maintaining fibre properties. Recommended pulping conditions are a pulping consistency of 4–15%, a pulping time of 10–20 min, at 20°C and consistent pH values in the pulper and paper machine headbox. Pulp is finally stored in a stock tank for 30–40 min to revitalise the fibres. The DIP is mainly used for newsprint, but also for magazine and tissue papers. WEPA Papierfabrik Paul Krengel GmbH and Co KG, Germany, produces market DIP for its highly demanding customers precisely in accordance with their specific product requirements (Berger, 2002). Market DIP has been produced by Ponderosa Fibers of America (Cox, 1992; O'Brien, 1993) since the 1960s, mostly for use in tissue products.

5.4 RCF Stock Preparation for Fine Paper System

Fine papers are high-quality printing/writing and cover papers with excellent surface characteristics for writing. This term is usually used in contrast to coarse/industrial and/or packaging papers. Fine papers have some of the highest quality requirements like printability and runnability. The paper must have good brightness, brightness stability, whiteness, cleanliness, opacity, suitable smoothness, compressibility, ink penetration and sufficient strength for the printing operations. A minimum opacity is normally required and this becomes critical as sheet weights are reduced. So, fine papers use some of the most elaborate and extensive system designs for deinking (Dick, 1992; Maier, 1993; Schwarz, 2000). Many have multiple screening, cleaning, dispersion, flotation and washing steps, as well as fairly elaborate bleaching sequences. Fractionation has also been proposed to produce higher grades of paper (Ackermann, Putz, & Gottsching, 1992). Many of these systems are located at integrated sites, which also have a pulp mill and paper machines making a variety of speciality grades as well as uncoated free sheets (Morrison, 1995; Rubio, 1991). The deinking system size has varied from 50 to over 300 tonnes/day. Systems were installed in the late 1980s and early 1990s to meet growing consumer demand for products with recycled fibre (Jensen, 1990). Also, mills have anticipated potential legislation that requires a certain level of secondary fibre in paper grades. Most systems were designed to recycle MOW or SMOW (S, sorted), being equipped to handle a variety of contaminants as well as TAPPI dirt levels in the incoming stock in the 1000–2500 ppm range.

System design for fine paper has varied. Most, but not all, systems include flotation and kneading or dispersion for reduction and removal of visible ink levels, which can be quite high owing to the high percentage of toner or electrostatic inks in the waste paper (Seifert & Gilkey, 1997). Systems in general can be categorised by the relationship between the flotation (F) and dispersion and/or kneading (D) steps. In general, various F and D sequences have been installed, including DF, FDF, DFDF, etc. FD-type sequences have been seen in some ONP systems, but many of these omit use of the dispersion part (Marquat, 1994). One of the major advantages of using the DF sequence is that the D step helps shift the particle size into a smaller range, where the F step has a greater chance of removing the ink particle (Galland, Julien, Armand, & Vernac, 1993). Also, there is evidence that the dispersion or kneading step helps to separate the fibres from the toner ink particles, making the ink particles more readily floatable (Vidotti, Johnson, & Thompson, 1997). The disadvantage of having the D step before the F is dispersion or size reduction of the sticky material, which may reagglomerate later in the system. Most North American MOW systems have been of the FDF or DFDF type. Examples are the Union Camp facility in Franklin, Virginia (Ferguson, 1995), and the Boise Cascade mill in Jackson, Alabama (Ferguson & McBride, 1993). The Compagne de L'Essone in France (Platier, 1995) and Stora Magle-Molle Papirfabrik in Denmark (Nielsen, 1992) are examples of fine paper deinking systems in Europe. At least one system in the UK has been of the FDFDF arrangement. Compagnie Papetiere de l'Essonne is using 100% waste paper, with no virgin pulp, for the manufacture of a wide range of printing and writing papers: offset printing paper with three levels of brightness 63/68/76; ecocopy paper for photocopying, ecopynature with two levels of brightness 70 and 78; paper for schools; computer printout paper, and Ecopress a 40 gsm paper for newspapers (Platier, 1995). The processing of waste paper to pulp is 85% efficient and the line speed is 580–620 m/min. The M-real's recycled fibre plant at

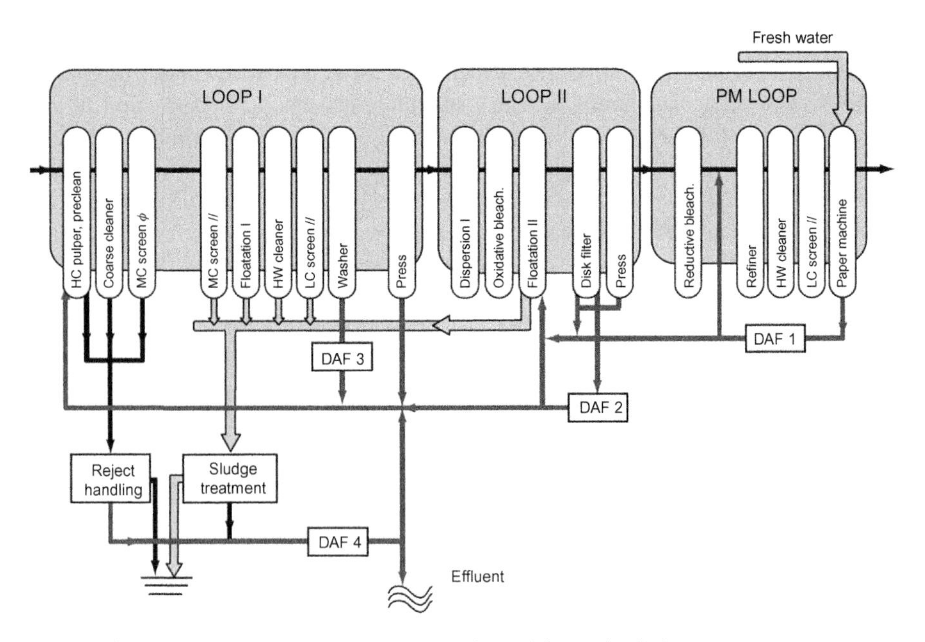

Figure 5.3 RCF stock preparation for high-quality writing and printing paper.
Source: Reproduced from Schwarz (2000) with permission from Fapet Oy, Finland.

Kemsley, Kent, UK is producing high-quality DIP from MOW (Bajpai, 2006). According to the company, the recycled fibre plant produces pulp that is brighter, whiter and cleaner than any other recycled fibres and can be used to produce the brightest, whitest and strongest recycled paper in Europe. It has the most favourable cost structure compared with other DIP plants. Being near to its source of raw material, and with 75% of customers within a 3 mile radius, the company can supply wet rather than dry pulp and thus remove drying and transport costs.

Figure 5.3 shows the arrangement of process stages used for the production of high-quality writing and printing grades (Schwarz, 2000). It has two processing loops. For optimum brightness and more efficient de-ashing, a washer replaces the disk filter at the end of the first loop. This allows controlled reduction of fines and fillers. Stock is then recovered from the filtrate of this washer in a separate dissolved air flotation (DAF 3). For high-grade writing and printing, bleaching is conducted in two stages. Peroxide bleaching not only increases brightness but also reduces the fibre mottling effect by brightening individual brown fibres. Reductive bleaching with dithionite or formamidine sulphinic acid (FAS) then contributes to an additional increase in brightness and colour stripping.

Dispersing uses a high specific energy demand of more than 100 kWh/tonne. A refining stage is also necessary after reductive bleaching. This post-refining reduces the 14 and 30 mesh fraction and adapts coarse TMP fibres in the newsprint furnish to the more refined TMP stock for manufacturing SC and LWC papers. This greatly improves smoothness and especially printability (Bergmann, 1988; Burger, 1993;

Demel & Möbius, 1987; Novak & Malesic, 1992; von Raven & von Troll, 1991). Owing to the chemical fibre content, DIP prepared in this way has significantly better strength characteristics than TMP or groundwood. For lightweight SC and LWC grades made exclusively of virgin pulp stock, minimum strength requirements must occur by adding chemical fibres to the TMP. When using DIP as an additional component, the chemical fibres contained therein largely make adding virgin chemical fibres unnecessary. Total raw material costs can therefore decrease by optimising the furnish.

5.5 RCF Stock Preparation for Tissue

Tissue is a collective term for papers of a grammage of less than 30 gsm that differ in application and composition but have the common feature of being thin. The extremely thin Japanese tissue papers are sometimes produced in grammages as small as 6–8 g. Examples of different types of tissue paper include sanitary grades such as toilet, facial, napkin, towels, wipes and special sanitary papers. Desirable characteristics in these types of tissue paper are softness, strength and freedom from lint. Other examples of tissue papers are decorative and laminated tissue papers and crepe papers, often used in gift wrapping and to decorate. Desirable characteristics here are appearance, strength and durability. Tissue papers are divided into three major categories: at-home (or consumer), away-from-home (or commercial and industrial) and specialty. Some tissues are made from 100% secondary fibres, others from 100% virgin pulp, with a spectrum of mixes also available. This reflects the reality that, although there are difficulties, acceptable product quality can be achieved when using secondary fibre raw material (Berger, 1990; French, 1992; Linck, 1995; Linck, Nyffeler, & Kern, 1994; Sector, 1992; Siewert, 1988). In North America, most tissue producers have integrated, often proprietary, deinking plants. Recycled deinked tissue is made from appropriate waste paper grades, depending on the final tissue quality desired. The type of fibre required to make the end product is the most significant variable in selecting waste paper input. Hardwood or softwood chemical pulp, or groundwood, are the major grades found. Often, brightness or final dirt count also dictate the raw material sourcing, depending on the capabilities of the existing deinking and bleach plants. Paper machine runnability may also force use of certain raw material pulps. The second general variable is the degree of contamination of the raw material by ink, debris and, most importantly, stickies. Tolerance to these materials depends to a high degree on the cleaning and deinking equipment available, the chemicals used and operational variables of the deink plant. Finally, ash content must be mentioned because hygienic tissues are restricted to about 1.0% of ash, or less. Ash can be removed from the raw material by washing. The main goal in tissue production from RCF is to reduce the filler content of the furnish. Ash removal also involves fibre fines removal. At high system de-ashing rates, ash removal is less selective, i.e. the inorganic part of the sludge removed can be below 50%. Total stock yield may then fall to 60% or even less. For cost-effective tissue stock preparation lines, the ash content of the finished stock should therefore

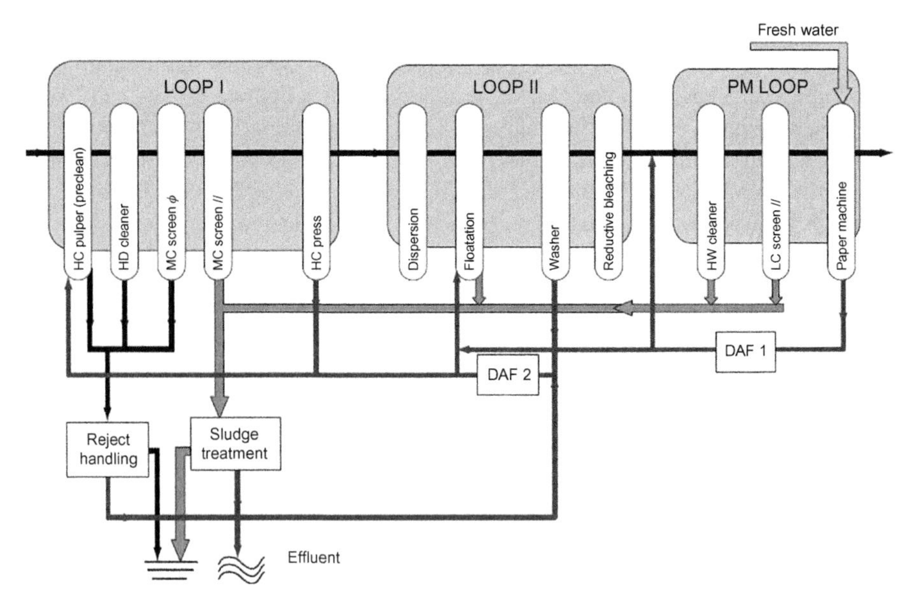

Figure 5.4 RCF stock preparation for tissue – standard and wood-free furnish.
Source: Reproduced from Schwarz (2000) with permission from Fapet Oy, Finland.

only be decreased as far as absolutely essential. Figure 5.4 shows the total system for a tissue stock preparation line (Schwarz, 2000). Because the furnish comprises wood-free recovered paper grades, the content of offset printed paper is low. Only a deinking flotation stage is necessary. After high consistency screening (MC) hole screening, slotted screening is also included in the MC section. As no low consistency screening (LC) screening stage exists, only small-size machinery is necessary for thickening to 30% before dispersing. This reliably breaks down toner particles. After flotation under neutral conditions, a washer is also part of the line. This removes fillers while also functioning as a thickener. At 8–10% stock consistency, reductive bleaching can easily be integrated afterwards. Stock is recovered from the washer filtrate of the DAF 2. The washer sludge is thickened with the fine rejects and flotation sludge (Kleuser, 1990; Siewert & Horsch, 1984). The microflotation stage of DAF 1 does not perform cleaning but merely recovers fibres from white water II. For this reason, the floating material normally returns to the mixing chest. Because microflotation requires no auxiliary filtering stage, the flotation sludge exclusively comprises solids from the white water. Using the DAF 1 stage to control ash content is therefore also possible. For this purpose, the microflotation sludge is treated partly or totally with the other sludge. In the example shown in Figure 5.5 the stock consistency in loop I after MC hole screening decreases. LC slotted screening with extremely narrow slots follows the cleaners. The advantages of this arrangement are highly efficient stock cleaning at the beginning of the line and the possibility of including a washer in loop I because of low stock consistency (Schwarz, 2000). With the second washer in loop II, this results in an ash removal with this system higher

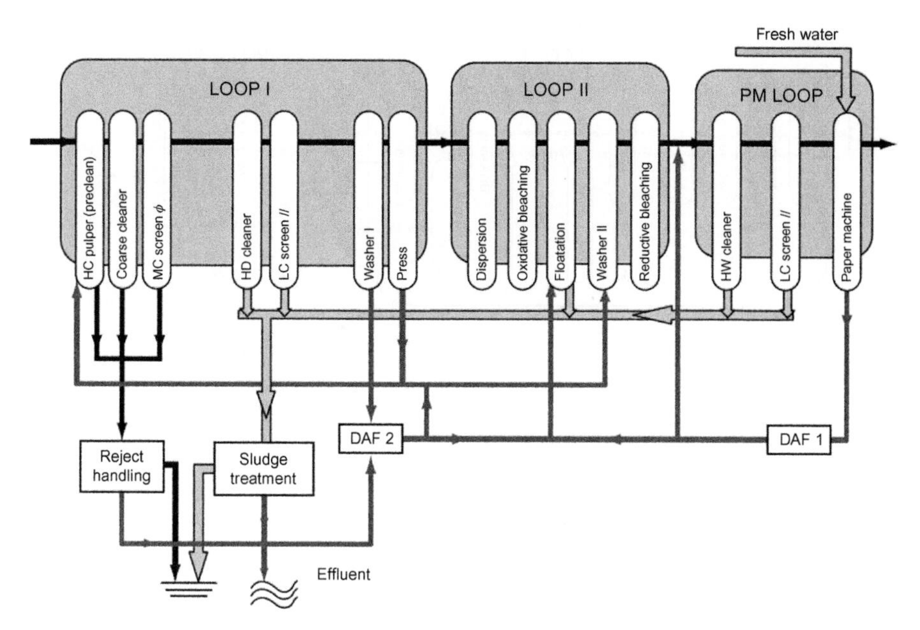

Figure 5.5 RCF stock preparation for tissue – high-grade wood-free furnish.
Source: Reproduced from Schwarz (2000) with permission from Fapet Oy, Finland.

than that of the standard process described above. For optimum brightness, two-stage bleaching (oxidative and reductive) is best. If the furnish also contains mechanical fibres, two flotation stages should be provided.

Tissues are very low weight papers, generally in the range 15–25 gsm for toilet and facial tissues. Towelling grades may range up to 45 gsm. These grades are often creped to enhance bulk, softness and absorbency, and require low final filler content. Generally, these tissues are readily made from DIP, and meet the requirements of strength, brightness, colour and cleanliness (Fischer, 1993). Although the dirt content of the DIP is not as critical in tissue grades as for printing grades, low stickies content is very important. Residual stickies deposit on paper machine wires and felts, and cause holes in the paper, which lead to breaks. The desired ash content for hygienic tissues is generally less than 1%. With raw material ash content in the pulper as high as 15%, strong washing steps are required in the deinking plant itself (Seifert, 1992; Siewert, 1988) and/or by using the natural washing action of the tissue paper machine (Guss, 1995). Microbiological purity of tissues for toilet and kitchen use has been addressed, by assessing bacteria and fungi in tissue samples. Microbe counts were found to be the same in both recycled and virgin tissue. Fungi were not found at all (Jokinen, Siren, & Osmonen, 1995). The odour of the product, when new or after storage, does not appear to be a problem when secondary fibres are used. Any odour change seemed to stem from aeration and oxidation, not from microbiological activity (Jokinen et al., 1995). Sometimes, market forces determine the content of recycled pulp in tissue, to the point that some manufacturers are

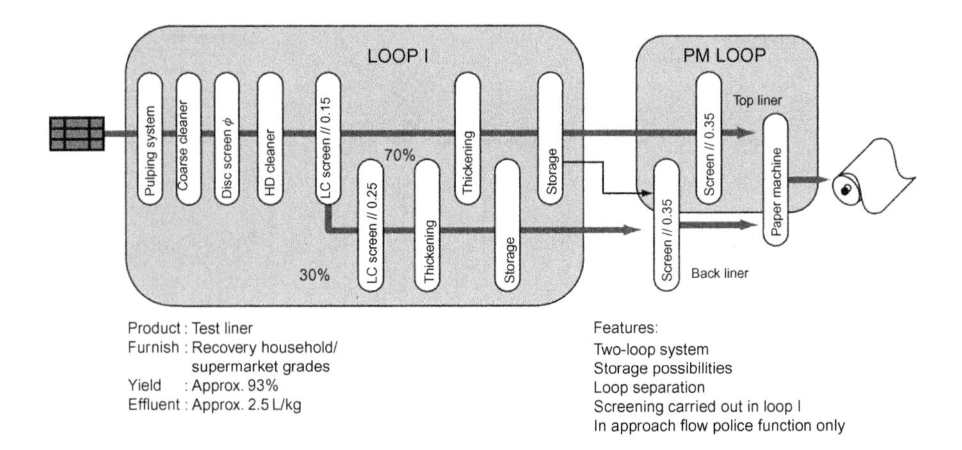

Figure 5.6 RCF stock preparation for test liner.

enhancing marketability by stressing high deink pulp content. For packaging tissues the situation is different, as the final desired ash content is high, to generate good opacity. Strong washing action is therefore not advocated.

5.6 RCF Stock Preparation for Test Liner

Test liner is mainly produced from waste paper used as even facing for corrugated board or as liner of solid board. It is often produced as duplex (two-layer) paper. The grammage is higher than 125 gsm. Figure 5.6 describes in detail a process example for recovered paper processing for the production of liner grades (Schwarz, 2000). The processes for the production of corrugating medium are very similar to those for liner. The critical differences are the less stringent demands put on the optical cleanliness of the medium. This essentially results in a lower screening requirement. The slot widths in the fine screening stage can be adjusted wider than in the production of test liner. Available raw material is low-quality recovered paper from households and supermarkets as post-consumer material. The system illustrated as an overview in Figure 5.6 relates to Central European conditions where the strength potential of the untreated stock is low. The process design is simpler and contains fewer process stages than the stock preparation systems for the production of graphic paper grades. Basically, technically sophisticated processes are also desirable in packaging paper and board production, but the requirements of the final product do not allow sufficient leeway for integration of additional fibre treatment and cleaning stages. Each individual machine must therefore be examined for its economic profitability even more closely than with graphic paper grades. Compared with writing and printing paper grades, the required optical properties for packaging paper grades are less demanding. Integrating a biological treatment into the process without impairing product or process technology is possible with packaging grades. Using a biological such as a 'system kidney' for dissolved organic substances results in reasonable loop closure down to zero effluent without running into trouble with micro-stickies and dissolved substances. Several mills around the world are producing high-quality corrugated medium and test liner from recycled furnish (Francin, Gray, & Wiest, 2005; Jussila, Welt, & Wirth, 2005).

References

Ackermann, C., Putz, H. J., & Gottsching, L. (1992). Waste paper treatment for the production of high quality graphic papers. *Wochenblatt Für Papierfabrikation*, *120*(11/12), 433.

Anonymous (1992). Largest US newsprint producer adds deinking system. *Resource Recycling*, *11*(2), 25.

Anonymous (1993). Deink market pulp properties – A mill survey. *Progress in Paper Recycling*, *2*(2), 54.

Anonymous (1997). Examination of the future of deinked pulp. *Allgemeine Papier-Rundschau*, *121*(14), 333.

Bajpai, P. (2006). *Advances in recycling and deinking*. UK: Smithers Pira. (pp. 180).

Berger, H. (1990). West German environmentally friendly tissue production with secondary fibres. *Fifth international papermakers tissue conference*. Toronto.

Berger, H. (2002). Systematic expansion of a recovered paper stock preparation line for tissue and market dip production. *Twogether*, *14*, 15.

Bergmann, W. (1988). Quality requirements of coated printing paper. *Wochenblatt für Papierfabrikation*, *116*(18), 788.

Burger, G. (1993). Removal of fines and fillers in the production of LWC papers. *Papier*, *47*(5), 233.

Chorley L. M.(1990). Deinking options for newsprint. *CPPA symposium on waste paper in newsprint and printing and writing grades*. Montreal.

Cox, J. (1992). Persistence, common sense are the qualities behind Ponderosa's success. *American Papermaker*, *55*(6), 16.

Demel, J., & Möbius, C. H. (1987). Korrosion in papier fabriken. *Allgemeine Papier-Rundschau*, *111*(24/25), 746.

Dick, D. (1992). Sequence of unit operations in deinking plant design. *TAPPI pulping conference*. Boston, MA.

Drier, T., & Schuelke, T. (1992). Calendering of newsprint with DIP. *TAPPI finishing and converting conference*. Nashville, TN.

European Commission, (2001). *Integrated Pollution Prevention and Control (IPPC). Reference document on best available techniques in the pulp and paper industry*. Seville: Institute for Prospective Technological Studies.

Ferguson, K. H. (1995). Union camp sets deinking standard with modern Franklin pulp faculty. *Pulp and Paper*, *69*(5), 51.

Ferguson, L., & McBride, D. (1993). Deinking sorted office waste. *CPPA 79th annual meeting*. Montreal.

Ferguson, L., Shaw, R., DeBerry, R., & Henriksson, C. (2001). Can deinked pulp dare to compete with virgin pulp? *Sixth PAPTAC research forum on recycling*. Quebec.

Fetterly N. W. (1991). Deinking ONP for directory: The low cost approach. *TAPPI pulping conference*. Orlando, FL.

Fischer J. (1993). Modern recycled fibre lines for tissue production. *Tissue World*. Nice.
Francin, S., Gray, H., & Wiest, F. (2005). Les Papeteries De Champagne: As Delicious the Name, Just as Innovative the Paper Mill. *Twogether*, *29*, 29.
French K. M. (1992). Difficulties manufacturing tissue utilizing secondary fibre. *TAPPI tissue runnability seminar*. Nashville, TN.
Frostman T. M., Henning T., Harding R., & Laube D. (1996). Storm water quality management at a particleboard and door manufacturer. *TAPPI international environmental conference*. Orlando, FL.
Galland G., Julien Saint Armand F., & Vernac Y. (1993). Upgrading of recycled pulp by kneading. *25th EUCEPA conference on pulp and paper*. Vienna.
Gray, H. (2003). Deinking at Stora Enso's Langerbrugge Mill: Voith supplies key components for the New PM4 DIP Plant. *Internationale Papwirtschaft*, *9*, 20.
Guss D. B.(1995). The use of flotation clarifiers on tissue machines. *TAPPI course on tissue runnability*. Appleton, WI.
Jensen, J. (1990). If you're not buying recycled products, you're not recycling. *PIMA Journal*, *72*(6), 49.
Jensen, J. (2005). Voith: More innovations in de-inking. *Papermaking and Distribution*, *15*(4), 18.
Jokinen, K., Siren, K., & Osmonen, R. M. (1995). Quality aspects of recycled fibre. *World Paper*, *220*(10), 31.
Jussila, T., Welt, T., & Wirth, B. (2005). Stock preparation at Palm Worth's PM6. *Solutions*, *37*, 37.
Kleuser J. (1990). Deinking systems for tissue, writing and printing papers. *Appita annual conference*. Rotorua.
Krauthauf E. (1992). Preparation of household collected papers for manufacture of newsprint. *PTS symposium*. Munich.
Linck, E., Nyffeler, U., & Kern, H. R. (1994). Waste paper preparation with integrated residue incineration – An economic concept for tissue production. *Wochenblatt für Papierfabrikation*, *122*(9/10), 360.
Linck, E. (1995). Secondary fiber for high quality tissue. *Papier Carton Cellulose*, *44*(7/8), 34.
Maier, J. (1993). High-class printings made from waste – A necessity and opportunity for paper producers. *Papier*, *47*(2), 56.
Marquat G. (1994). Lightweight printing papers with deinked pulp content. *PTS symposium*. Munich.
Matzke, W. (1995). Deinking for LWC. *Asia Pacific Papermaker*, *5*(1), 32.
Morrison, J. B. (1995). Woodfree printing and writing grade papers requirements of RCF. *Papier Carton Cellulose*, *44*(7/8), 24.
Nielsen, P. B. (1992). Success on paper. *Grafiske Fag*, *6*, 26.
Novak, G., & Malesic, J. (1992). Base paper as one of the most important factors of coated paper quality. *Papier*, *46*(3), 109.
O'Brien, J. (1993). Ponderosa fibers set to harvest the urban forest. *Paper Age*, *109*(2), 8.
Pfitzner, T. (2003). Ready in a flash: For almost a decade stora enso sachsen mill has successfully produced flash-dried market dip from old magazines and newspaper. *Pulp and Paper International*, *45*(4), 18.
Pimley, J. (1991). Washing the waste. *Paper*, *215*(7), 24.
Platier, C. (1995). Printing and writing papers from 100% of waste paper: The Compagnie Papetiere De L'essonne Challenge. *Papeterie*, *187*, 18.
Rankin P. (2005). Trends in 100% DIP newsprint quality. *91st annual meeting of the Pulp and Paper Technical Association of Canada*. Montreal.

Rodden, G. (2005). Norske skog's risk is paying off. *Pulp and Paper International*, *47*(12), 14.

Rubio OA. (1991). Recycled Vs. Virgin Fine Papers:. *A Comparison, Tappi Paper Recycling Symposium*. Atlanta GA.

Sauzedde, C. (1993). Deinking. *Asia Pacific Papermaker*, *3*(1), 33.

Schmid, H. (1994). Deinking system for processing household collected papers, at neutral pH values. *PTS symposium*. Munich.

Schwarz, M. (2000). Design of recycled fiber processes for different paper and board grades. In L. D. Gottsching & H. Pakarinen (Eds.), *Papermaking science and technology* (Vol. 7, pp. 210), Helsinki, Finland: Fapet Oy.

Sector R. F. (1992). Contaminant removal in deinking for tissue systems. *TAPPI tissue runnability seminar*. Nashville, TN.

Seifert P. (1992). Understanding washing of secondary fibre. *TAPPI pulping conference*. Boston, MA.

Seifert, P., & Gilkey, M. (1997). Deinking: A literature review (p. 139), Leatherhead, Surrey, U.K: Pira International.

Siewert W. H. (1988).The use of waste paper in tissue production. *TAPPI pulping conference*. New Orleans, LA.

Siewert W. H., & Horsch G. R. (1984). Operating results and experiences with the Escher Wyss Variosplit. *TAPPI pulping conference*. Atlanta, GA.

Stawicki B., & Barry B (2009). The future of paper recycling in Europe. *Opportunities and limitations*. <www.cost-e48.net/10_9_publications/thebook.pdf> accessed January, 2012.

Stockler A., & Grossmann H. (1994). Haindl increases waste paper inputs in graphic printing papers. *PTS deinking symposium*. Munich.

Vidotti, R. M., Johnson, D. A., & Thompson, E. V. (1997). Influence of toner detachment during mixed office waste paper repulping on flotation efficiency. Part I: Particle fractionation. *Pulp and Paper Canada*, *98*(4), 55.

von Raven, A., & von Troll, S. (1991). Future development of magazine printing papers in Europe and in the USA. *Wochenblatt für Papierfabrikation*, *119*(22), 891.

Watanabe, N. (1991). New inverse polymer emulsion papermaking chemicals. *Japanese Journal of Paper Technology*, *7*, 6.

6 Effects of Recycling on Pulp Quality*

6.1 Introduction

There were very few investigations into the effect of recycling on sheet properties until the late 1960s. From then until the late 1970s, a considerable amount of work was done to identify the effects of recycling on pulp properties and the cause of these effects (Nazhad, 1993, 2005). In the late 1980s and early 1990s, recycling issues emerged stronger than before owing to the higher cost of landfills in developed countries and an evolution in people's awareness. The findings of the early 1970s on recycling effects have since been confirmed, although attempts to trace the cause of these effects are still not resolved (Howard & Bichard, 1992). Recycling significantly reduces the papermaking potential of fibres. This is mainly due to the loss of bonding capacity, which is related to reduced fibre swelling. The surface properties of the fibre also appear to be important. Substantial work has been done to identify the effects of recycling on pulp properties. Several researchers have examined fundamental problems in recycling: how fibres are affected by the recycling processes and the effects on paper properties.

Recycled fibres have lower strength and higher drainage resistance than virgin fibres. The mechanical properties of fibres as well as their ability to swell are diminished after they are exposed to the pulping and drying conditions imposed during the papermaking cycle. The reduction in swelling and the loss of fibre flexibility after drying reduce the strength potential of recovered fibres. Contamination and age degradation also contribute to the reduced strength of secondary fibre (Bhatt, Heitmann, & Joyce, 1991). When a fibre is dried, physical discontinuities in the cell wall are collapsed by high surface tension forces that pull the surfaces together. These surfaces become hydrogen bonded, which reduces swelling in the next cycle. In subsequent beating stages, the recycled fibre will not be able to delaminate and swell as well as virgin fibre. This mechanism is confirmed by experimental work where hydroxyl groups on the cellulose fibre were blocked by derivatisation, reducing the amount of irreversible shrinkage during drying. With lower irreversible shrinkage, there were fewer differences between the properties of sheets made from once-dried and never-dried pulp (De Ruvo, Farstrand, Hagen, & Haglund, 1986).

For equivalent beating times, a sheet containing recycled fibre is less dense and usually more absorptive than virgin fibre stock. The fines created when secondary fibres are beaten consist largely of microfibrils that were strongly coupled to each

*Some excerpts taken from Bajpai, 2006 with kind permission from Pira International, UK. Freeness reduction during beating is much faster for secondary fibres.

Recycling and Deinking of Recovered Paper. DOI: http://dx.doi.org/10.1016/B978-0-12-416998-2.00006-4

other when they were originally dried on the paper machine. When liberated during refining, they increase the specific surface area of suspension more than the swelling potential. They start to behave as fillers, with a small effect on strength but a large effect on the drainage properties (De Ruvo et al., 1986).

In general, the greater the degree of refining of the virgin fibres, the lower the recovery potential of sheet properties that are a direct function of fibre bonding, for example, burst strength and tensile strength (McKee, 1971). Folding endurance of recycled paper is also considerably lower than for sheets made from virgin stock. Sheet density decreases each time the fibres are recycled (Bajpai, 2006). These strength losses may be the result of a loss in bonding potential, either in the strength of the interfibre bonds or in their number (Guest & Weston, 1986). Sheets made from re-pulped fibre have slightly lower brightness than virgin fibre sheets. Strength loss can generally be regained by refining (Chase, 1975). Unfortunately, this usually reduces drainage and production capacity. Increased refining also limits the amount of strength that can be regained by refining in future cycles.

6.2 General Effects of Recycling on Papermaking Properties

An excellent review on this topic has been published by Howard (1990, 1995). Use of recycled fibres in commodity grades such as newsprint and packaging paper and board does not cause noticeable deterioration in product quality and performance at the present utilisation rate (Čabalová, Kačík, & Sivák, 2009). The expected increase in recovery rates of used paper products will require a considerable consumption increase of recycled fibres in higher-quality grades such as office paper and magazine paper. To promote expanded use of recovered paper, an understanding of the fundamental nature of recycled fibres and the differences from virgin fibres is necessary. Essentially, recycled fibres are contaminated, used fibres. Recycled pulp quality is, therefore, directly affected by the history of the fibres, that is, by the origins, processes and treatments that these fibres have experienced. McKinney (1995) classified the history into five periods: fibre furnish and pulp history; papermaking process history; printing and converting history; consumer and collection history; recycling process history. To identity changes in fibre properties, many recycling studies have been conducted in different laboratories. Realistically repeating all the stages of the recycling chain is difficult, especially when including printing and deinking. Some insight into changes in fibre structure, cell-wall properties and bonding ability is possible from investigations using various recycling procedures, testing methods and furnishes. Mechanical pulp is chemically and physically different from chemical pulp, so the recycling effect on those furnishes is also different. When chemical fibres undergo repeated drying and rewetting, they are hornified and can significantly lose their originally high bonding potential (Bouchard & Douek, 1994; da Silva, Mocchiutti, Zanuttini, & Ramos, 2007; Kato & Cameron, 1999; Khantayanuwong, Toshiharu, & Fumihiko, 2002; Khantayanuwong, 2003; Somwand, Enomae, & Onabe, 2002; Song & Law, 2010; Zanuttini, McDonough, Courchene, & Mocchiutti, 2007). The degree of hornification can be measured by water retention value (WRV)

(Kim, Oh, & Jo, 2000). In contrast to the chemical pulps, originally weaker mechanical pulps do not deteriorate but somewhat even improve bonding potential during a corresponding treatment. Several studies (Ackerman, Göttsching, & Pakarinen, 2000; Maloney, Todorovic, & Paulapuro, 1998; Weise & Paulapuro, 1998) have shown good recyclability of mechanical fibres. Adámková and Milichovský (2002) presented the dependence of beating degree (Schopper–Riegler (SR) degree) and WRV from the relative length of hardwood and softwood pulps. Their results showed that the WRV increase in dependence on the pulp length alteration is more rapid in hardwood pulp, but finally this value is higher in softwood pulps. Kim et al. (2000) determined the WRV decrease in softwood pulps with higher numbers of recycling. Use of secondary fibres to furnish at paper production decreases the initial need for woody raw material (fewer trees) but the paper quality is not significantly worse.

For low-yield fibres such as sulphate and sulphite pulps, recycling causes a major reduction in breaking length, burst and fold, with a lesser reduction in apparent density and stretch. Increases in tear, stiffness, scattering coefficient, opacity and air permeability are usually observed. These changes have been largely ascribed to decreased swelling capacity and flexibility of the fibres, which lead to a loss of bonding potential. It has also been speculated that loss of bonding potential could also be due to changes in surface properties during recycling. The first recycling causes the greatest change in any property, regardless of whether the virgin fibre was wet or dry. The loss of intrinsic fibre properties such as bonding capacity, flexibility and swelling potential during papermaking is associated with irreversible hardening or hornification of fibres during drying (Laivins & Scallan, 1993). During drying of low-yield chemical pulps, hydrogen bonds are formed between cellulose chains in the cell wall, and some of these bonds remain unbroken on rewetting. Hornification reduces interfibre bonding, which lowers the paper's density, tensile strength and burst strength (Howard, 1995; Howard & Bichard, 1991; Valade, Law, & Peng, 1994). However, experience with sulphate pulps show that, unlike beaten pulps, unbeaten pulps exhibit increases in tensile strength during the drying/rewetting process (Howard, 1995; Howard & Bichard, 1991; Valade et al., 1994). The increase in tensile strength of the unbeaten kraft fibres may be explained by the stress-releasing effect (Valade et al., 1994) or by decurling (Howard & Bichard, 1991; Howard, 1995). This implies that the effect of hornification on paper properties is determined not only by the chemical nature of pulp fibres but also by their physical state. The amount of lignin in the pulp also influences the recycling characteristics (Klungness & Caulfield, 1982; Laivins & Scallan, 1993). Bleached pulps showed greater hornification than unbleached pulps. Hornification appears to be a carbohydrate phenomenon (Klungness & Caulfield, 1982; Laivins & Scallan, 1993). Gurnagul, Ju, and Page (2001) demonstrated that the primary cause of tensile strength reduction on drying was the loss of interfibre bonding strength, with a minor reduction caused by a loss in interfibre bonded area.

At the repeated use of the secondary fibres, the paper properties alter because of fibre deterioration during the recycling, when many alterations are irreversible. The alteration depth depends on the cycle's number and the way the fibres are used. The main problem is the decrease in the mechanical properties of the secondary pulp with continuing recycling, mainly the paper strength (Garg & Singh, 2006; Geffertová, Geffert, & Čabalová, 2008; Hubbe & Zhang, 2005; Jahan, 2003; Khantayanuwong et al., 2002; Sutjipto, Li, Pongpattanasuegsa, & Nazhad, 2008). This decrease is an effect of many alterations, which can but need not arise in the secondary pulp during the recycling process. The recycling causes hornification of the cell walls, which results in the decline of some pulp properties. It is due to irreversible alterations in the structure of the cells during drying (Diniz, Gil, & Castro, 2004; Kim et al., 2000; Oksanen, Buchert, & Viikari, 1997). Worse properties of recycled fibres than primary fibres can be caused not only by hornification but also by the decrease in the hydrophilic properties of the surface of the fibres during drying because of the redistribution or migration of resin and fatty acids to the surface (Nazhad, 2005; Nazhad & Paszner, 1994). Okayama (2002) observed the enormous increase of the contact angle with water, which is related to fibre inactivation at recycling. This process is known as irreversible hornification.

Paper recycling saves the natural wood raw stock, decreases the operation and capital costs to the paper unit, decreases water consumption and finally, yet importantly, it gives rise to the preservation of the environment. A key issue in paper recycling is the impact of energy use in manufacturing. Processing waste paper for paper and board manufacture requires energy that is usually derived from fossil fuels, such as oil and coal. In contrast to the production of virgin-fibre-based chemical pulp, waste paper processing does not yield a thermal surplus and thus thermal energy must be supplied to dry the paper web. If, however, the waste paper is recovered for energy purposes, the need for fossil fuel would be reduced and this reduction would have a favourable impact on the carbon dioxide balance and the greenhouse effect. Moreover, pulp production based on virgin fibres requires consumption of round wood and causes emissions of air-polluting compounds, as does the collection of waste paper. For better paper utilisation, an interactive model, the optimal fibre flow model, considers both a quality (age) and an environmental measure of waste paper recycling (Byström & Lönnstedt, 1997).

Very few recycling studies have been conducted on mechanical and ultra-high-yield pulps (Ferguson, 1992). Mechanical pulps recycle in an entirely different way from chemical pulps (Chatterjee, Roy, & Whiting, 1992; Howard, 1995; Howard & Bichard, 1991). Refined chemical pulps lose density and tensile strength, but mechanical pulps show small gains in strength and density (Howard, 1995; Howard & Bichard, 1991). There are two reasons for this different behaviour. The first is that, unlike chemical pulps, the walls of mechanical pulp fibres are not extensively delaminated in the wet state. Hornification during drying is therefore limited and will have little, if any, effect on interfibre bonding. The reason for the increase in strength and density is believed to be progressive flattening and flexibilising of the stiff, uncollapsed fibres during each successive papermaking and re-slushing cycle. The flatter, more flexible fibres bond better and give a thinner, denser sheet.

Mechanical and ultra-high-yield pulps show little or no irreversible hornification because of the presence of lignohemicellulose gel in fibre walls that prevents any direct contact between cellulosic surfaces during drying (Laivins & Scallan, 1993). However, results (Howard, 1995; Howard & Bichard, 1991) have shown increases in fibre collapse, fibre bonding, sheet density and air resistance of spruce mechanical pulps despite the reduction in fibre saturation point. If the nature of hornification is due to the loss of swelling of cell-wall material, its influence alone cannot account for the drying effects on paper properties of unbeaten chemical and softwood mechanical pulps. Another report (Bouchard & Douek, 1994) indicated that there is no direct correlation between the changes in chemical composition of fibres during recycling and variations of strength properties.

It appears that drying of lignocellulosic fibres is a complex phenomenon and requires further investigation. Different yields of chemithermomechanical pulps (CTMPs) corresponding to the extent of chemical pre-treatment influence development of the sheet properties during recycling. Law (1996) recycled thermomechanical pulp (TMP), CTMP and chemimechanical pulp (CMP) made from spruce and aspen without deinking chemicals but with the retention of fines. Compared with CTMP, the CMP process had stronger sodium sulphite liquor, higher temperature and longer treatment time; it gave yields of 85% for spruce CMP and 80% for aspen CMP. Spruce and aspen CTMP had yields of 92% and 90%, respectively. The influences of repeated drying and re-slushing depended on the yield of pulps and wood species.

TMP and CTMP spruce pulps showed some increase in sheet density and tensile strength. Recycling of CMP exhibited a decrease of 15% in density and 30% in tensile strength. When the raw material was aspen, CTMP already showed a declining tendency in those properties. These results indicate that the CMP suffered noticeable reduction of bonding potential in the recycling process, but mildly treated CTMP hardly suffered at all. Chemical pre-treatment to a yield of 80–90% removes lignohemicellulosic gel from cell-wall lamellae to such an extent that it leaves the structures or some structure exposed to an irreversible hornification during drying. The exact yield boundary depends on the wood species. Differences in recycling behaviour probably exist between pulps made of different wood species. In a study of TMP and CTMP made of spruce, pine, poplar and birch, the pulps were recycled with a broke-like procedure without fines recovery (Bayer, 1996). Fines were lost, and the tensile and burst strengths dropped in all the pulps. The drop was lowest in spruce pulps, slightly more in pine pulps and highest in hardwood pulps. The conclusion is that the bonding ability of fibre fractions in TMP and CTMP made of hardwood suffered more in recycling than those of softwood. Recycling studies conducted by Liebe (1995) with various unbleached and bleached kraft pulps showed very similar development of sheet properties independent of bleaching method (elemental chlorine free or totally chlorine free), with hardwood pulps being more prone to hornification than softwood. Sheet properties that depend on bonding ability such as modulus of elasticity and Scott bond (delamination resistance) have, very logically, responded in recycling experiments. They have increased with increasing sheet bonding and vice versa. The stiffness of sheets has very consistently declined with all pulp types in recycling trials. According to paper physics, the resistance to bending action increases with growing modulus of elasticity or thickness of the sheet. The reason for less rigid sheets in recycled low-yield pulps is probably the loss in fibre bonding (E modulus). For mechanical pulps, the reason is the loss of sheet thickness owing to flattening

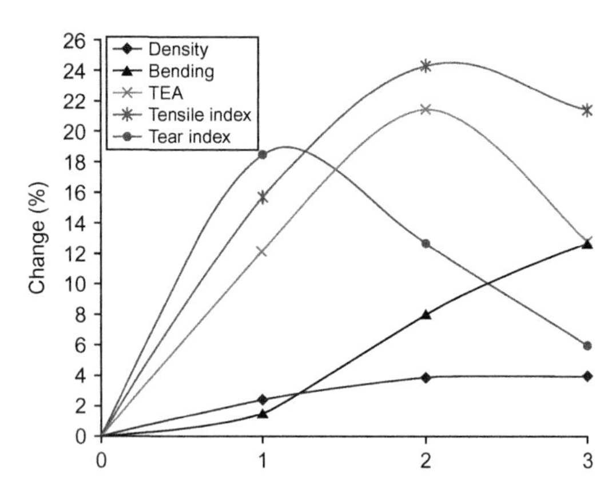

Figure 6.1 Strength properties of unbeaten softwood market pulp.
Source: Sutjipto et al. (2008); reproduced with permission.

of fibres. A dried (hornified) chemical pulp provides better dimensional stability for the sheet than undried pulp. Drying actually decreases the hygroexpansion ability of fibres. Transmission electron microscopy was used by Brandstrom, Joseleau, Cochaux, Giraud-Telme, and Ruel (2005) to investigate the ultrastructure of recycled pulp fibres originating from household refuse collection plants and intended for packaging paper production. It assessed three recovered paper grades – 1.02 sorted mixed papers and board, 5.02 mixed packaging and 1.05 old corrugated container (OCC) – and recycling processes that included pulping, screening, cleaning and refining, with emphasis on surface and internal fibrillation as well as xylan localisation. A large heterogeneity in fibre ultrastructure was observed within and between the grades. Screening and cleaning steps had no detectable effects but refining clearly increased cell-wall delamination and surface fibrillation.

To study the changes in fibre properties in the process of recycling, different chemical pulps (i.e. hardwood and softwood market pulps as well as laboratory pulps) and CTMP pulp were recycled three times, and some of their physical properties were determined and analysed (Sutjipto et al., 2008). It was observed that the recycling effect on chemical pulps (kraft pulps) is similar regardless of the wood species. The rate of strength loss was more pronounced for the laboratory pulps than the market pulp. In general, the rate of tensile strength loss was always twice that of the density regardless of wood species. Recycling CTMP suggested that the gain in density was almost negligible compared with the gain in tensile strength. Recycled chemical pulps showed a very high bending stiffness compared with virgin pulp. Bending stiffness of CTMP pulp also benefited from recycling (Figures 6.1–6.5). It was inferred that recycled pulp could be a better choice for some grades of paper and board, a point that has been overlooked by papermakers.

Immunolabelling of xylans showed that they were distributed rather evenly across cell walls. They were also present on fines. Two different mechanisms for

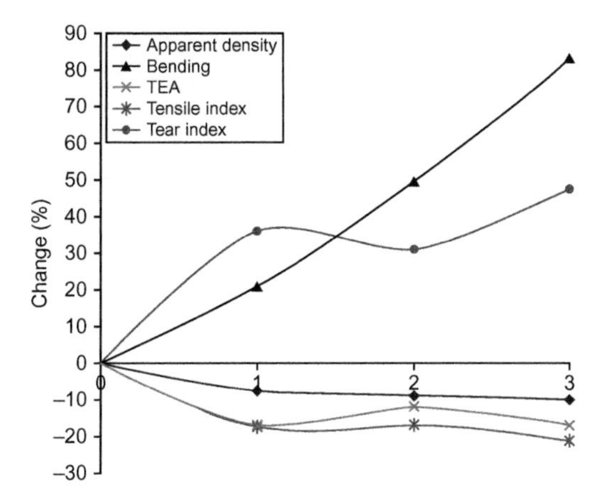

Figure 6.2 Effect of recycling on market softwood pulp (beaten).
Source: Sutjipto et al. (2008); reproduced with permission.

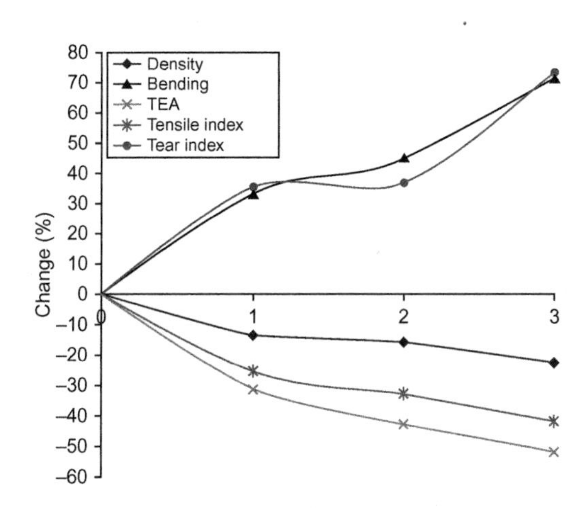

Figure 6.3 Properties of laboratory scale softwood pulp.
Source: Sutjipto et al. (2008); reproduced with permission.

fibre delamination and surface fibrillation were found. The first implies that internal and external fibrillation takes place simultaneously across the cell wall. The second implies successive peeling of layers or sublayers from the outside towards the inside. It is suggested that recycled fibres of chemical pulp origin follow the first mechanism and that recycled fibres containing lignin binding the cell-wall matrix follow the second mechanism. Because several recycled fibres were severely delaminated and almost fractured, it was suggested that, to produce quality packaging paper, the recycled pulp needs to contain a significant proportion of fibres with intrinsic strength.

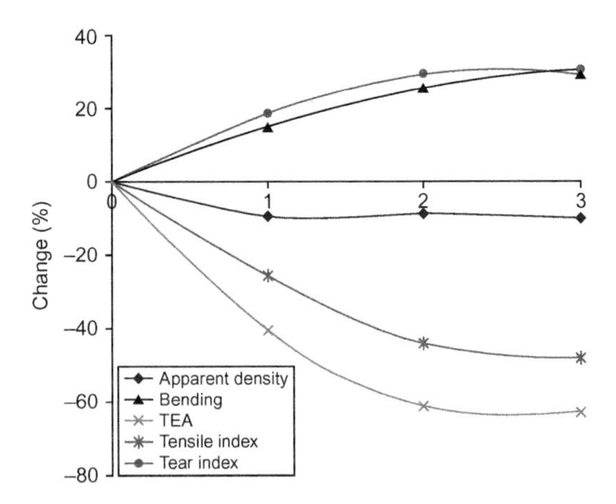

Figure 6.4 Strength properties of laboratory hardwood pulp.
Source: Sutjipto et al. (2008); reproduced with permission.

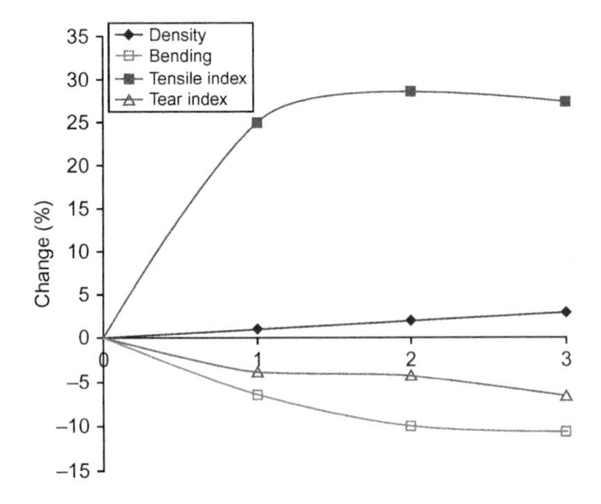

Figure 6.5 Strength properties of market CTMP.
Source: Sutjipto et al. (2008); reproduced with permission.

6.3 Factors Influencing Recycling

Recycled pulp quality is influenced by the following factors (Howard, 1995):

- Pulp type
- Processes of papermaking, converting, use and storage
- Recycling operations

Types of pulp have different responses to recycling. Chemical pulps that are not refined do not behave in the same way as refined chemical pulps. The actual behaviour depends on whether or not the pulps have been dried. Never-dried chemical

Table 6.1 Effect of Recycling on Sulphate Pulp Properties (Bleached Beaten Pulp)

	0	1	2	3	4	5
Density (g/cm^3)	0.704	0.658	0.645	0.637	0.613	0.621
Breaking length (km)	10.81	10.29	9.91	9.30	9.35	9.58
Burst index (kPa m^2/g)	8.97	8.29	8.14	7.79	7.39	7.45
Scott bond (J/m^2)	469	347	298	260	242	248
Tear index (mN m^2/g)	8.80	9.69	10.14	10.41	11.43	11.24
Scattering coefficient (m^2/kg)	21.64	26.10	27.96	27.90	28.66	28.95
Air resistance (s/100 mL)	936.7	527.2	457.4	217.5	217.5	224.1
MIT double folds	1868	1550	1634	1494	1895	1781

Source: Based on Howard and Bichard (1992, 1993)

pulps, even in the unrefined state, are quite swollen and will therefore make a strong virgin sheet. However, these pulps will lose bonding potential owing to hornification during drying, so they will form a weaker, bulkier sheet after re-pulping. On the other hand, dry pulps have already been hornified by the pulp drying process. Further drying during virgin papermaking does not significantly reduce the already low level of swelling, so the recycled pulp would not be expected to lose much bonding potential. Yet the fibres of dry pulps, especially bleached pulps, are frequently curly, which reduces the pulp strength potential (Mohlin & Alfredsson, 1990; Page, 1985).

Howard and Bichard (1992, 1993), while studying the recycling potential of 11 Canadian pulps using standard laboratory procedures, found that different pulp types showed very different recycling effects. Mechanical pulp fibres became flatter and more flexible, giving a denser, stronger sheet. Beaten chemical pulp fibres hornified, resulting in a bulkier, weaker sheet. Unbeaten chemical pulp fibres were initially curly; recycling removed the curl. A blend of mechanical and chemical pulp revealed that these effects occur at different rates. No pulp showed evidence of fibre strength loss or fibre embrittlement. In these laboratory experiments, fines loss during sheet-making affected the magnitude of the sheet properties but not the trends. CTMP (once dried) behaves like stone groundwood and TMP, but with the derailed difference that almost all the pulp property development occurred in the first cycle, whereas for the stone groundwood and TMP, the properties changed progressively over several cycles. This is attributed to a reduction in fibre wall rigidity arising from the chemical pre-treatment. Tables 6.1 and 6.2 show the effect of recycling on sulphate and sulphite pulp properties (Howard and Bichard, 1992, 1993).

Not much information is available about the recycling behaviour of hardwood and the non-wood pulps (Mansito, Agnero, & Sosa, 1992; Yamagishi & Oye, 1981). A bleached bagasse soda pulp lost bonding capacity owing to swelling reductions. Hornification was less pronounced in an unbleached bagasse CTMP owing to the lignin content. The greater the initial degree of chemical pulp refining, the greater the change of pulp quality on recycling (Higgins & MacKenzie, 1963; Lundberg & De Ruvo, 1978; McKee, 1971). This loss of recycling potential corresponds to the loss of fibre wall swelling. Compared with the unrefined pulp, refined pulp is more

Table 6.2 Effect of Recycling on Sulphite Pulp Properties (Bleached Beaten Pulp)

	0	1	2	3	4	5
Density (g/cm^3)	0.662	0.599	0.602	0.585	0.581	0.585
Breaking length (km)	5.29	4.48	4.36	4.19	3.93	4.01
Burst index (kPa m^2/g)	2.88	2.46	2.31	2.29	2.33	2.29
Scott bond (J/m^2)	320	171	157	148	156	148
Tear index (mN m^2/g)	5.28	5.74	5.90	6.04	6.07	5.99
Scattering coefficient (m^2/kg)	36.85	44.45	44.49	45.59	44.35	43.52
Air resistance (s/100 mL)	129.9	75.3	66.5	55.3	56.8	53.5
MIT double folds	29	12	9	10	10	11

Source: Based on Howard and Bichard (1992, 1993).

internally delaminated. These delaminations close up during the drying process. The greater the extent of initial swelling, the greater the change in pulp properties after drying. Fines are generated during refining; they will be present when the recycled fibres are reused unless they are lost from the system during papermaking or subsequent recycling. Compared with the virgin pulp, the freeness of recycled pulp is usually lower. This has been attributed to the generation of fines (Bovin, Hartler, & Teder, 1973; De Ruvo & Htun, 1981; De Ruvo, Htun, & Ehrnrooth, 1978). Recycled fines fill the sheet but do not improve the bonding (Szwarcsztajn & Przybysz, 1978). However, another study reported that recycled fines could make a modest contribution to the recycled pulp strength, but this contribution is much less than could be obtained from fines created by subsequent refining (Hawes & Doshi, 1986). Stock preparation also reduces the fibre length in addition to creating fines. So, compared with virgin pulp, the average fibre length in the recycled pulp will be shorter unless the shorter material is lost during recycling. The fines content will also be higher. So there will be significant loss of swelling, a reduction in fibre length and an increase in fines content in case of refined recycled chemical pulp. This will reduce the recycling potential.

Not much information is available about the effect of wet pressing (Carlsson & Lindstrom, 1984; Pycraft & Howarth, 1980). It seems that pressing has a detrimental effect on the recycling potential; this is due to the 'healing' of internal delaminations, the same process that was proposed to explain irreversible hornification. In fact, pressing helps bring together the delaminations in the fibre, producing a more hornified fibre after drying (Howard &Bichard, 1991). Nevertheless, it seems that pressing has a less detrimental effect than drying on the recycling potential of fibres. Pycraft and Howarth (1980) showed that the recycling potential of the more highly wet-pressed paper was lower than the recycling potential of the less highly pressed paper. For everyday papermaking, this affects only the recyclability of dry paper and not of wet broke until there is a high level of express solids.

In mechanical pulps, WRVs dropped by less than 5%, even when pressed to solids contents as high as 73% (Carlsson & Lindstrom, 1984). The fibre walls of mechanical pulps do not delaminate extensively, even if mildly refined at low consistency, so pressing would not be expected to cause significant reduction in swelling, and hence recycling potential. Pressing might be expected to flatten mechanical pulp fibres and make them more flexible, increasing the recycled pulp strength rather than reducing it (Chatterjee et al., 1992; Howard, 1990, 1995).

Papermaking always involves the use of chemical additives. Paper additives and manufacturing aids used in paper production from virgin pulps re-enter the paper cycle together with the waste paper, at least in certain proportions. They affect the waste paper treatment process and the effectiveness of the chemicals added in the manufacture of recycled fibre-based papers (retention aids, sizing agents and strength agents). It was found that rosin and alum sizing in the original paper caused a further loss of recycling potential beyond what would have been expected anyway (Eastwood & Clarke, 1977; Guest & Voss, 1983; Horn, 1975). It is likely that the sized fibres retain their hydrophobic surfaces and somewhat inhibit bonding in the recycled sheet (Guest & Voss, 1983). The neutral sized paper (using an alkyl ketene dimer (AKD) size) retained more of its original burst strength, but less of its breaking length (Forester, 1985). The recycling potential of paper sized with AKD was considerably worse than for paper sized with rosin and alum (Guest & Voss, 1983).

Lindstrom and Carlsson (1982) have shown how much the chemical environment affects the fibre swelling of virgin chemical pulp and the strength of paper made from it. Unbleached pulps dried under acid conditions in the first making showed considerably less swelling, hence strength, than pulps initially dried under alkaline conditions. The authors were able to identify the importance of the acid group content of the pulp (and its ionic form) to the initial swelling. Bleached pulps are unaffected by initial pH because of their low acid group content. All these studies were conducted on chemical pulps. McComb and Williams (1981) checked paper acidity, which may be present because of the manufacturing processes or chemicals and showed that acid penetrated the open amorphous regions of the fibre-cut carbon–oxygen glycosidic bonds, lowering the overall degree of polymerisation (acid hydrolysis). On the whole, the chemical environment during the first sheet-making controls the swelling of the pulp after drying and recycling. Drying is not only a problem for recyclers but also for papermakers using virgin pulps. Indeed, a huge difference exists between never-dried fibres from integrated mills and dried fibres from non-integrated mills (Guest & Weston, 1986). It is well known that paper manufactured with never-dried fibres has higher mechanical strength than paper made with dried fibres. The strength decrease is greatest after the first time the pulp is used. The drying conditions typically prevailing in a mill environment must be expected to have the most negative and lasting impact on the fibre properties. The major effects on the fibre, and hence its recycling potential, occur during drying. Much work has been done in this area. Bawden and Kibblewhite (1997) examined four handsheet formations, four drying and rewetting cycles and the associated interstage refining treatments to see how they affected the fibre walls of unbleached market kraft pulp from radiata pine. The initial drying of the fibres caused the greatest change in fibre dimensions. Further drying treatments continued to reduce fibre size and increased collapse, but to a lesser extent. Fibre walls became more dense as the fibre size and wall area decreased. Water accessibility was reduced by drying. The refining of dried fibres caused them to change shape and size. Handsheet tensile strength was lost when the fibres were first dried. The unbleached radiata pine pulp was resilient to the drying, rewetting and refining treatments. Jayme (1944) observed a drop in WRV after air drying or drying at 70°C for an unbleached kraft pulp.

Treiber and Abrahamson (1972) noted a substantial drop in fibre saturation point when drying a dissolving pulp at 120°C instead of 25°C. Okayama, Okada, and Oye (1982) reported that freeze drying and microwave drying produced their own distinct effect on the WRV of the recycled pulp. After studying the effects of recycling on the properties of masson pine pulp, Liying, Beihai, Yinggang, and Huaiyu (2005) reported that more damage to fibre quality occurs with a severe drying process. Sturmer and Gottsching (1979) studied the effect of drying history on the hygroscopic behaviour of recycled fibres. A recycled kraft pulp sheet that was dried rapidly absorbs significantly less water from humid air than a recycled sheet slowly dried at room temperature. More intense drying brings still more hornification that further reduces ingress of water into the cell-wall matrix by irreversibly closing the micropores. This prevents fibre swelling. In contrast, the initially lower hygroexpansive ability of mechanical fibres hardly changes during recycling (Hoppe & Baumgarten, 1997). As a fibre dries, the internal pore structure generated during preparation collapses and leads to a partly irreversible hornification, which is most marked for chemical pulps (Kibblewhite & Dell Bawden, 1995). Strong hydrogen bonds are formed within the fibre walls and between the fibres, and they are too strong to be broken simply by soaking (Phipps, 1994). The thickness of the fibre decreases with drying, which implies shrinkage of the internal fibre volume. Weise and Paulapuro (1998) have studied the relation between fibre shrinkage and hornification. They observed three phenomena: deformation of the fibre cross-section, wrinkling of the fibre surface and progressing shrinkage. This effect is even more marked when the drying temperature is high (Weston & Guest, 1985).

The effect of recycling on fibre properties depends on the pulping or papermaking history. However, the effects fade away after the fifth cycle. Some recycled pulps at worst loose 10–30% of their original strength after the fifth cycle, so, they are still a valuable raw material. The cited literature suggests that paper strength is a consequence of many different interactions. Therefore, attributing the strength loss or gain to a single factor might be an oversimplification of the whole issue. However, contemporary observations overall support the fact that the specific bond strength plays a major role in these developments (Nazhad, 2005). Based on this fact, the loss in paper strength could be restored by increasing the surface bonding agents of the fibres.

Pycraft and Howarth (1980) found that the drying cylinder temperatures on a pilot paper machine affected the recycling potential. Lundberg and De Ruvo (1978), De Ruvo et al. (1978) and De Ruvo (1980) reported that increased drying temperatures during the drying of Formette Dynamique prepared sheets gave rise to reduced swelling and restrained drying reduced the swelling a little more. They used a commercial bleached kraft pulp. These researchers noted that pulp prepared from paper dried at the higher temperature could not regain the WRV of the initial pulp, even after prolonged beating. Chemical pulps were used in all these cases and hornification of the fibre wall accounts for the basic phenomenon. Drying procedures have been shown to affect more than just bonding. Calendering also has an important effect on recycling potential but its effects have not yet been extensively studied. Gottsching and Stiirmer (1978) showed that calendering (steel-on-steel nips) and supercalendering (steel-on-paper nips) shorten fibres and reduce the breaking length and tear strength of paper after recycling. The higher the load, the greater the effect and the greater the loss in mechanical properties; even zero-span tensile is dramatic. Gratton (1991) studied the effect of different calendering methods and showed that in temperature-gradient calendering and extreme calendering treatment at a temperature over 200°C, the rheology of the fibre surface is altered and fibres are permanently deformed and flattened. The surface of a handsheet from extreme calendering treatment shows fibre breaking under the electron microscope. Calendering also has a significant impact on the sheet's initial strength properties. In temperature-gradient and extreme calendering treatment, the fibres are generally permanently deformed and flattened. Damage done in calendering is not reversible by re-slushing and recycling. Calendering significantly reduces the elastic modulus of all handsheets, probably because of bond breakage and fibre damage.

Printing operations mainly affect the optical properties of recycled pulp. It is not just the amount of residual ink after deinking that controls the recycled pulp brightness but also the particle size distribution. Ink particles smaller than 50 µm cannot be resolved individually by the human eye, but their presence lowers the brightness and lends a grey tinge to the pulp. Above 50 µm, ink particles appear as specks. Instruments are now available for determining the particle size and the ink. With these instruments, the recycler can produce brighter, cleaner pulp.

The quality of recycled pulp is significantly influenced by events before the recycling process, including a variety of unit operations such as mechanical, chemical and thermal treatment. It is now well known that mechanical action at higher consistencies will impart significant curl and microcompression into fibres. This effect increases with consistency, resulting in a weaker, bulkier, more stretchy pulp (de Grace & Page, 1976). In recycling operations, high-consistency mechanical action can occur at the pulper, in dispersion units and in bleaching. The permanence of the curling effect will depend on the fibre furnish, as well as on the mechanical treatments and temperatures that follow the high-consistency stage, which may tend to remove the imposed curl (Page, Barbe, Seth, & Jordan, 1984).

Not much information is available about consumer effects on recycling potential. Apart from adding further contamination to an already contaminated fibre, a very significant post-consumer effect could be ageing. If the newsprint is stored for a longer time before recycling, the strength properties will be lowered (Andrews, 1990). The quality of recycled pulp is also influenced by consumers and collectors as totally different grades of paper are mixed. The quality of the recycled pulp then falls somewhere between the quality of the best fibre in the waste paper and the worst. The important implication is that beyond any loss of quality arising from the reuse of a particular paper grade, there will be a further change due to 'dilution' by other paper grades.

Processing of waste paper involves the use of chemicals. For example, during the re-pulping of old newspapers and magazines, typical additives at the pulper will include sodium hydroxide, hydrogen peroxide, sodium silicate, surfactants (such as fatty acid soaps) and chelating agents. Although sodium hydroxide may promote pulp strength, an opposite result could arise from the presence of surfactants. If surfactants are not washed out and carry over into the paper machine, they may lower the surface tension and reduce bond strength and paper strength (Springer, Dullforce, & Wegner, 1986).

6.4 Techniques to Enhance Strength Properties of Recycled Fibres

Several techniques have been used to enhance the strength properties of recycled fibres (Bhardwaj, Bajpai and Bajpai, 1997; Bhatt et al., 1991; Howard, 1995; Kessel & Westenbroek, 2004; Sarkar, 1996; Sharma, Rao, & Singh, 1998). Strength loss generally can be regained by refining (Chase, 1975). Unfortunately, this usually reduces drainage and production capacity. Increased refining also limits the amount of strength that can be regained by refining in future cycles The use of chemical additives, which improve the strength properties without changing the re-pulping requirement, can provide an alternative method to refining (Chan, 1976). Two resins often used are an anionic polymer (Chan, 1976), which is capable of facilitating hydrogen bonding, and a cationic polymer, which is capable of forming strong electrostatic bonds between fibres and fines. These resins improve the dry strength of paper by increasing both the strength and the area of the interfibre bonds (Linke, 1968). Treatment of waste paper with sodium hydroxide increases the freeness and the strength properties of recycled fibre (Eastwood & Clarke, 1977). Sodium hydroxide treatment promotes fibre swelling, thereby increasing fibre flexibility and surface conformability. Both alkaline treatment and delignification can improve the papermaking potential of recycled fibres.

Oxygen–alkali delignification has been studied as a means of improving strength properties in OCC recycled pulp (De Ruvo et al., 1986). The delignification treatment was found to improve bonding and strength characteristics, probably because of softening, swelling and lignin removal. The strength improvement in the fibre is especially noticeable in the higher burst value and strain-to-failure value at a given drainage rate (Markham & Courchene, 1988)

Bhatt et al. (1991) examined the effect of several techniques for enhancing the strength of secondary fibre. Re-pulping under alkaline conditions and refining are the most commonly used methods to improve the strength of secondary fibres. High-shear-field (HSF) treatment produces an effect similar to refining while producing less fines. The best results are obtained using a combination method of alkali treatment followed by HSF treatment. Strength properties are higher than refining and comparable to virgin pulp in some cases.

Moderate low-consistency refining can be used to improve the papermaking potential of recycled fibres and reduce paper manufacturing costs. Nevertheless, incorrect refining produces negative effects such as increased water resistance, and reduced fibre length, strength and bulk (Lumiainen, 1992, 1994); the fibre potential of secondary fibre is not comparable to that of virgin fibre. Total energy consumption in low-consistency refining of recycled fibres with Conflo refiners is typically 30–60 kW h/tonne. In some cases, when refining strong kraft waste, it can exceed 100 kW h/tonne. High-consistency refining has been examined as a method of developing pulp properties without undue fines generation and drainage rate reduction (Fellers, Htun, Kolman, & De Ruvo, 1978). However, the energy consumption is much higher and the pulp properties are different from those developed by low-consistency treatments. Beneficial strength improvements are achieved with some dispersion treatments that use high consistencies (Rangamannar & Silveri, 1989); these processes may see more development in the future.

Chemical additives that improve strength properties without changing the repulping requirement can provide an alternative method to refining (Chan, 1976). Two resins often used are an anionic polymer (Chan, 1976), which can facilitate hydrogen bonding, and a cationic polymer, which can form strong electrostatic bonds between fibres and fines. These resins improve the dry strength of paper by increasing the strength and the area of the interfibre bonds (Linke, 1968). Treatment of waste paper with sodium hydroxide increases the freeness and the strength properties of recycled fibre (Eastwood & Clarke, 1977). Sodium hydroxide treatment promotes fibre swelling, which increases fibre flexibility and surface conformability. Both alkaline treatment and delignification can improve the papermaking potential of recycled fibres. Oxygen–alkali delignification has recently been studied as a means of improving strength properties in OCC recycled pulp (De Ruvo et al., 1986). Cationic starch is the most common additive for improving the strength of recycled fibres. Howard & Jowsey (1989) studied the mechanism by which the addition of cationic starch at the wet end increases the tensile strength of paper. The long-fibre fraction of an unbeaten Scandinavian kraft dry-lap pine pulp was used, together with a proprietary quaternary ammonium cationic starch, to give sheets of low initial bond strength. Measurements were made of formation and fibre collapse. It was found that cationic starch increases bond strength per unit bonded area of the sheet with only marginal effects on the relative bonded area. The initial 0.5% starch addition had the greatest effect, which was compared with starch bonding to low-grammage fibre networks of glass. HSF treatment, in a pulp consistency range of 10–20%, can be used to produce an effect similar to refining. The fibre wall structure is modified by the brushing and bending action, which increases the bonding area. The HSF treatment produces fewer fines than refining, hence less freeness loss. A combination of alkali and HSF treatment may be a better alternative to obtain high product quality from secondary fibre. The strength properties of the recycled paper obtained by the combination alkali/HSF treatment are higher than those obtained by refining and are sometimes comparable to virgin pulp. The combination treatment seems most effective in restoring ring crush and Concora flat crush strength. This treatment offers a potentially valuable, practical method of increasing the use of secondary fibre in boxboard as well as corrugating medium.

Enzymes can be used to increase the freeness of the secondary fibre without affecting the quality of the final product. A preparation of cellulase and hemicellulase at 0.2% enzyme concentration, 30min, 10% pulp consistency, pH 5 and 45°C was the most economical and practical level for the pulps investigated (Bhatt et al., 1991). Using monocomponent enzymes, Kessel and Westenbroek (2004) found positive results on fibre properties important for papermaking. Relative bonded area, flexibility and fibrillation of the treated fibres increased whereas fibre length remained constant. Trials using industrial pulp with the monocomponent enzymes showed that drainability of the pulp and porosity of the final paper increased. Flexibility and relative bonded area were also affected. Lower doses of enzymes were found to be effective compared with the dosage used under laboratory conditions.

Several chemicals, including modified acrylamides, modified starches and an enzyme mixture of cellulase and hemicellulase (Pergalase), were studied (Bhardwaj et al., 1997) to see how effectively they increased the strength and drainage properties of corrugated kraft cuttings and boxes containing reclaimed fibres. An anionic polyacrylamide gave the best results, but all the chemical additives gave significant improvements in drainage and pulp strength. The Pergalase enzyme improved drainage by 40% but had no effect on pulp strength. In another study (Sarkar, 1996), six mill trials were established using cellulase and hemicellulase enzymes in conjunction with synthetic drainage aid polymers to improve the strength and drainage properties of recycled fibres. Improvements to machine speed as a result of fibre freeness were noted. Improved strength was achieved by modifying the fibre surface and the activity of the drainage aid polymer.

The blending of high-yield recycled pulp with kraft recycled pulp can also improve strength properties. This may have some attractive economic advantages. Initial studies are needed to find the optimum blending proportion of recycled high-yield and kraft pulps and to examine possible combinations with refining, HSF treatment, chemical addition, alkali treatment or oxygen–alkali treatment for improved pulp qualities.

Fractionation is quite a common technique in board manufacture for upgrading the recycled pulp performance. This technique separates fibres of different origin and therefore reduces the harmful effects of pulping (Kessel & Westenbroek, 2004; Yu, Defoe, & Crossley, 1994). Fractionation trials have shown that fractionation of recycled pulp can improve and control pulp and paper characteristics by redistribution of fines and ash in the pulp. Putz, Torok, and Gottsching (1989) have reported that in many German board mills which produce test liner and corrugating medium using a waste-based, kraft-containing furnish, the stock is separated by screening and the long-fibre fraction is beaten separately. It is then reused according to one of the following schemes. The long and short fibre fractions may be converted to paper either on two paper machines producing two different paper grades such as test liner and medium, or as different plies in one product. Remixing of the long-fibre fraction after treatment with the untreated short fibre fraction produced only marginally better characteristics.

References

Ackerman, C. H., Göttsching, L., & Pakarinen, H. (2000). Papermaking potential of recycled fiber. In L. Göttsching & H. Pakarinen (Eds.), *Recycled fiber and deinking*. Finland: Papermaking Science and Technology. Chapter 10, pp (book 7).

Adámková, G., & Milichovský, M. (2002). Beating of mixtures hardwood and softwood pulps. *Papír a celulóza*, *57*(8), 250–254.

Andrews, W. C. (1990). Contaminant removal, timely use vital to quality ONP fiber yield. *Pulp and Paper*, *64*(9), 126.

Bajpai, P. (2006). *Advances in Recycling and deinking* (180 pp.). U.K: Smithers Pira.

Bawden, A. D., & Kibblewhite, R. P. (1997). Effects of multiple drying treatments on kraft fiber walls. *Journal of Pulp and Paper Science*, *23*(7), J340.

Bayer, R. (1996). Einfluß Des mehrfachen recyclings Auf Ausgewählte eigenschaften von TMP Und CTMP. *Allgemeine Papier-Rundschau*, *120*(27), 739.

Bhardwaj, N. K., Bajpai, P., & Bajpai, P. K. (1997). Enhancement of strength and drainage of secondary fibres. *Appita Journal*, *50*(3), 230.

Bhatt, G., Heitmann, J. A., & Joyce, T. W. (1991). Novel techniques for enhancing the strength of secondary fiber. *TAPPI Journal*, *74*(9), 151.

Bouchard, J., & Douek, M. (1994). The effects of recycling on the chemical properties of pulps. *Journal of Pulp and Paper Science*, *20*(5), J131.

Bovin, A., Hartler, N., & Teder, A. (1973). Change in pulp quality due to repeated papermaking. *Paper Technology*, *14*(5), 261.

Brandstrom, J., Joseleau, J. P., Cochaux, A., Giraud-Telme, N., & Ruel, K. (2005). Ultrastructure of commercial recycled pulp fibers in the production of packaging paper. *Holzforschung*, *59*(6), 675.

Byström, S., & Lönnstedt, L. (1997). Paper recycling: Environmental and economic impact. *Resources, Conservation and Recycling*, *21*, 109–127.

Čabalová, I., Kačík, F., & Sivák, J. (2009). Changes of molecular weight distribution of cellulose during pulp recycling. *Acta Facultatis Xylologiae Zvolen*, *51*(1), 11–17.

Carlsson, G., & Lindstrom, T. (1984). Hornification of cellulosic fibers during wet processing. *Svensk Papperstidning*, *87*(15), 119.

Chan, L. (1976). Dry strength resins: Useful tools for papermaking. *Pulp and Paper Canada*, *77*(6), 43.

Chase, R. (1975). Supplementing kraft linerboard furnish with old corrugated. *TAPPI Journal*, *58*(4), 90.

Chatterjee, A., Roy, D. N., & Whiting, P. (1992). Effect of recycling on strength, optical and surface properties of handsheets. *CPPA 78th annual meeting*. Montreal.

da Silva, T. A., Mocchiutti, P., Zanuttini, M. A., & Ramos, L. P. (2007). Chemical characterization of pulp components in unbleached softwood kraft fibers recycled with the assistance of a laccase/HTB system. *BioResources*, *2*(4), 616–629.

de Grace, J. H., & Page, D. H. (1976). The extensional behaviour of commercial softwood bleached kraft pulps. *TAPPI Journal*, *59*(7), 98.

De Ruvo, A. (1980). Fundamental and practical aspects of papermaking with recycled fibres. *Industria della Carta*, *18*(6), 287.

De Ruvo, A., & Htun, M. (1981). *Fundamental and practical aspects of papermaking with recycled fibres*. London: Mechanical Engineering Publications. p. 195.

De Ruvo, A., Farstrand, P., Hagen, N., & Haglund, N. (1986). Upgrading of pulp from corrugated containers by oxygen delignification. *TAPPI Journal*, *69*(6), 100.

De Ruvo, A., Htun, M., & Ehrnrooth, E. (1978). Fundamental aspects on the maintaining properties of paper made from recycled fiber. *EUCEPA symposium*. Warsaw.

Diniz, J. M. B. F., Gil, M. H., & Castro, J. A. A. M. (2004). Hornification – Its origin and interpretation in wood pulps. *Wood Science Technology*, *37*, 489–494.

Eastwood, F. G., & Clarke, B. (1977). Laboratory and pilot scale machine upgrading of mixed waste paper. *Paper Technology Industry*, *18*(5), 155.

Fellers, C., Htun, M., Kolman, M., & De Ruvo, A. (1978). The effect of beating strategy in the manufacture of board from recycled fibres. *Svensk Papperstidning, 81*(14), 443.

Ferguson, L. D. (1992). Effects of recycling on strength properties. *Paper Technology, 33*(10), 14.

Forester, W. K. (1985). *TAPPI pulping conferences.* Atlanta, Georgia, USA: TAPPI Press. Bk 1, p. 141.

Garg, M., & Singh, S. P. (2006). Reason of strength loss in recycled pulp. *Appita Journal, 59*(4), 274–279.

Geffertová, J., Geffert, A., & Čabalová, I. (2008). Hardwood sulphate pulp in the recycling process. *Acta Facultatis Xylologiae Zvolen, L, 1*, 73–81.

Gottsching, L., & Stiirmer, L. (1978). Fibre–water interactions in papermaking. *Journal of the British Paper and Board Industry Federation, 2*, 877.

Gratton, M. F.(1991). The recycling potential of calendered newsprint fibres. *First CPPA research forum on recycling.* Toronto.

Guest, D., & Weston, J. (1986). Fibre–water interactions: Modified by recycling. *TAPPI pulping conference.* Toronto.

Guest, D. A., & Voss, G. P. (1983). Improving the quality of recycling fibre. *Paper week '83.* London.

Gurnagul, N., Ju, S., & Page, D. H. (2001). Fibre–fibre bond strength of once-dried pulps. *Journal of Pulp and Paper Science, 27*(3), 88.

Hawes, J. M., & Doshi, M. (1986), *TAPPI pulping conference.* Atlanta, GA.

Higgins, A. G., & MacKenzie, A. W. (1963). The future of paper in the electronic age – Paper on the way to the third millennium, part 2. *Appita Journal, 16*(5), 145.

Hoppe, J., & Baumgarten, H. L. (1997). The future of paper in the electronic age – Paper on the way to the third millennium, part 1. *Wochenblatt für Papierfabrikation, 125*(18), 860.

Horn, R. A. (1975). What are the effects of recycling on fibre and paper properties? *Paper Trade Journal, 159*(7/8), 78.

Howard, R. C. (1990). The effects of recycling on paper quality. *Journal of Pulp and Paper Science, 16*(5), J143.

Howard, R. C. (1995). The effects of recycling on pulp quality. In R. W. J. Mckinney (Ed.), *Technology of paper recycling.* New York: Blackie. Chapter 6.

Howard, R. C., & Bichard, W. (1991).The basic effects of recycling on pulp properties. *First CPPA research forum on recycling.* Toronto.

Howard, R. C., & Bichard, W. (1992). The basic effects of recycling on pulp properties. *Journal of Pulp and Paper Science, 18*(4), J151.

Howard, R. C., & Bichard, W. (1993). The basic effects of recycling on pulp properties. *Journal of Pulp and Paper Science, 19*(2), J57.

Howard, R. C., & Jowsey, C. J. (1989). The effect of cationic starch on the tensile strength of paper. *Journal of Pulp and Paper Science, 15*(6), J225.

Hubbe, M. A., & Zhang, M. (2005). Recovered kraft fibers and wet-end dry-strength polymers. *Proceedings of TAPPI 2005 practical papermakers conference.* TAPPI Press, Atlanta.

Jahan, M. S. (2003). Changes of paper properties of nonwood pulp on recycling. *TAPPI Journal, 2*(7), 9–12.

Jayme, G. (1944). Mikro-Quellungsmessungen an Zellstoffen. *Wochenblatt für Papierfabrikation, 6*, 187.

Kato, K. L., & Cameron, R. E. (1999). A review of the relationship between thermally accelerated ageing of paper and hornification. *Cellulose, 6*, 23–40.

Kessel, L., & Westenbroek, A.(2004). Aims and technologies for fibre upgrading. *Eighth Pira international conference on paper recycling technology.* Prague.

Khantayanuwong, S. (2003). Determination of the effect of recycling treatment on pulp fiber properties by principal component analysis. *Kasetsart Journal (Nat. Sci.), 37*, 219–223.

Khantayanuwong, S., Toshiharu, E., & Fumihiko, O. (2002). Effect of fiber hornification in recycling on bonding potential at interfiber crossings: Confocal laser-scanning microscopy (CLSM). *Japan TAPPI Journal*, *56*(2), 239–245.

Kibblewhite, R. P., & Dell Bawden, A. (1995), Effects of multiple drying treatments on kraft fibre walls. *Third CPPA research forum on recycling*. Vancouver.

Kim, H. J., Oh, J. S., & Jo, B. M. (2000). Hornification behaviour of cellulosic fibres by recycling. *Applied Chemistry*, *4*(1), 363–366.

Klungness, J. H., & Caulfield, D. (1982). Mechanisms affecting fiber bonding during drying and aging of pulps. *TAPPI Journal*, *65*(12), 94.

Laivins, G. V., & Scallan, M. (1993).The mechanism of hornification of wood pulps. *Tenth Pira international fundamental research symposium*. Oxford.

Law, K. N. (1996). Study on cyclic reslushing of mechanical pulps. *Progress in Paper Recycling*, *6*(1), 32.

Liebe, H. (1995). *Einfluss Des Papierhersteliungsprosesses Auf Das Festigkeitspotential Von ECFUnd TCF-Zellstoff*, MSc Thesis, Darmstadt.

Lindstrom, T., & Carlsson, G. (1982). The effect of carboxyl groups and their ionic form during drying on the hornification of cellulose fibers. *Svensk Papperstidning*, *85*(15), R146.

Linke, W. (1968). Retention and bonding of synthetic dry strength resins. *TAPPI Journal*, *51*(11), 59A.

Liying, Q., Beihai, H., Yinggang, H., & Huaiyu, X. (2005). Effects of recycling on masson pine pulp properties. *Progress in Paper Recycling*, *14*(3), 19.

Lumiainen, J. (1992). Refining recycled fibers: Advantages and disadvantages. *TAPPI Journal*, *75*(8), 92.

Lumiainen, J. (1994). Refining – A key to upgrading the papermaking potential of recycled fibre. *Paper Technology*, *35*(7), 41.

Lundberg, R., & De Ruvo, A. (1978). The influence of defibration and beating conditions on the paper-making potential of recycled paper. *Svensk Papperstidning*, *81*(12), 383.

Maloney, T. C., Todorovic, A., & Paulapuro, H. (1998). The effect of fiber swelling in press dewatering. *Nordic Pulp and Paper Research Journal*, *13*(4), 285–291.

Mansito, O., Agnero, C., & Sosa, M. E. (1992). Recycling of bagasse pulps. *Papel*, *29*(June–July) p. 69.

Markham, L., & Courchene, C. E. (1988). Oxygen bleaching of secondary fiber grades. *TAPPI Journal*, *71*(12), 168.

McComb, R. E., & Williams, J. C. (1981). The value of alkaline papers for recycling. *TAPPI Journal*, *64*(4), 93.

McKee, R. C. (1971). Effect of repulping on sheet properties and fiber characteristics. *Paper Trade Journal*, *155*(21), 34.

McKinney, R. W. J. (1995). *Technology of paper recycling*. New York: Blackie A and P. 0-7514-0017-3, pp. 401, 12.03.2011. Available from: <http://cgi.ebay.com/Technology-Paper-Recycling-NEW-R-W-J-McKinney-/130395108560>.

Mohlin, U. B., & Alfredsson, C. (1990). Fibre deformation and its implications in pulp characterization. *Nordic Pulp and Paper Research Journal*, *5*(4), 172–179.

Nazhad, M. M. (2005). Recycled fibre quality – A review. *Journal of Industrial and Engineering Chemistry, Korean Journal*, *11*(3), 314.

Nazhad, M. M., & Paszner, L. (1994). Fundamentals of strength loss in recycled paper. *TAPPI Journal*, *77*(9), 171–179.

Okayama, T. (2002). The effect of recycling on pulp and paper properties. *Japan TAPPI Journal*, *56*(7), 62–68.

Okayama, T., Okada, Y., & Oye, R. (1982). Einfluß des Recycling auf Zellstofffasern. IV. Einfluß der Entwässerungsbedingungen. *Japan TAPPI Journal*, *36*(3), 388–399.

Oksanen, T., Buchert, J., & Viikari, L. (1997). The role of hemicelluloses in the hornification of bleached kraft pulps. *Holzforschung*, *51*, 355–360.

Page, D. H. (1985). The mechanism of strength development of dried pulps by beating. *Svensk Papperstidning*, *88*(3), R30.

Page, D. H., Barbe, M. C., Seth, R. S., & Jordan, B. D. (1984). The mechanism of curl creation, removal and retention in pulp. *Journal of Pulp and Paper Science*, *10*(5), J74.

Phipps, J. (1994). The effects of recycling on the strength properties of paper. *Paper Technology*, *35*(6), 34.

Putz, H. J., Torok, I., & Gottsching, L. (1989). Making high quality board from low quality waste paper. *Paper Technology*, *30*(6), 14.

Pycraft, C. J. H., & Howarth, P. (1980). Does better paper mean worse waste paper? *Paper Technology and Industry*, *21*(10), 321.

Rangamannar, G., & Silveri, L. (1989). Diskpersion – An effective secondary fiber treatment process for high quality deinked pulp. *TAPPI pulping conference*. Seattle, WA.

Sarkar, J. M.(1996). Recycle paper mill trials using enzyme and polymer for upgrading recycled fiber. *Appita 50th annual general conference*. Auckland.

Sharma, C., Rao, N. J., & Singh, S. P. (1998). Effect of recycling on fibre characteristics. *IPPTA convention issue* (p. 69), December.

Somwand, K., Enomae, T., & Onabe, F. (2002). Effect of fiber hornification in recycling on bonding potential at interfiber crossings, confocal laser scanning microscopy. *Japan TAPPI Journal*, *56*(2), 239–245.

Song, X., & Law, K. N. (2010). Kraft pulp oxidation and its influence of recycling characteristics of fibres. *Cellulose Chemistry and Technology*, *44*(7–8), 265–270.

Springer, A. M., Dullforce, J. P., & Wegner, T. H. (1986). Mechanisms by which white water system contaminants affect the strength of paper produced from secondary fiber. *TAPPI Journal*, *69*(4), 106.

Sturmer, L., & Gottsching, L. (1979). Physical properties of secondary fibres under the influence of their prehistory, part V: The effects of converting (corrugating). *Wochenblatt für Papierfabrikation*, *107*(3), 6.

Sutjipto, E. R., Li, K., Pongpattanasuegsa, S., & Nazhad, M. M. (2008). Effect of recycling on paper properties, TAPPSA (technical articles), 07.02.2011. Available from: <http://www.tappsa.co.za/archive3/index.html>.

Szwarcsztajn, E., & Przybysz, K. (1978). Fibre–water interactions in papermaking. *Transactions of the BPBIF symposium* (Vol. 2, p. 999).

Treiber, E., & Abrahamson, B. (1972). Beitrag zur Beurteilung der Zellstoffverhornung bei verschiedenen Trocknungsprozessen. *Holzforschung und Holzverwertung*, *24*(3), 54.

Valade, J. L., Law, K. N., & Peng, Y. X. (1994). Influence of blending virgin pulp on the papermaking potential of reslushed fibres. *Progress in Paper Recycling*, *3*(2), 60.

Weise, U., & Paulapuro, H. (1998). Relation between fiber shrinkage and hornification. *Progress in Paper Recycling*, *7*(3), 14.

Weston, J. D. W., & Guest, D. A. (1985). The importance of cell wall structure in recycling fibres. *Paper Technology and Industry*, *26*(7), 309.

Yamagishi, Y., & Oye, R. (1981). Influence of recycling on wood pulp fibres – Changes in properties of wood pulp fibres with recycling. *Japan TAPPI Journal*, *35*(9), 33.

Yu, C. J., Defoe, R. J., & Crossley, B. R. (1994). Fractionation technology and its applications. *TAPPI pulping conference*. San Diego, CA.

Zanuttini, M. A., McDonough, T. J., Courchene, C. E., & Mocchiutti, P. (2007). Upgrading OCC and recycled liner pulps by medium-consistency ozone treatment. *TAPPI Journal*, *6*(2), 3–8.

7 Chemicals Used in Deinking and Their Function*

7.1 Introduction

A wide variety of recovered papers are deinked. Each of these recovered paper grades has differences both in fibre and in filler characteristics. A variety of different deinking processes are used based on the combination of recovered paper being used and the desired final pulp requirements. Flotation deinking tends to be more selective than wash deinking and thus results in higher yields. The selectivity in both flotation and wash deinking can be dramatically enhanced by using one or more specialist chemicals. It is well understood that flotation and wash deinking are dominated by different physico-chemical properties and as such the two processes usually require different types of chemical.

The proper selection of deinking chemistry is often a compromise between costs and performance. Some mills do not use deinking chemicals but on the other hand use a more expensive and higher-quality recovered paper to reach a certain target. Other mills save costs with the recovered paper by using a lower-quality furnish and then use deinking chemicals. One of the large production costs for a deinking mill come from the recovered paper. The costs for pulping chemicals are often around €10–20 per tonne, and bleaching may be the same or up to double that amount depending on bleaching dosages and sequences. Recently, owing to increasing chemical, energy and pulp prices, mills have been exploring new ways of saving costs using new concepts for recycled fibre processing. Some chemical suppliers offer neutral deinking where chemical costs in the pulper can be saved in addition to other process and chemical benefits. Owing to the complexity of the entire recycling process and the inter-dependency of each process step, it is not sufficient to look only at the pulper or flotation stage but rather the entire process.

The deinking process involves various steps like slushing of waste paper, coarse screening, pulping, fine screening, flotation in one or more stages, thickening, bleaching (oxidative and/or reductive), etc. as shown in Figure 7.1. The major chemicals used in flotation deinking with their dose levels and usual points of addition are presented in Table 7.1. To be removed from the fibre/water mixture in the flotation cell, an ink particle must adhere to an air bubble and float. It will do this best when it is approximately 10–150 μm in diameter, has a mainly hydrophobic surface and is not stuck to a fibre. Deinking chemicals added to the pulper help the removal

*Some excerpts taken from Bajpai, 2006 with kind permission from Pira International, UK.

Recycling and Deinking of Recovered Paper. DOI: http://dx.doi.org/10.1016/B978-0-12-416998-2.00007-6

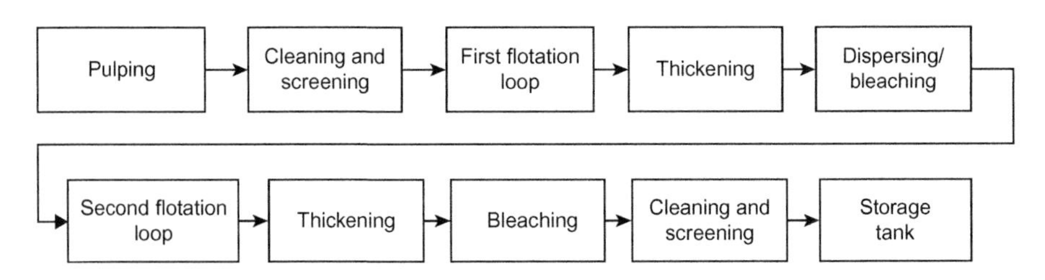

Figure 7.1 Schematic of a deinking process for the production of deinked pulp.
Source: Based on McKinney (1995) and Hannuksela and Rosencrance (2006).

Table 7.1 Principal Deinking Chemicals

Chemical	Application
Sodium hydroxide	Pulper, bleach tower, disperser
Hydrogen peroxide	Pulper, bleach tower, disperser
Sodium silicate	Pulper, bleach tower
Chelating agents	Pulper, flotation cells
Soap	Pulper, flotation cells
Surfactants	Pulper, flotation
Talc	Pulper, flotation cells, final stock
Calcium salts	Flotation cells
Sulphuric acid	Final stock
Sodium hydrosulphite	Final stock
Sodium hypochlorite	Final stock
Flocculating chemicals	Backwater

Source: Based on Ferguson (1992a,b), Lassus (2000), Turvey (1995) and Zhao et al. (2004)

of the print from the fibre surface. Chemicals added to the pulper and to the flotation cell increase the hydrophobic nature of the surface of the detached print particles (Ferguson, 1992a,b; Turvey, 1995). The role of different chemicals in deinking system is described below and has been summarised in Table 7.2 (Bajpai, 2006; Ferguson, 1992a,b; Lassus, 2000; Seifert & Gilkey, 1997; Turvey, 1995; Zhao, Deng, & Zhu, 2004). Table 7.3 provides typical use levels of deinking chemicals found at recovered paper mills (Dingman & Perry, 1999; European Commission, 1999; Hamilton & Leopold, 1987).

7.2 Alkali

Sodium hydroxide (NaOH or lye) is used to adjust the pH towards the alkaline region of 9–10. NaOH is added for reasons such as ink detachment and ink dispersion. It is commonly believed that ink detachment is improved using NaOH both because of fibre swelling effects and chemical hydrolysis of some bonds between

Table 7.2 Function of the Common Deinking Chemicals

Sodium hydroxide
Fibre swelling, breaks down ink vehicle by saponification or hydrolysis, ink dispersion
Sodium silicates
Source of alkalinity and pH buffering agent, wetting, ink dispersion, peroxide stabilisation
Surfactants
Wetting, ink removal, ink dispersion, emulsification
Hydrogen peroxide
Bleaching, ink degradation, anti-yellowing
Collector soap
Ink collector (renders the ink hydrophobic and facilitates its attachment to the air bubble)
Calcium salts
Reacts with the collector soap to form small insoluble calcium soap particles that adhere to the ink particle; calcium helps agglomerate the ink particles into large hydrophobic clusters
Chelating agents
The role of the chelant is to form soluble complexes with heavy metal ions. The complexates prevent these ions from decomposing the hydrogen peroxide

Source: Based on Ferguson (1992a,b) and Lassus (2000)

Table 7.3 Typical Formulation Range of the Pulping Chemicals for Wood-Containing and Wood-Free ONPs

Chemicals	Wood-Containing ONPs	Wood-Free ONPs
Chelant	0.15–0.4%	–
Sodium silicate	1.0–3.0%	–
Sodium hydroxide	0.8–1.5%	1.0–1.5%
Hydrogen peroxide	0.5–2.0%	–
Surfactant/collector	0.25–1.5%	0.25–1.5%
Temperature	45–55°C	50–60°C
pH	9.5–10.5	10–11
Consistency	5–15%	5–15%
Time	4–60 min	4–60 min

Source: Based on Ferguson (1992a)

the substrate and some ink species. The pH at the exit of the pulper is often between 8.5 and 10.5 when deinking a newspaper/magazine furnish. The swelling of the fibre caused by the lye can help the mechanical release of the stiffer ink from the surface of the fibre. The alkaline environment also saponifies the fatty acids and hydrolyses the ink resins. The NaOH causes an ionisation of the carboxylic groups of the cellulose fibres and saponification of some ink binder and acid resins in the wood. The release of the ink is due to ionisation of fibres and the ink's surface groups, generation of electrostatic repulsive forces, mechanical stress at the ink/fibre interface after swelling and release of fatty acid soaps.

With the increase in the alkali concentration, the swelling of cellulosic fibres increases and, thus, the deinking effectiveness increases (Ferguson, 1992a; Turvey, 1995). The adsorption of hydroxide ion is thought to increase the electrostatic repulsion between the fibres and the ink particles, thus resulting in greater ink detachment. The increase in alkali concentration increases the pH of the system. Higher pH values cause yellowing of wood-containing pulps, which is counteracted by the addition of hydrogen peroxide (H_2O_2) (Bajpai, 2006). It also induces the dissolution of soluble components in recovered papers (increasing the chemical oxygen demand and biological oxygen demand of process waters). When re-pulping under alkaline conditions, high pH values of 9–10 are normally limited to the re-pulping stage and in the following deinking stages, the pH is progressively lowered to 8–9 in the flotation stage and 7.5 in the second process water loop. The increased ink detachment effect of NaOH during re-pulping has been well demonstrated, especially through brightness and effective residual ink concentration measurements on whole and hyperwashed pulps, and the direct measurement of the ink particle distribution by image analysis (Ferguson, 1992a,b; Pelach, Puig, Vilaseca, & Mutje, 2001; Turvey, 1995).

The pigment in ink is released by breaking down the oil-based vehicle carrying the pigment. NaOH reacts with oil-based inks giving out soap and alcohol through saponification. However, inks based on synthetic resins and special inks cannot be removed by alkali. Milder alkali like sodium carbonate (Na_2CO_3) may partly replace NaOH. The alkali level is decided based on experimentation.

The effect of caustic soda on the detachment of the ink is attributed to mechanical stress at the ink/fibre interface generated by fibre preferential swelling, ionisation at the surface of both of fibres and of ink particles, generation of electrostatic repulsion forces and release of fatty acid soaps (saponification). In the specific case of toner particles, a relevant decrease in fragmentation, caused by the lubricating action of NaOH and fibre swelling, and a corresponding increase in contamination by specks, were observed when re-pulping under alkaline conditions (Azevedo, Drelich, & Miller, 1999).

Deinking without addition of NaOH has been successfully applied to raw materials of high quality (wood-free office paper), whereas with wood-containing household papers, a significant drop in brightness and an insufficient disintegration of wet strength papers were observed.

Few studies have investigated the use of alternative alkaline sources. One was in Australia at the University of Tasmania. Starting by testing several alkalis such as MgO, $Mg(OH)_2$, CaO, $Ca(OH)_2$ and NH_4OH, the research then focused on magnesium oxide and the best conditions to improve its efficiency in the deinking process. It appeared that the highest deinking efficiency was achieved when MgO was used with soap and hydrogen peroxide and that this efficiency is further increased when MgO is hydrated and added as a slurry. This suggests that $Mg(OH)_2$ is actually the most efficient alternative alkali source. In addition it was demonstrated that the $Mg(OH)_2$-based systems generate the lowest chemical oxygen demand and dissolved solids in effluent. Calais and Blanc (2008) investigated low alkalinity deinking using $Mg(OH)_2$ as an alternative alkali source in the deinking of an old newspaper (ONP)-rich recovered paper mixture (70% ONP/30% old magazine, OMG), compared with NaOH-based chemistry. $Mg(OH)_2$-based deinking proved to have great advantages compared with NaOH-based deinking in many aspects: intrinsic higher deinking performances were obtained after flotation in terms of optical properties (at least a 1 point brightness increase); near-neutral conditions obtained (pulping pH>>8.5) induced a drastic reduction in chemical oxygen demand (−20% to −50%); chemical consumption decreases (lower silicate and peroxide charges necessary); and hydrosulphite post-bleaching response is unchanged (leading to potential additional decreases in chemical consumption). The process was found to be economically attractive. The use of $Mg(OH)_2$-based deinking resulted in a chemical cost saving of €3–5 per tonne of deinked pulp produced.

7.3 Stabilisers

The role that sodium silicate plays in a newsprint deinking system has been studied by several workers (Ali, McLellan, Adiwinata, May, & Evans, 1994; Borchardt, 1995, 1997; Ferguson 1991, 1992a,b; Mahagaonkar, Banham, & Stack, 1996, 1997; Mathur, 1991; McCormick, 1990, 1991; Pauck & Marsh, 2002; Read, 1986, 1991; Renders, Chauveheid, & Dionne, 1996; Santos, Carre, & Roring, 1996). It is apparent that sodium silicate's role seems to be multi-faceted, and can be summarised as follows: its main function is to stabilise the hydrogen peroxide used to bleach the pulp in the pulper. In addition, sodium silicate has buffering and saponification properties (Ferguson, 1991, 1992a). Sodium silicate has been reported to assist in the dispersion of the ink particles and influence their size (Ali et al., 1994; Mahagaonkar et al., 1997; Read, 1986). Also, it appears to act as an ink collector (Santos et al., 1996), and it reduces fibre losses and suppresses the flotation of fillers (Liphard, Hornfeck, & Schreck, 1991; Mathur, 1994; Turvey, 1990). However, several authors refute these findings and claim that sodium silicate has no influence on the final brightness (Mak & Stevens, 1993; Mathur, 1994; Zabala & McCool, 1988). The Merebank mill's own experience has shown that sodium silicate plays a vital role in the deinking process. The elimination of sodium silicate from the process resulted in an immediate deterioration in deinking performance.

An attempt was made to understand the multiple roles of sodium silicate in a deinking system by investigating the dispersing power of sodium silicate for a typical newsprint ink, in the presence of the other main non-fibrous components of a newsprint deinking system, namely calcium ions and fatty acid soap. This work was performed using model ink systems. It was found that sodium silicate had a poor dispersing action for a typical hydrophobic newsprint ink, but the effects of calcium on ink particle size and morphology could be considerably modified by the presence of sodium silicate. This was demonstrated to be due to the chelation of the calcium ion by sodium silicate, which resulted in increased levels of soap in solution. It was postulated that this would lead to increased dispersion of ink particles and thus better ink removal from the fibre surface during pulping. Consequently, this would have better flotation performance. This led to the hypothesis that the chelating effect that sodium silicate has on calcium has an indirect but significant effect on the performance of a deinking system (Pauck, 2002; Pauck & Marsh, 2002).

Sodium silicate is added to the pulper. The reason for the addition of sodium silicate is to stabilise the hydrogen peroxide. It deactivates metal ions which catalytically decompose peroxide (Ferguson, 1992a,b; Lassus, 2000; Turvey, 1995). By inactivating the metal ions in the process causing breakdown of hydrogen peroxide, and by maintaining a stable pH, an optimisation of the effect of the hydrogen peroxide is acquired. There are also other minor effects of the sodium silicate such as improved dispersion of ink, owing to attachment to colloidal particles. It also buffers the system at around the pH where peroxide works best. However, the presence of sodium silicate in the pulper increases the detachment and dispersion of the ink in alkaline conditions. Renders (1992) and Renders et al. (1996) have reported that sodium silicate prevents ink re-deposition on the fibres and agglomerates small ink particles to form large ones (10 μm) so that they have a more suitable size for flotation. Sodium silicate increases the pH of the flotation cell.

At low concentrations, silicates are good emulsifiers. Sodium silicate can also help fibre wetting and dispersion. It can be added with chelating agents for better performance. It is known to have a powerful effect on deinking performance, particularly in newsprint deinking. Sodium silicate comes in a variety of SiO_2/Na_2O ratios; in deinking a common ratio that is used is 3.3/1.

7.4 Hydrogen Peroxide

Hydrogen peroxide is used in flotation deinking where it is commonly added to the pulper (Ferguson, 1992a,b; Lassus, 2000; Renders, 1992; Turvey, 1995). The main reason the hydrogen peroxide is added to the process is to prevent the yellowing of the paper that occurs with the addition of NaOH. The hydrogen peroxide forms a perhydroxyl anion in water which attacks the groups causing yellowing in the lignin called chromophores. The hydrogen peroxide can also break the chemical cross-linkage that is formed between alkyl binders of the ink when stored and dried.

According to Ferguson (1992a), it is not efficient to use peroxide as a bleaching agent in the pulper. The ink and contraries load present in the pulper reduce the bleaching efficiency of the peroxide.

Hydrogen peroxide is also sometimes added to the bleach towers between the pulper and the cells. Addition to bleach towers after deinking is practiced and in some mills addition to dispersing equipment is also performed. The initial reason for adding hydrogen peroxide to deinking plants was certainly to act as bleach and to increase the brightness of the stock. It is known to be effective at stopping the darkening of mechanical fibres carried by NaOH. Now, it has been also suggested that peroxide plays a role in deinking in addition to the bleaching, which undoubtedly occurs. Peroxide helps in ink removal by penetrating into the fibre and subsequent decomposition. Peroxide addition to pulper can give higher final stock brightness than peroxide addition to a post-flotation bleach tower. One suggestion to explain this is that peroxide destroys material that is possibly alkali-extracted from fibres, which may give print particles hydrophilic surfaces and stop them sticking to air bubbles during flotation. It is also suggested that peroxide breaks bonds in print networks, which can help detach print from fibres and create smaller print particles. These smaller particles may float better, which will decrease speckness. The action of dispersers can sometimes be improved by addition of peroxide.

When peroxide reacts with caustic soda, the perhydroxyl anion (HOO^-) is formed, which is the active bleaching agent. To get the best use of the peroxide, it is important to maximise the amount of the perhydroxyl anion. The options available are to raise the pH by increasing the caustic level, raise the temperature, reduce the competing side reactions and increase the amount of peroxide. The competing reactions are those that can decompose peroxide such as the presence of heavy metal ions like manganese, copper and iron, enzymes such as catalase, and high pH and temperature. The peroxide decomposition products and conditions have been identified as contributing to the loss of brightness in wood-containing virgin pulps. It is reasonable to expect a similar effect with recycled pulps. The decomposition of peroxide can be reduced by the addition of stabilising agents such as chelants and sodium silicate. These chemicals do not stabilise the hydrogen peroxide itself, but stabilise the environment within which the peroxide works. Hydrogen peroxide is also used as a post-bleaching agent. The balance between how much peroxide should be added in the pulper against how much is used in the bleaching stage must be optimised for each furnish. It should be remembered that the peroxide is added to the pulper simply to offset the formation of chromophores created by the alkaline pH.

The problem that occurs with the use of peroxide is catalase decomposition. Catalase is an enzyme generated by bacteria to protect themselves against peroxide attack. It catalytically destroys peroxide very rapidly. The effect in a deink plant is sharp reduction in brightness. It can be easily identified by measuring how fast a backwater sample destroys peroxide; if it is rapid, then either catalase or metal ions are present. To check which species is present, a sample of backwater is boiled and then cooled. If this treated water does not destroy peroxide, then catalase is present (boiling denatures and destroys the enzyme, whereas metal ions are unaffected by such treatment). It is difficult to free a system from catalase. Stressing the bacteria by either raising or lowering the temperature sometimes works, and adding large amounts of peroxide can also resolve the problem. The use of peracetic acid has been suggested as a way to clean the system, as has hypochlorite addition. Draining the system and cleaning mechanically also helps, whereas the use of biocides offers little help, as most of these work best in acid or neutral environments.

7.5 Chelating Agents

The most commonly used chelant is diethylenetriaminepentaacetic acid (DTPA) although ethylenediaminetetraacetic acid (EDTA) is also used (Ferguson, 1992a). The role of the chelant is to form soluble complexes with heavy metal ions. The complexates prevent these ions from decomposing the hydrogen peroxide. Some deinking mills have found that their metals content is low enough to preclude the use of a chelant. It is worth mentioning that chelants like DTPA and EDTA have been banned in some countries, for example Sweden and Norway, when the effluent stream discharges into a water system because of the disruptive effects of the chemicals on aquatic life. The USA and Canada allow free use of chelants. Magnesium sulphate, commonly known as epsom salts, has also been found to be an effective chelating agent in virgin pulps, but is rarely used in deinking applications. It is believed that magnesium works by halting the peroxide decomposition reaction rather than by deactivating the metal ions.

It has been shown that magnesium will not work in the presence of iron and copper and will catalyse peroxide decomposition in the presence of manganese. A paper reported a synergistic effect between high doses of epsom salts and DTPA, but effective recovery of the peroxide residual is necessary to make this cost effective. It has been reported that magnesium sulphate works in conjunction with sodium silicate to inactivate metal ions. The maximum stabilisation effect requires magnesium or calcium ions in a concentration of 50 ppm, which is usually supplied by normal mill operations (Ferguson, 1992a).

7.6 Surfactants

Surfactants are surface-active agents. These chemically unique species have a dual character. This consists of hydrophilic and hydrophobic portions of the chemical structure. In an aqueous environment, the hydrophilic portion is water-loving and relatively polar, whereas the hydrophobic part of the surfactant is water-hating and relatively non-polar. Surfactants can be non-ionic, anionic or cationic. For the anionic and cationic species, the hydrophobic part normally includes a hydrocarbon chain ($-CH_2-CH_2-$) and the hydrophilic part includes a variety of chemical functionalities. These can include entities such as an amine group ($-NH_3$), a carboxyl ($-COOH$), sulphonate ($-SO_3$), or sulphate ($-SO_4$) group among others. Non-ionic synthetic surfactants generally contain blocks or units of various alkoxylates, most commonly derived from either ethylene or propylene oxides.

The surfactants react on the surface of the fibre to release the ink particles and help to disperse them in water so that they are not re-deposited on the fibres (Ferguson, 1992a,b; Lassus, 2000; Turvey, 1995). In general, surfactants are applied with NaOH, Na_2SiO_3 and H_2O_2. A surfactant has two distinct parts; one part will dissolve in water whereas the other part will not. The latter part usually has an affinity for oils and similar materials and prefers air to water. Put into water, surfactants go to surfaces; hence, with a slight change of spelling, their name surface-active agents.

Surfactants have solubilities in the medium in which they are distributed. Typically, if surfactants are sufficiently soluble a concentration can be reached, which is referred to as the critical micelle concentration (CMC). The kraft point is the temperature at which the solubility of a given surfactant equals the CMC of that surfactant. Once the CMC is achieved, a series of new physical chemical phenomena occur and as such lead to a variety of unique and valuable attributes. The effect of the CMC is rooted in minimisation of the surfactants' free energy in a given system. The CMC depends on the temperature and composition of the aqueous system and is unique for each surfactant and system. Concepts such as micro-solubilisation are directly linked to micelle formation.

Surfactants used in flotation deinking cannot work in the way that surfactants probably work in wash deinking. Larger print particles with hydrophobic surfaces are the particles that float. Several surfactants have been identified which improve flotation deinking. They usually contain ethoxylated chains as the hydrophilic end, with a variety of different structures as the hydrophobic end. The ratio of the hydrophobic to the hydrophilic part of a surfactant molecule can be given a numerical value, which is called the hydrophilic–lipophilic balance (HLB) value. The higher the HLB value, the more soluble is the surfactant. Optimum flotation deinking effectiveness around HLB values of 15 has been reported. In general, surfactants play three roles in flotation deinking: (i) as a dispersant to separate the ink particles from the fibre surface and prevent the re-deposition of separated particles on fibres during flotation deinking; (ii) as a collector to agglomerate small ink particles to large ones and change the surface of particles from hydrophilic to hydrophobic; and (iii) as a frother to generate a foam layer at the top of a flotation cell for ink removal (Zhao et al., 2004). Although surfactants play important roles in deinking, they can also have some adverse effects on ink removal, fibre quality and water reuse. For example, the adsorption of dispersant and frother (Epple, Schmidt, & Berg, 1994; Panek & Pfromm, 1996) on fibre surfaces may reduce fibre–fibre bonding and create foaming problems in paper machines. Not all types of surfactant are needed in flotation deinking. For instance, no collector is necessary in deinking hydrophobic inks, such as photocopy toner. The dispersant may also be unnecessary if the ink particles can be separated from fibres by other chemicals, such as sodium silicate, NaOH and enzyme, or by mechanical actions such as magnetic and electrical fields and ultrasonic irradiation. However, a frother must be used to obtain a stable foam layer for ink particle removal in flotation deinking (Zhao et al., 2004). Although HLB value, cloud point, CMC and detail structure are useful tools for characterising surfactants, they should not be used as the primary factors in selecting a surfactant for flotation deinking. Deinking conditions, such as furnish, water hardness, pH and temperature, are different in each paper recycling mill. As a result, a surfactant formulation that works well at one mill may not work at another.

7.7 Dispersion Agents

Non-ionic surfactants are used today in flotation mainly as dispersion agents. There are many surfactants that have been produced especially for the flotation deinking process and are used in the concentration range of 0.001–0.01% on dry fibre. The surfactants used in flotation deinking are non-ionic owing to their low sensitivity towards the hardness of the water. Ionic surfactants do not improve the result of the flotation and can lead to high yield losses in the flotation stage. Non-ionic surfactants are also preferred when working at neutral flotation deinking conditions where the use of calcium soaps is difficult because of the low alkalinity. The dispersion of surfactants is caused by the adsorption of the hydrophobic ink particles, thereby creating a hydrophilic surface and thereby increasing detachment of ink. The non-ionic surfactants used in deinking often, to some part, consist of ethylene oxide

Table 7.4 Some of the Surfactants Used in Deinking Chemistry

Chemical Name
Ethylated fatty alcohol
Ethylated fatty acid
Propylene oxide/ethylene oxide
Block copolymers

Source: Based on Svensson (2011)

and propylene oxide polymers. According to Theander and Pugh (2004), many patents indicate that the optimal performance of the surfactant is with a hydrocarbon chain length of 16–18. The ethylene oxide:propylene oxide ratio should be between 1:2 and 4:117. These types of non-ionic surfactant are often produced as a block co-polymer, but there are indications in patents that random co-polymers are more effective, giving higher flotation efficiency of newsprint. Table 7.4 gives some of the surfactants used in deinking chemistry.

The HLB of a non-ionic surfactant affects the flotation efficiency to some extent. HLB values describe how the surfactant interacts with water-soluble and water-repellent substances. In deinking flotation, it has been suggested that non-ionic surfactants with an HLB value of 14–15 are optimal for deinking of paper and magazine mixtures. HLB values are theoretical and calculated from the chemical groups that the non-ionic surfactants are composed of.

The HLB number is not an unambiguous indication when looking at the behaviour of non-ionic surfactants. It does not account for the temperature sensitivity of a non-ionic surfactant (Svensson, 2011). This phenomenon is called the cloud point. It is an important factor for non-ionic surfactants and therefore affects the choice of surfactants in the process. The ethylene oxide/propylene oxide surfactants become soluble in water owing to hydration of the ether oxygen group in the polyoxyether. Increases in temperature break this bond between the water and ether oxygen group, thus resulting in a decrease in water solubility of the non-ionic surfactant in the water. The system separates into two phases and because of this a turbid appearance is observed, giving the name cloud point. Deinking efficiency has been generally correlated with the cloud point. The highest deinked sheet brightness was obtained when the process temperature was within 5°C of the surfactant cloud point. There are several positive and negative aspects of using non-ionic surfactants in the process, according to Johansson and Johansson (2000). Some of these are listed in Table 7.5.

Non-ionic surfactants, such as polyethyleneoxide alkyl ether and modified compounds of this surfactant type, are capable of assisting the detachment of coldset offset ink particles from newsprint (Rao & Stenius, 1998; Pirttinen & Stenius, 2000). Dispersants can solubilise the detached ink particles and create a stable emulsion that does not readily re-deposit onto the fibre. Because dispersants provide the ink particles with a solubilising surface chemistry as the surfactant adsorbs, this kind of chemical can adversely affect favourable interactions with hydrophobic calcium soap complexes and result in poor flotation efficiency. Therefore, dispersants should

Table 7.5 Positive and Negative Aspects of Non-Ionic Surfactants in the Process

Positive Aspect
Reduced ink re-deposition
Decreased air bubble coalescence
Reduced fibre losses
Increased foaming in the flotation cell
Negative Aspect
Increased ink fragmentation (dispersion)
Reduced ink agglomeration
Reduced collection of ink by air bubbles
Increased foaming in the flotation cell

Source: Based on Johansson and Johansson (2000)

be used with caution in flotation systems that use fatty acid soap as the collector. One of the most commonly used surfactants in the flotation deinking process is fatty acid soaps. Generally, fatty acids react with calcium ions in the system to form calcium soaps, which can adsorb onto the ink surface and provide the collector action (Somasundaran, Zhang, Krishnakumar, & Slepetys, 1999). The fatty acid soap system is not a good dispersant in general. It can function as a dispersant only if there are free fatty acids in the pulp suspension.

The Stephenson Group has created a new cost-saving innovation for the paper recycling industry that reduces the amount of chemicals used in the process (Anonymous, 2009). Serfax SLR effectively deinks paper by sticking to the printed ink and removing it. It delivers optimum performance even in low quantities and requires minimal investment in new application equipment. The chemical is suitable for a range of paper products including newsprint, tissue, linerboard and white grades. After an effective trial in a leading German recycling mill, Serfax SLR is now being marketed to the paper recycling industry around the world. Jamie Bentley, chief executive of the Stephenson Group, said, 'Serfax SLR will dramatically reduce the amount of chemicals used in the deinking process, which will be a major cost benefit to companies'.

A new range of additives based on polysiloxanes developed by Nopco Paper Technology GmbH offers high levels of ink and dirt particle separation under neutral conditions (Nellessen, 2006). The effectiveness of the new additives was tested in laboratory trials on papers with different coloured and black and white print. Samples were dissolved in a laboratory kneader with ink separation determined by hyperwashing, with standard washing used to assess washing behaviour, and flotation measured on a laboratory flotation tester. The results were evaluated by measuring the effect on brightness and the number of dirt specks, as well as flotation losses. The trials showed that the silicone additives had a significant effect on ink separation and dispersal, with improved ink removal in the standard washing process. Flotation trials were done under neutral conditions and in the presence of 0.15% non-ionic agglomeration agents.

7.8 Collecting Chemicals

Fatty acids are often used as a collector owing to their ability to form ink-affinitive soaps with calcium ions. The function of a collector is to aggregate very small ink particles that have been released from the fibres by the pulping action. The particles are then more efficiently removed by the uprising air bubbles injected by the flotation unit. The optimal particle size range for flotation deinking is 10–100 μm. However, before collectors are added, most ink particles are much smaller than 10 μm. Collectors can be made from naturally occurring materials, such as fatty acid soaps, synthetics such as polyethylene oxide and polypropylene oxide copolymers, and blends such as ethoxylated fatty acids. It is always necessary to balance the effects of a collector with those of a dispersant in flotation deinking, if dispersant is used.

Fatty acids have been used to produce fatty acid salts (soaps) for centuries. These soaps have been used to remove dirt from different surfaces. Fatty acids consist of a long hydrocarbon chain ($–CH_2–CH_2–$) with a carboxyl group, typically at the terminus of the molecule. The hydrocarbon chain can be saturated or unsaturated (containing double bonds) depending on the origin of the fatty acid. Saturated fatty acids, such as stearic acid (18 carbons) or palmitic acid (16 carbons) are solid at room temperature whereas unsaturated fatty acids, like linoleic acid (18 carbons with two double bonds at the 9 and 12 positions) can be liquid (Table 7.6). Fatty acid products are normally mixtures of several fatty acids with different structures. Vegetable-based fatty acids normally consist in large part of mixtures of unsaturated fatty acids (oleic, linoleic and linolenic acid) whereas animal-based (tallow) fatty acids generally consist in large part of mixtures of various saturated (stearic, palmitic acid) and unsaturated (oleic acids) fatty acids.

Fatty acids, being anionic salts of carboxylic acids, demonstrate different properties depending on pH and the presence of various other ionic species present in the solution. In acid form, the fatty acids are nearly insoluble in water but upon addition of alkalinity to elevate the pH the fatty acid transitions to form an essentially water-soluble fatty acid soap. If calcium ions are present, the fatty acid soaps form relatively water-insoluble calcium–fatty acids soaps. Several metal–fatty acid complexes

Table 7.6 The Structure of Some Fatty Acids Normally Found in Deinking Aids

Fatty Acid	Structure (Short Name)	Melting Point (°C)
Palmitic acid	$CH_3(CH_2)_{14}CO_2H$ (C16:0)	63
Stearic acid	$CH_3(CH_2)_{16}CO_2H$ (C18:0)	69
Palmitoleic acid	$CH_3(CH_2)_5CH{=}CH(CH_2)_7CO_2H$ (C16:1)	0
Oleic acid	$CH_3(CH_2)_7CH{=}CH(CH_2)_7CO_2H$ (C18:1)	13
Linoleic acid	$CH_3(CH_2)_4CH{=}CHCH_2CH{=}CH(CH_2)_7CO_2H$ (C18:2)	−5
Linolenic acid	$CH_3CH_2CH{=}CHCH_2CH{=}CHCH_2CH{=}CH(CH_2)_7$ CO_2H (C18:3)	−11

Source: Based on Hannuksela and Rosencrance (2006)

and speciations can exist, depending on the overall composition of the system. The calcium soaps are good ink collectors in flotation deinking processes. These benefits are the result of the calcium soaps simultaneously impacting both the surface energy and the size of ink-containing agglomerates.

Fatty acid soaps are often formed by pre-neutralisation of the fatty acid in the presence of alkalinity before introduction into the re-pulping system. It is important also to evaluate the entire pulping and papermaking process to ensure that no unfavourable metal soaps, such as calcium–fatty acid soap, are contributing to deposit problems. Fatty acid dosages to the pulper normally range between 3 and 7 kg/tonne or between 2 and 4 kg/tonne to the flotation cell. In processes where high amounts of newspaper are used, like in the USA, mills tend to use more synthetic surfactants or blends rather than pure fatty acids. This is in part because the synthetic surfactants allow for more controlled and enhanced modification of surface properties, especially in the areas of ink detachment and ink dispersion.

Currently, fatty acid soaps are widely used in ONP–OMG deinking. The calcium–fatty acid soap formulation is considered to be the most prevalent collector system in the industry today, and most commercially available deinking agents are founded upon this chemistry. Although fatty acid soap formulation is widely used in flotation deinking, it still has some unfavourable effects on deinking performance. Generally, calcium ions need to be added to convert the fatty acid soaps to the calcium soaps. However, the calcium ions are believed to cause scaling and deposition problems on paper machines and other equipment in a deinking plant.

7.9 Frothing Agent

The function of the frother is to generate a foam layer at the top of the flotation cell for ink removal. A frother must be used to obtain a stable foam layer to remove ink particles. Non-ionic surfactants are widely used as frothers in flotation deinking because they have excellent foamability and function independently of water hardness. Froth stability is critical for ink removal. Ink removal efficiency increases with an increase in froth stability. A fatty acid system could serve not only as the collector but also as the frother and dispersant in a flotation deinking operation only when there are free fatty acids in the pulp suspension.

Ink removal efficiency depends on several factors such as the ability to separate the ink particles from the fibres, the collision probability between ink particles and air bubbles, the interfacial energy between ink particles and the air bubble surface, the specific contact surface area between ink particles and air bubbles, the stability of the froth for final ink removal, etc. It is well known that surface chemistry plays a key role in flotation deinking. It has also been shown that froth stability is critical for ink removal. Ink removal efficiency increases with an increase in froth stability. Unfortunately, the increase in frother concentration in the pulp suspension may increase the adsorption of surfactant by ink particles, resulting in a reduction in surface hydrophobicity of the ink particles and therefore low ink removal (Epple et al., 1994). Therefore, there is an optimum frother concentration for efficient ink removal.

7.10 Defoamer

The addition of defoamer to the flotation pulp slurry is required sometimes to control froth stability, froth structure and froth dynamics, which are critical to ink removal and fibre and water losses. Also, addition of defoamer suppresses the formation of foams during the papermaking process. Low HLB surfactant and finely divided hydrophobic silica particles dispersed in silicone oil are effective defoamers (Brandt, Teasley, & Anderson-Noms, 1996; Hendriks & Barnett, 1997).

7.11 Emulsions

Owing to the difficulty in handling some fatty acids, which can be solids or semi-solid at common atmospheric temperatures of interest, formations of liquid emulsions and/or dispersions are often made. In many cases the fatty acid is saponified and in all cases the products contain at least one emulsifying/dispersing agent. These agents are commonly non-ionic surfactants. Fatty acid levels in these products can range from 10% to 50% by weight. Emulsions/dispersions may be used in smaller deinking mills that do not have a saponification unit.

7.12 Modified Inorganic Particle

To improve collection in flotation deinking systems, a new technology has been developed (Rosencrance, Horacek, & Hale, 2005). This is based on introducing a hydrophobically modified inorganic particle to the pulper. The particle will collect hydrophobic substances like ink and not only improve flotation deinking selectivity but also significantly reduce ink re-deposition. This technology is new and has shown very promising results in mill-scale trials where very low attached ink values have been observed.

7.13 Calcium Salts

Calcium salts are added to flotation deinking systems that have low calcium ions to make fatty acid soaps function as deinkers. The most commonly added salts are $CaCl_2$ and $Ca(OH)_2$, which are added to hardness levels of 180 ppm and above and are typically added to the first flotation cell. Calcium ions form a complex with the print particles, which then sticks to the fibres. This makes them more hydrophobic and, therefore, they float.

In recent years, suppliers have dramatically improved the efficiency and performance of surfactants and other recycling chemistry (Patrick, 2001). As a result, today's plants are able to stabilise foam in the deinking segment with generally less ash content. In this regard, some ONP plants, for example, are now finding they can reduce the addition of OMG in the fibre stream (coating mineral content in these grades provides ash for the deinking segment) and experience a gain in yield. Plants can get valuable kraft fibre content from OMG, but the disadvantage is the drop in yield when high percentages of the mineral coating are lost in subsequent washing stages. Some ONP plants are now operating with no OMG addition at all, because of recent improvements in deinking chemistry.

Experts predict high growth rates for deinking chemicals in China, as a consequence of the rising consumption of waste paper for newsprint production (Anonymous, 2006). Historically, deinking agents have grown 23% in value and 17% in volume over the past 10 years. As the consumption of waste paper increases at the expense of virgin pulp, the consumption of newsprint grows with the economy, and customer demand for rising quality emphasises deinking. The Chinese government are reportedly backing the increased importation of waste paper. The expectation of survey respondents was that recycling and deinking agents would grow by 10.25% per year in the production of newsprint, and by 5.86% and 8.48% per year respectively in printing and writing papers/tissues.

References

Ali, T., McLellan, F., Adiwinata, J., May, M., & Evans, T. (1994). Functional and performance characteristics of soluble silicates in deinking. Part I: Alkaline deinking of newsprint/magazine. *Journal of Pulp and Paper Science*, *20*(1), J3–J8.

Anonymous, (2006). Recycling and deinking chemicals. *China Paper Industry*, *1*(6), 5.

Anonymous. (2009). Stephenson group launches new deinking chemical for paper recycling industry. UK, 29 July. <http://www.risiinfo.com/technologyarchives/chemicals/Stephenson-launches-new-deinking-chemical-for-paper-recycling-industry.html>. Accessed January 2012.

Azevedo, M. A. D., Drelich, J., & Miller, J. D. (1999). The effect of pH on pulping and flotation of mixed office wastepaper. *Journal of Pulp and Paper Science*, *25*(9), J317.

Bajpai, P. (2006). *Advances in Recycling and deinking* (180 pp.). U.K: PIRA International.

Borchardt, J. K. (1995). Chemistry of unit operations in paper deinking mills: *Plastics, rubber and paper recycling*. Washington, USA: American Chemical Society. (pp. 323–341; Chapter 27).

Borchardt, J. K. (1997). The use of surfactants in deinking paper for paper recycling. *Current Opinion in Colloid and Interface Science*, *2*(4), 402–408.

Brandt, C. S., Teasley, J. G., & Anderson-Noms, A. (1990). Water-based silicone defoamers: New generation of defoamers. *Paper Age*, *112*(10), 24.

Calais, C., & Blanc, J. (2008). Keeping up to speed – Study on alkaline deinking. Arkema's Rhône-Alpes Research Center. Pulp and Paper International. [Online] <http://www.risiinfo.com/magazines/September/2008/PPI/PPIMagSeptember-Keepingup-to-speed.html>. Accessed February 2012.

Dingman, D. J., & Perry, C. D. (1999). Caustic-free repulping for newsprint production, part I. *TAPPI pulping conference*. Orlando, FL.

Epple, M., Schmidt, D. C., & Berg, J. C. (1994). The effect of froth stability and wettability on the flotation of a xerographic toner. *Colloid and Polymer Science*, *272*, J1264.

European Commission. (1999). Integrated Pollution Prevention and Control (IPPC) Act. Draft reference document on best available techniques in the pulp and paper industry.

Ferguson, L. D. (1991). The role of pulper chemistry in deinking: *TAPPI 1991 pulping conference proceedings*. Atlanta, GA: TAPPI Press.

Ferguson, L. D. (1992a). Deinking chemistry: Part 1. *TAPPI Journal*, *75*(7), J75.

Ferguson, L. D. (1992b). Deinking chemistry: Part 2. *TAPPI Journal*, *75*(8), J49.

Hamilton, F., & Leopold, B. (1987). Pulp and paper manufacture. In M. J. Kocurek (Ed.), *Secondary fibers and non-wood pulping* (Vol. 3). Canada: Joint Textbook Committee of the Paper Industry.

Hannuksela, T., & Rosencrance, S. (2006). Deinking chemistry, Cost E46 report. <www.cost-e46.eu.org/.../Deinking%20primer/Deinking%20Chemistry-FINAL>.

Hendriks, W., & Barnett, D. (1997). Antifoam compositions for aqueous systems. US Patent No. 2,154,387.

Johansson, B., & Johansson, M. (2000). Agglomeration of ink particles using a mixture of a fatty acid sodium salt and a non-ionic surfactant. *Nordic Pulp and Paper Research Journal*, *15*(3), 243–248.

Lassus, A. (2000). Deinking chemistry. In L. Göttsching & H. Pakarinen (Eds.), *Papermaking science and technology, book 7: Recycled fiber and deinking*. Jyväskylä, Finland: Fapet Oy.

Liphard, M., Hornfeck K., & Schreck, B. (1991). The surface chemical aspects of filler flotation in waste paper recycling. *Proceedings of the first research forum on recycling* (pp. 55–64). Toronto, 29–31 October. Vancouver: Omni Continental.

Mahagaonkar, M., Banham, P., & Stack, K. (1997). The effects of different furnishes and flotation conditions on the deinking of newsprint. *Progress in Paper Recycling, February*, 50–57.

Mak, N., & Stevens, J. S. (1993). Characteristics of fatty acid as an effective flotation deinking collector. *Proceedings of the second research forum on recycling* (pp. 145–152). Ste-Adele, 5–7 October. Montreal: CPPA.

Mathur I. (1991). Chelant optimisation in deinking formulation. *First research forum on recycling* (pp. 115–123). Toronto, October. Vancouver: Omni Continental.

Mathur, I. (1994). Preferred method of removal of filler from deinked pulp: *Proceedings of the 1994 recycling symposium* (pp. 53–57.). Atlanta, GA: TAPPI Press.

Mckinney, R. W. J. (1995). Waste paper preparation and contaminant removal. In R. W. J. Mckinney (Ed.), *Technology of paper recycling* (Chapter 3, p. 48). London, U.K.: Blackie.

Nellessen, B. (2006). New types of additives for dirt dispersion. *Allgemeine Papier-Rundschau*, *130*(1), 24.

Panek, J., & Pfromm, P. (1996). Interfacial properties of toner in flotation deinking. *Progress in Paper Recycling*, *5*(2), 49.

Patrick, K. (2001). Advances in paper recycling technologies. *Paper Age*, *117*(7), 16.

Pauck, W. J. (2002). *The role of sodium silicate in flotation deinking*, MSc Dissertation, University of Natal, Durban.

Pauck, W. J., & Marsh, J. (2002). The role of sodium silicate in the flotation deinking of newsprint at Mondi Merebank. *TAPPSA Journal, January*, 20–25.

Pelach, M. A., Puig, J., Vilaseca, F., & Mutje, P. (2001). Influence of chemicals on deinkability of wood-free fully coated fine paper. *Journal of Pulp and Paper Science*, *27*(10), 353.

Pirttinen, E., & Stenius, P. (2000). The effects of chemical conditions on newsprint ink detachment and fragmentation. *TAPPI Journal*, *83*(11), 72.

Rao, R. N., & Stenius, P. (1998). Mechanisms of ink release from model surfaces and fibre. *Journal of Pulp and Paper Science*, *24*(6), 183.

Read, B. R. (1986). Deinking chemicals and their effects. *Stephenson process chemicals, technical report*. Bradford.

Read, B. R. (1991). The chemistry of flotation deinking: *Proceedings of the 1991 TAPPI pulping conference*. Atlanta, GA: TAPPI Press.

Renders, A. (1992). Hydrogen peroxide and related chemical additives in deinking processes. *TAPPI pulping conference*. Boston, MA.

Renders, A., Chauveheid, E., & Dionne, P. Y. (1996). The use of chemical additives in deinking. *Paper Technology, March*, 39–47.

Rosencrance, S., Horacek, B., & Hale, K. (2005). A unique new ONP/OMG "trueneutral deinking technology". *EXFOR secondary fibre conference*. Montreal.

Santos, A., Carre, B., & Roring, A. (1996). Contribution to a better understanding of the basic mechanisms involved in the pulping and flotation of offset ink particles: *Proceedings of the 1996 recycling symposium* (pp. 339–347) . Atlanta, GA: TAPPI Press.

Seifert, P., & Gilkey, M. (1997). *Deinking: A literature review*. Leatherhead, Surrey, U.K.: Pira International. (p. 139).

Somasundaran, P., Zhang, L., Krishnakumar, S., & Slepetys, R. (1999). Flotation deinking – A review of the principles and techniques. *Progress in Paper Recycling*, *8*(3), 22.

Svensson R. (2011). *The influence of surfactant chemistry on flotation deinking*, MSc Thesis, Material and Nanotechnology, Chalmers University of Technology, Goteberg.

Theander, K., & Pugh, R. J. (2004). Surface chemicals concepts of flotation deinking. *Colloids and Surfaces A: Physicochemical and Engineering Aspects*, *240*, 111–130.

Turvey, R. W. (1990). The role of calcium ions in flotation deinking. *PITA annual conference 1990 – Technology of secondary fibre* (pp. 47–54). Manchester, 27–28 March. Leatherhead, Surrey: Pira.

Turvey, R. W. (1995). Chemical use in recycling. In R. W. J. McKinney (Ed.), *Technology of paper recycling* (p. 130). London, U.K.: Blackie.

Zabala, J. M., & McCool, M. A. (1988). Deinking at Papelera Peninsular and the philosophy of deinking system design. *TAPPI Journal*, *71*(8), 62–68.

Zhao, Y., Deng, Y., & Zhu, J. Y. (2004). Roles of surfactants in flotation deinking. *Progress in Paper Recycling*, *14*(1), 41.

8 Deinking with Enzymes*

8.1 Introduction

During the past decade, environmental, economic and market pressures have led to a consistent increase in the utilisation rate of recovered paper and board. Recovered paper is an important source of raw material for the pulp and paper industry. Indeed, utilisation of these secondary fibres is increasing all over the world, the deinking being an important step in the recycling process for white grade papers. In traditional deinking, large quantities of chemicals are used, which makes the method expensive, environmentally damaging and increases the release of contaminants. Many conventional deinking processes require large quantities of chemicals, resulting in high wastewater treatment costs to meet environmental regulations. Deinking processes create substantial amounts of solid and liquid waste. Disposal is a problem, and deinking plants would benefit from more effective and less polluting processes. Recently enzymes have appeared as an alternative deinking aid. Enzymes have been found to be promising in the deinking of mixed office prints under neutral pH, thus contributing to the avoidance of some deinking chemicals and hence lower environmental impact (Anonymous, 2010; Bajpai, 1999, 2006, chap. 6, 2012; Bajpai & Bajpai, 1998; Ma & Jian, 2002; Mohammed, 2010; Puneet, Bhardwaj, & Singh, 2010). Introducing enzymatic deinking technology in a mill-scale operation requires extensive customisation of the enzyme formulation and process variables to achieve optimal effectiveness.

8.2 Enzymes Used in Deinking and Their Mechanism

The enzymes used for deinking include hemicellulases, cellulases, lipases, esterases, amylases, pectinases and ligninolytic enzymes (Bajpai & Bajpai, 1998). Most published literature on deinking deals with cellulases and hemicellulases. Many patents for the use of enzymes in deinking have been granted or applied for. Deinking involves dislodging ink particles from the fibre surfaces then separating the dispersed ink from the fibre suspension by washing or flotation. Enzymatic approaches involve attacking the fibre surfaces or the ink. Lipases and esterases degrade vegetable-oil-based inks. Pectinases, hemicellulases, cellulases and ligninolytic enzymes alter the fibre surface or bonds in the vicinity of the ink particles, thereby freeing ink for removal

*Some excerpts taken from Bajpai (2006) with kind permission from Pira International, UK and Bajpai (2012) with kind permission from Springer Science & Business media.

Recycling and Deinking of Recovered Paper. DOI: http://dx.doi.org/10.1016/B978-0-12-416998-2.00008-8

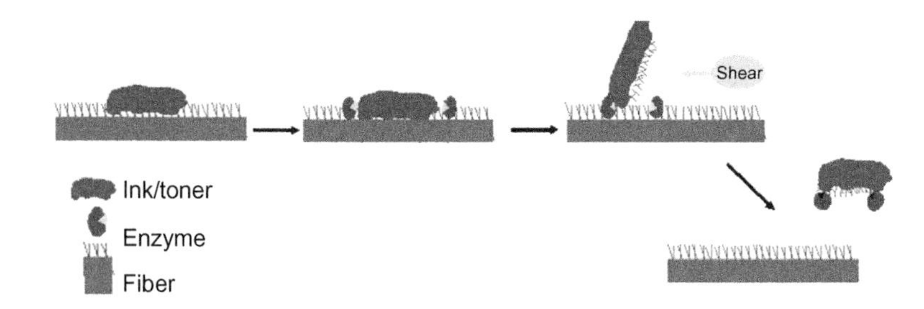

Figure 8.1 The mechanism of cellulase enzyme on fibre.
Source: Reproduced with permission from Mohammed (2010).

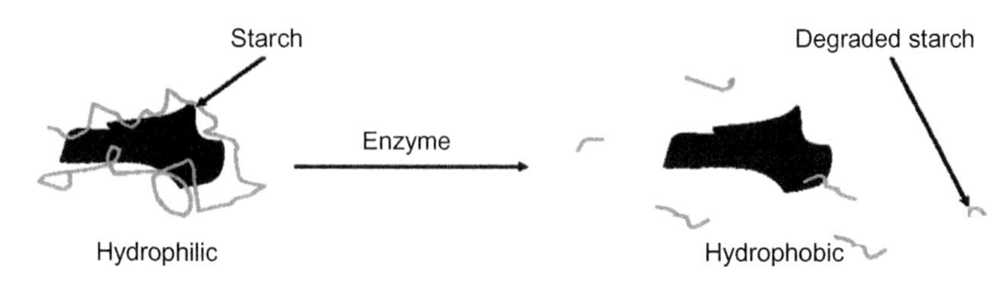

Figure 8.2 The mechanism of amylase enzyme on starch in the raw material.
Source: Reproduced with permission from Mohammed (2010).

by washing or flotation (Welt & Dinus, 1995). Several patents specify the use of cellulases, particularly alkaline cellulases for deinking. Few patents claim that esterases can be used whereas others specify the use of lipases or pectinases. One patent application mentions the use of laccase enzyme from white rot fungi. The mechanism of deinking by these enzymes has not been completely elucidated, although some hypotheses have been described (Eom & Ow, 1990; Jeffries, Klungness, Sykes, & Rutledge-Cropsey, 1994; Kim, Ow, & Eom, 1991; Mohammed, 2010; Putz, Renne, Gottsching, & Jokinen, 1994; Woodward, Stephan, Koran, Wong, & Saddler, 1994; Zeyer, Joyce, Heitmann, & Rucker, 1994). The cellulose and hemicellulose hydrolysis on the surface of the fibres leads to a removal of small fibrils, a phenomenon known as 'peeling-off fibres', which facilitates ink detachment from the surface. It has also been suggested that the alteration in hydrophobicity of the ink particles because of the removal of small fibrils enhances their separation (fibre/ink separation) in the flotation/washing step. α-Amylase influences the degradation of the starch layer on the surface of the surface sized paper in mixed office waste (MOW). Toner particles adhering to the paper surface that are released by enzymatic treatment undergo subsequent separation from the pulp suspension by flotation. A probable mechanism of cellulose and amylase action on fibre is presented in Figures 8.1 and 8.2 (Mohammed, 2010).

Interest in the use of laccases, alone or in combination with cellulases and/or hemicellulases, for deinking secondary fibres has greatly increased in recent years. Laccases in the presence of redox mediators, the so-called laccase-mediator system described two decades ago, has been largely used to delignify, and bleach, different types of pulp (Call & Strittmatter, 1992). This fact offers the possibility of deinking secondary fibres rich in lignin, such as fibres based on mechanical pulps. With the removal of lignin, the bonds between the fibre and ink particles become loose, facilitating ink detachment. Moreover, successful transformation of recalcitrant dyes by the laccase-mediator system has been recently described, which has suggested a new method of secondary fibre deinking by direct decolourisation of the inks.

8.3 Developments in Enzymatic Deinking

Extensive work has examined the potential of enzymes in deinking different types of waste paper (Bajpai, 2006, chap. 2, 2012; Bajpai & Bajpai, 1998; Baret, Leclerc, & Lamort, 1991; Chezick et al., 2004; Eriksson and Adolphson, 1997; Floccia, 1988; Franks & Munk, 1995; Gallo, Mosele, & Elegir, 2004; Gu, You, Yong, & Yu, 2004; Haynes, 2000; Heise et al., 1996; Heitmann, Joyce, & Prasad, 1992; Jeffries et al., 1994; Jeffries, Sykes, Cropsey, Klungness, & Abubakr, 1995; Kim et al., 1991; Leenen & Tausche, 2004; Ma & Jian, 2002; Magnin, Lantto, & Delpech, 2001; Ow, Park, Han, Srebotnik, & Messner, 1996; Paik & Park, 1993; Prasad, 1993; Prasad, Heitman, & Joyce, 1992a,b; Putz et al., 1994; Rushing, Joyce, & Heitmann, 1993; Rutledge-Cropsey, Jeffries, Klungness, & Sykes, 1994; Spiridon & de Andrade, 2005; Spiridon & Belgacem, 2004; Sykes, Klungness, Abubakr, & Rutledge-Cropsey, 1995; Tausche, 2002, 2005, 2007; Wang & Kim, 2005; Woodward et al., 1994; Xu et al., 2005; Yang, Ma, Eriksson, Srebotnik, & Messner, 1995; Zeyer et al., 1994; Zeyer, Heitmann, Joyce, & Rucker, 1995; Zhang & Hu, 2004; Zhang, Renaud, & Paice, 2008; Zuo & Saville, 2005).

Spiridon and de Andrade (2005) studied the effects of three enzymatic preparations – mixed cellulase and xylanase, cellulose alone and lipase – on the properties of old newsprint (ONP) that was 7 months old before deinking using a conventional flotation technique. The treatments using cellulase/xylanase and lipase produced the best properties. There was a considerable improvement in drainage rates for the treated fibre suspensions. The enzymatic treatments affected the handsheet mechanical properties. Pulps treated with cellulase/xylanase and pulps treated with lipase had constant tensile and burst indices, but their tear indices decreased. All showed improved optical properties: opacity, brightness and effective residual ink content (ERIC).

Prasad et al. (1992a) and Prasad (1993) evaluated low-pH cellulase and hemicellulase mixtures for deinking of letterpress and colour offset-printed newsprint at pH 5.5. The highest brightness increase for letterpress paper was obtained with a hemicellulase plus cellulase preparation (Table 8.1). However, the lowest residual ink areas as measured by image analysis were achieved with a cellulase preparation. For coloured offset papers, the best brightness was obtained with a mixture of cellulases and hemicellulases. These researchers also used similar enzymes to deink flexographic-printed newspaper (Heitmann et al., 1992; Prasad et al., 1992b). Enzyme treatment and flotation removed the water-based ink with ease, resulting in brightness levels well above those obtained with conventional deinking. When the hemicellulases from *Aspergillus niger* and cellulases from *Trichoderma virdie* were evaluated for deinking, brightness increased with increasing enzyme dosage and reaction time (Paik & Park, 1993).

Table 8.1 Effect of Enzymatic Deinking on Brightness Improvement for ONP

Type of Printing	Enzyme	Brightness, % ISO	
		Enzyme Treatment	Control
Coloured flexo	Cellulase	51	46
Black-and-white flexo	Cellulase	55	51
Black-and-white letterpress	Cellulase + hemicellulase	58	53

Based on data from Prasad et al. (1992a,b)

Table 8.2 Effect of Cellulose Enzyme Treatment on MOW Pulp Quality

Treatment	Brightness, % ISO	Dirt Count, mm^2/m^2	Freeness, mL CSF	Breaking Length, km	Tear Index, $mN\,m^2/g$	Burst Index, $kPa\,m^2/g$
Blank	82.1	4200	400	6.5	8.5	4.55
Control	84.2	1200	440	6.2	7.9	4.38
Enzyme	88.3	70	490	6.8	8.0	4.75

Based on data from Prasad (1993)

Treatment with a pure alkaline cellulase significantly improved brightness levels of photocopied and laser-printed papers relative to pulping in water without enzymes (Prasad et al., 1993). A brightness improvement of four ISO units was observed, as seen in Table 8.2. Residual ink area (dirt count) was reduced by 94%. Enzyme treatment also affected fibre length distributions. Such results might be expected, because these papers typically contain bleached softwood chemical pulp, and cellulases are more likely to affect fibre distributions of chemical pulps. Enzyme-treated pulps showed a similar increase in freeness and in strength properties (breaking length and burst index) relative to control pulps, as seen in Table 8.2.

Mixtures of lipase, cellulase and xylanase for deinking ONP were used by Gu et al. (2004). An equal mixture of cellulase and xylanase, itself mixed at a ratio of 60:40 with lipase, gave the best deinking performance. The breaking length, the burst index and the tear index of handsheets from the deinked pulp were increased by 3.2%, 7.4% and 7.1%, respectively, compared with pulp deinked with the cellulase/xylanase mixture alone. Higher pulp yield and improved pulp drainage were also obtained. Zhang and Hu (2004) studied the enzymatic deinking of post-consumer printing paper using cellulase and compared the deinking efficiency of the enzymatic process with the conventional process using chemicals. The enzymatically deinked pulp showed superior drainage, improved physical properties and better bleachability than the chemically deinked pulp.

Cellulases and hemicellulases have a significant effect on the enzymatic deinking of ONP, improving deinking efficiency and fibre modification (Wang & Kim, 2005). Compared with deinked pulps from conventional chemical materials, the enzymatically deinked pulps exhibit better bleachability. The enzymatically bleached pulp showed a brightness of 59.1% ISO, which was 9% higher than unbleached pulp.

The efficacy of immobilised cellulase for deinking of MOW was studied by Zuo and Saville (2005). When they used immobilised enzyme treatment, they found the residual ink levels were lower than with soluble enzyme treatment. Their results suggest that immobilised cellulase could be useful for deinking MOW. The effectiveness of enzymes in deinking office papers was studied by Spiridon and Belgacem (2004). They took recycled fibres from office papers and gave them enzymatic pretreatment using cellulase alone or a mixture of cellulase and xylanase. Then they observed the effects on freeness, sheet strength, sheet optical and paper surface properties. The fibre suspensions showed a significant improvement in drainage rates, and the mechanical properties of the handsheet showed a substantial increase in burst strength. The tensile strength remained almost constant for pulps treated with the mixture of cellulase and xylanase. For all treatments, the tear index decreased significantly, but the brightness and ERIC improved. The mixture of cellulase and xylanase was the most suitable treatment for laser-printed paper.

Treatment of 100% multiprints furnishes with cellulase and amylase enzymes at pH 7–7.5 improved the pulp brightness by two ISO points in the laboratory investigation as a part of EUREKA Enzyrecypaper Project (Gill et al., 2007). The ink particles released on treating with amylase enzymes appeared to be more hydrophobic than those released on treating with cellulose. During the mill trial using highly specific amylases, the brightness was significantly improved up to eight points. The ash content also greatly reduced after flotation and washing, resulting in a change of the final pulp characteristics.

Deinked pulp obtained after deinking sorted office waste with hydrolytic enzymes showed higher brightness (1.0–1.6 points), whiteness (2.7–3.0 points) and lower residual ink than chemically deinked pulp (Bajpai, Mishra, Mishra, Kumar, & Bajpai, 2004). It was possible to obtain pulp with <10 parts per million dirt count with a combination of cellulase and α-amylase enzymes, resulting in reduced chemical consumption. Chemical oxygen demand (COD) and colour loads were lower in effluents generated during enzymatic deinking (Tables 8.3–8.5).

MOW often contains a large variety of dyed papers. For this reason, it is a frequently underutilised source of waste papers. Usually, several chemical bleaching agents like ozone, oxygen, hydrogen peroxide or sodium hydrosulphite have been used to bleach secondary fibres. Now, there is an alternative colour-stripping process for secondary fibres: the laccase-mediator system. In a study by Arjona, Vidal, Roncero, and Torres (2007), a bleaching sequence included an enzyme stage called the laccase-mediator system stage (L), a hydrogen peroxide stage (P) and a sodium hydrosulphite stage (Y) on a mixture of different coloured writing and printing papers. After application of the L–P–Y sequence, a pulp with optical properties near to eucalyptus totally bleached pulp was obtained. The L–P–Y sequence provided a colour removal of 90% and saved chemicals in the final stages.

Table 8.3 Deinking of Sorted Office Waste with Different Hydrolytic Enzymes Alone and in Combination

Deinking Enzyme	Brightness, % ISO	CIE Whiteness	Dirt count, parts per million	Yield, %	Ash, %
Hemicellulase	81.5	71.4	15	70.6	6.0
α-Amylase	80.2	69.1	19	70.8	6.1
Cellulase	81.6	71.5	12	71.0	6.1
Cellulase + α-amylase	81.7	71.5	8	70.9	6.2
Cellulase + hemicellulase	81.3	71.7	13	71.2	6.2
Cellulase + hemicellulase + α-amylase	81.5	71.5	7	70.8	6.2
Chemicals (control)	80.1	69.2	16	71.1	6.3

Initial brightness, 61.1% ISO; initial whiteness, 53.6%; initial ash content, 14.20%; dirt count in sorted office waste, 2997 parts per million
Source: Based on data from Bajpai et al. (2004)

Table 8.4 Generation of Pollutants in Different Stages of Deinking of Sorted Office Waste

Parameter	Enzymatic Deinking with Same Chemicals[a]	Enzymatic Deinking with Less Chemicals[b]	Chemical Deinking (Control)
COD, kg/tonne of pulp	39.80	36.80	40.70
TSS, kg/tonne of pulp	20.1	20.2	19.9
Colour, kg/ tonne of pulp	7.71	8.57	10.50

[a]NaOH, 1%; sodium silicate, 1%; H_2O_2, 0.5%; surfactant, 0.2%; soap, 0.001%.
[b]Surfactant, 0.2%; soap, 0.001%.
Source: Based on data from Bajpai et al. (2004)

Table 8.5 Chemical Consumption in Deinking Process

Chemicals, kg/tonne of pulp Sorted Office Waste	Chemical Deinking	Enzymatic Deinking
Enzyme	–	0.4
Sodium hydroxide	26.0	16.0 (−10)
Sodium silicate	10.0	Nil (−10)
Hydrogen peroxide	25.0	20.0 (−5)
Diethylenetriaminepentaacetic acid	2.0	2.0
Surfactant	2.0	2.0
Coagulating/flocculating agent	3.0	3.0

Source: Based on data from Bajpai et al. (2004)

Ibarra, Concepción Monte, Blanco, Martínez, and Martínez (2012) compared the use of cellulases/hemicellulases with the laccase-mediator system for deinking printed fibres from newspapers and magazines. For this, two commercial enzyme preparations with endoglucanase and endoxylanase activities (Viscozyme Wheat from *Aspergillus oryzae* and Ultraflo L from *Humicola insolens*, Novozymes) and a commercial laccase (NS51002 from *Trametes villosa*, Novozymes), the latter in the presence of synthetic or natural (lignin-related) mediators, were evaluated. The enzymatic treatments were studied at the laboratory scale using a standard chemical deinking sequence consisting of a pulping stage, an alkaline stage using NaOH, sodium silicate and fatty acid soap, and a bleaching stage using hydrogen peroxide. The handsheets were then prepared and their brightness, residual ink concentration and strength properties measured. Among the different enzymatic treatments assayed, both carbohydrate hydrolases were found to deink the secondary fibres more efficiently. Brightness increased up to 3–4% ISO on newspaper fibres, being Ultraflo 20% more efficient in the ink removal. Up to 2.5% ISO brightness increase was obtained when magazine fibres were used, being Viscozyme 9% more efficient in the ink removal. The laccase-mediator system, alone or combined with carbohydrate hydrolases, was ineffective in deinking newspaper and magazine fibres, resulting in pulps with worse brightness and residual ink concentration values. However, pulp deinking by the laccase-mediator system was displayed when secondary fibres from printed cardboard were used, with a 3% ISO brightness increase and lower residual ink concentrations.

A mill-scale enzymatic deinking project was begun in 2001 by Van Houtum Papier (VHP) of the Netherlands and Enzymatic Deinking Technologies (EDT) of the USA, with project subsidies from the Dutch government (Leenen & Tausche, 2004). EDT analysed the mill and the product development process. In a laboratory trial, a Blue Print analysis indicated poor dirt removal and brightness development, and incorrect dilution. A mixture of enzymes was tested on VHP recycled paper furnishes. A short mill trial of 2 weeks was then conducted in which furnish and stock preparation were optimised, giving a brightness gain of 2.7 points and furnish savings. In a long mill trial of 2 months, further optimisation led to reduced enzyme dosage and changes in stock preparation. So enzymatic deinking can have a significant effect on a mill's performance but it is necessary to customise the treatment to suit the mill's situation.

Xu, Qin, Shi, Zhang, and Xu (2004) studied the deinking of ONP using cellulase or hemicellulase with a laccase-mediator system. The synergistic use of the two enzymes led to the production of pulps with superior brightness and strength compared with those prepared using only one of the enzymes. ONP deinked with cellulase and the laccase-mediator system had a brightness after bleaching with hydrogen peroxide of 55.9% ISO, a breaking length of 2.13 km and a tear index of 6.43 mN m^2/g. The respective increases in brightness were 2.4% and 3.8% compared with the use of cellulase and the laccase system alone. The breaking length was 30% higher, the pulp brightness (after hydrogen peroxide bleaching) was 60.4% ISO, breaking length of 1.94 km and tear index 6.54 mN m^2/g. The respective increases in brightness were 2.7% and 8.3% compared with the use of hemicellulase and the

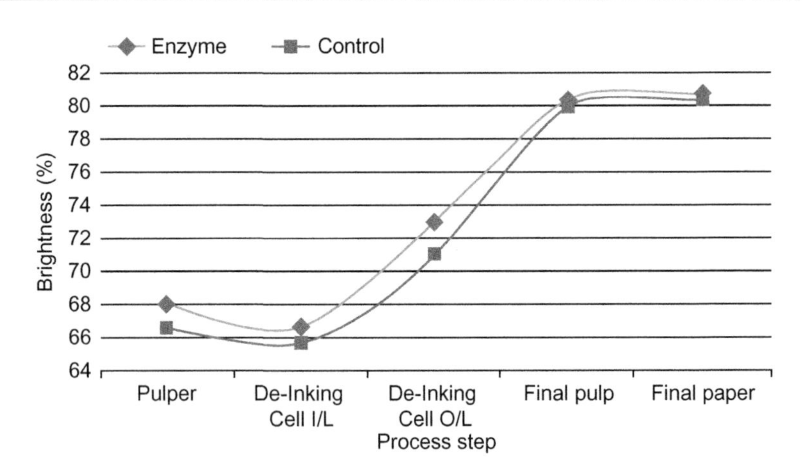

Figure 8.3 Effect of enzyme on brightness.
Source: Reproduced with permission from Mohammed (2010).

laccase system alone. The breaking length was 20% higher than that obtained using hemicellulase alone.

Mill-scale results show that enzymatic deinking gives a 50% reduction in visible and subvisible dirt (Tausche, 2002). Effective residual ink concentrations have been reduced by 35% in ONP/OMG (old magazine) mills using enzymatic deinking. Stickies reductions have reached 30–50% in mills that use tracking systems. There are also optical benefits of cleaner pulp for tissue and towel production. Yield improvements have averaged 2%. Some mills using waste paper mix have achieved a 15% decrease in furnish costs by using enzymatic deinking. Mohammed (2010) reported that in one Indian paper mill, use of enzymatic deinking on multigrade furnish like Coated Book Stock, MOW and ONP for producing writing and printing paper reduced the residual ink count, increased brightness, gave a cost saving of almost 50% in sodium hydroxide, a 37% saving in sodium silicate and complete elimination of hydrogen peroxide (Figures 8.3–8.5).

Magnin et al. (2001) conducted pilot-scale trials to compare enzymatic deinking with conventional alkaline deinking on typical wood-containing and wood-free paper compositions. The results were promising, particularly a reduction in the number and area of specks in the final deinked pulp. Full-scale enzymatic deinking was then performed at a mill producing wood-free deinked pulp from 100% printed coated wood-free papers. The results showed that good ink removal and lower specks contamination were obtained by enzymatic treatment in neutral conditions.

Yehia and Reheem (2012) studied the use of oleic acid and xylanase as surface modifiers of ink particles. The results showed that addition of enzyme alone to the ink suspension increases ink dispersion, because of its adsorption on ink particles. This facilitates separation of ink in a washing process. On the other hand, the addition of enzyme–oleic acid blend to ink suspension reduces the electronegativity of ink particles. This make the ink particles flocculate and facilitates their adhesion to

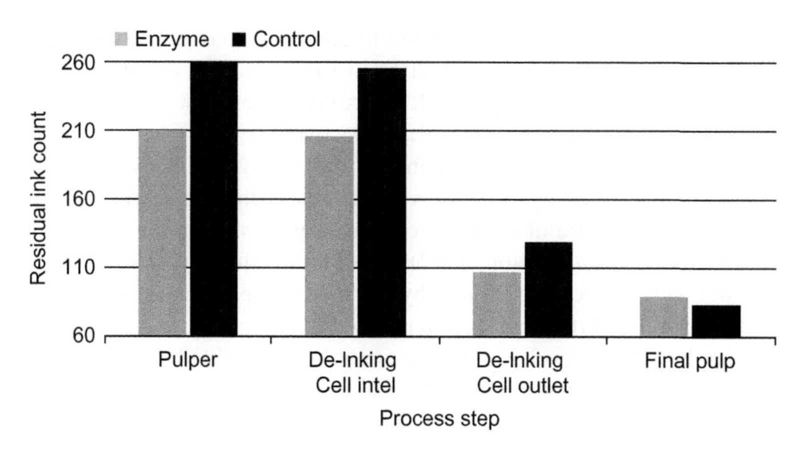

Figure 8.4 Effect of enzyme on residual ink count.
Source: Reproduced with permission from Mohammed (2010).

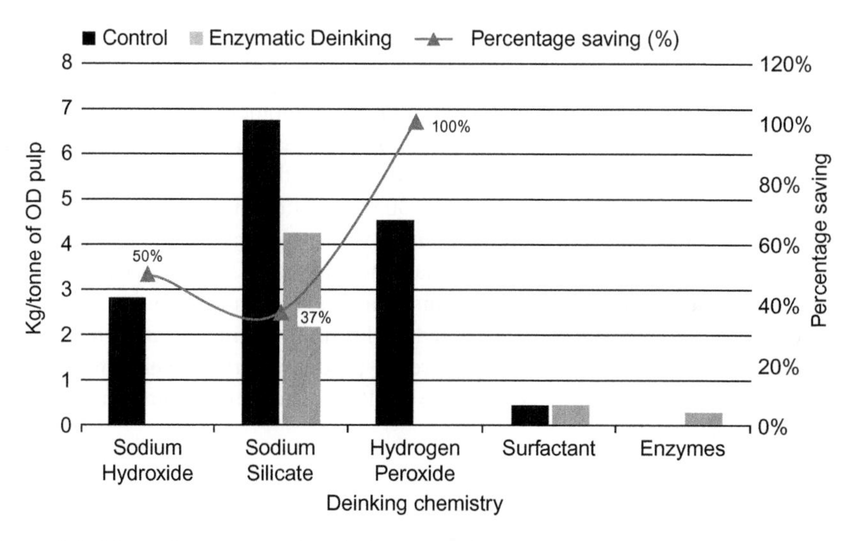

Figure 8.5 Effect of enzyme on chemical consumption.
Source: Reproduced with permission from Mohammed (2010).

the negatively charged air bubbles during flotation. It is expected that this fundamental knowledge will lead to improvements in the deinking of office waste paper.

Bio-deinking of sorted office paper by various enzymes – cellulase, xylanase, amylase and lipase – was investigated by Dutt, Tyagi, Singh, and Kumar (2012). The effects of various enzymes on pulp brightness, ERIC, deinkability factor based on brightness, deinkability factor based on ERIC, dirt counts, strength properties and effluent characteristics were compared with their respective controls. Also, the effects of different enzymes on fibre surface and fibre morphological changes during deinking were studied by atomic force microscopy and scanning electron microscopy. Results showed that, compared to control, a maximum improvement in brightness by 13.30% (ISO) and a reduction in ERIC and dirt counts by 68.18 and 88.04%, respectively, were achieved with a mixture of cellulase, xylanase, amylase, lipase at a dosage of 6, 3, 1.5 and 6 IU/mL, respectively.

Commercial use of enzymes for deinking has started in many countries. The Enzynk process developed by Eriksson's group has been commercialised by EDT. The process uses a mixture of enzymes combined with surfactants and a few other chemicals (Eriksson & Adolphson, 1997). The Stora Dalum deinking plant uses an enzymatic deinking process developed by EDT (Knudsen, Young, & Yang, 1998). Trials have shown that dirt specks can be reduced by up to 35% and stickies by up to 50%, and that brightness levels can be increased by 1.2% ISO before bleaching and by 2.2% ISO after bleaching. The use of certain chemicals was also reduced and the mill experienced a 1.8% higher yield by using enzymes. The average daily production of the mill increased by over 8 tonnes per day.

EDT is exhibiting its Enzynk technology for use in deinking mills to improve quality and reduce total production costs (Tausche, 2002, 2005a,b, 2007). EDT's mill-specific enzyme blend tailors a treatment based on the mill's furnish mix, deinking plant configuration, key operating conditions and desired results from the treatment. The company says that this patented technology provides superior ink and contaminant detachment from the fibres such that deinking plant equipment can be more efficient across flotation, cleaning and washing stages. In one case, a mill wishing to reduce furnish costs was able to achieve a 12–20% reduction in them. A second mill was able to achieve a 5% improvement in yield with a 30% reduction in sludge generation.

Enzymes for deinking have been used with a variety of strategies to reduce energy costs, both in the deink plant and on the machine. Tausche (2007) reported five such strategies to reduce energy costs per tonne: increased tonnage output, enhanced pulper defibrisation, reduced dispersion, reduced refining and reduced drying energy. Enzymatic deinking has helped mills increase production rates while maintaining good deinked pulp quality, overcoming the challenges resulting from running faster or at higher consistencies.

8.4 Effects of Enzymes on Fibre, Paper Quality and Pulp Yield

Enzymatically deinked pulp possess super physical properties, higher brightness and lower residual ink than chemically deinked recycled pulps. Enzymatic deinking works in neutral or slightly alkaline environments, which reduces overall chemical requirements and minimises yellowing of reclaimed paper after alkaline deinking. Deinking actually reduces the use of chemicals which actually results in lower waste treatment costs and reduced environmental impacts of pulp (Baret et al., 1991; Heise et al., 1996; Prasad et al.,1993; Yang et al.,1995). Enzyme-deinked pulp also has better runnability on paper machines. It is mainly the enhanced drainage and oxygen wet web strength that contribute to enhanced runnability. A hemicellulase preparation has shown the largest strength increase with the smallest improvement in freeness.

Yield results are inconclusive. Some yield reduction appears to arise from losses of fines and other small particles through the action of the applied enzyme. More precise control over enzyme doses and reaction times are expected to minimise these losses (Kim et al., 1991; Paik & Park, 1993).

8.5 Effects of Enzymes on Effluent Characteristics

COD is lower in enzymatic deinking than in alkaline deinking, which reduces the load on wastewater treatment systems (Yang et al., 1995). Wastewater effluent from enzymatic deinking was reported to have a 20–30% lower COD than wastewater from chemical deinking (Kim et al., 1991). There may be extra environmental advantages by avoiding the use of high alkalinity in the pulping stage (Jeffries et al., 1994). Another report indicated that the COD load after enzyme treatment was 50% lower than for conventional deinking (Putz et al., 1994).

Knudsen et al. (1998) reported that COD coming in to the biological treatment plant was not statistically different between chemical and enzymatic deinking. Bioplant outlet COD and changed. Sludge volume index was not impaired by the transition to enzymatic deinking. Ash content almost doubled, probably because of increased transfer of calcium ions from the deinking plant to the biological treatment plant. This changed the dewatering properties of the biosludge, necessitating a minor modification to the dewatering equipment. Thereafter a significantly higher solids content could be achieved compared with reference.

8.6 Benefits and Limitations

Enzyme-based deinking offers a potential means for reducing the amount of chemicals in the deinking process, hence reducing the load on wastewater treatment systems. Conventional methods are relatively ineffective in deinking MOW, which presents technical and economic challenges to the paper recycler. MOW contains a wide variety of fibres and contaminants as well as toners and other non-impact polymeric inks from laser printing, which are the most difficult to deal with. Toners and laser-printing inks do not disperse readily during a conventional repulping process and are not readily removed during flotation or washing. Conventional deinking uses surfactants to float toners away from fibres, high temperatures to make toner surfaces form aggregates, and vigorous high-intensity dispersion for size reduction. Fewer bleaching chemicals are usually needed for enzymatic deinking than for conventional chemical deinking. Less use of chemicals would reduce waste treatment costs and the impact on the environment. Lower bleaching costs and less pollution can also be anticipated, because enzymatically deinked pulps have proved easier to bleach and require fewer chemicals than pulps deinked by conventional methods.

Microbial enzymes enhance the release of toners from office waste papers. The size distribution and the shape of the ink removed can be effectively controlled using the enzymatic process to maximise the efficiency of the flotation process, which relies heavily on particle size. This can be accomplished by selectively varying enzyme composition, dose and residence time, and by varying other additives and the pH of the system to dislodge the normally large, flat and rigid ink particles effectively into much finer and non-platelet forms. Enzymes may also retard re-deposition of ink particles onto the fibres. The most promising implication of high deinking efficiency from enzyme-enhanced deinking is that the dewatering and dispersion steps – as well as subsequent re-flotation and washing – may not be essential. This should save capital expenses in construction of deinking plants while reducing consumption of electrical energy for dewatering and dispersion.

Enzymatically deinked pulp also displays improved drainage, superior physical properties, higher brightness and lower residual ink than chemically deinked recycled pulps. Improved drainage results in faster machine speed, which yields significant energy savings, hence overall cost savings. In addition, the use of recycled fibre reduces the need for virgin pulp. This brings great savings in the energy required for pulping, bleaching, refining, etc., which will also reduce pollution problems.

8.7 Future Prospects

Enzymes for deinking are now commercially available and at lower cost than in the past. Several pilot-plant and mill-scale trials have been done and promising results obtained. Several mills around the world are regularly using enzymes for deinking. EDT has been one of the most active companies. Increased use of and advances in fermentation technology are expected to lower the production costs of enzymes. Alternatively, genetic engineering techniques can be used to identify the gene for a specific enzyme and transfer it to another organism, for example *Escherichia coli*, that normally does not produce the enzyme. Transfer and expression of cellulase genes have also been accomplished and several firms are now producing individual cellulases. The future may also see development of biomimetic catalysis. These compounds are simpler and have lower molecular weights than enzymes but maintain function and specificity. Such synthetic catalysts would be more stable in commercial deinking environments, thereby allowing wider application.

References

Anonymous, (2010). Enzymes: The new approach to minimizing fiber costs? *Perini Journal*, *35*, 94–96. 99.

Arjona, I., Vidal, T., Roncero, M. B., & Torres, A. L. (2007). A new color stripping sequence for dyed secondary fibres. *Tenth international congress on biotechnology in the pulp and paper industry*, p. 127. Madison, WI, 10–15 June, PS LPA 3.2.

Bajpai, P. (1999). Application of enzymes in pulp & paper industry. *Biotechnology Progress*, *15*(2), 147–157.

Bajpai, P. (2006). *Advances in recycling and deinking* (pp. 75–88). Leatherhead, Surrey, U.K: Pira International.

Bajpai, P. (2012). *Biotechnology in pulp and paper processing* (412 pp). NewYork, USA: Springer US New York Inc.

Bajpai, P., & Bajpai, P. K. (1998). Deinking with enzymes: A review. *TAPPI Journal, 81*(12), 111.

Bajpai, P., Mishra, O. P., Mishra, S. P., Kumar, S., & Bajpai, P. K. (2004). Enzyme assisted deinking of sorted non-impact white office paper. *Ninth international conference on biotechnology in the pulp and paper industry*, p. 121. Durban, 10–14 October, 1.5.

Baret, J. L., Leclerc, M., & Lamort, J. P. (1991). Enzymatic deinking process. International Application No. PCT DK91/00090.

Call, H. P., & Strittmatter, G. (1992). Application of ligninolytic enzymes in the paper and pulp industry – Recent results. *Papier*, *46*(10A), V32.

Chezick, C., Allen, J., Hill, G., Lapierre, L., Dorris, G., Merza, J., et al. (2004). A 10-day mill trial of near-neutral sulfite deinking. Part I: Deinked pulp optical and physical properties. *Pulp and Paper Canada, 104*(4), 33–38.

Dutt, D., Tyagi, C. H., Singh, R. P., & Kumar, A. (2012). Effect of enzyme concoctions on fiber surface roughness and deinking efficiency of sorted office paper. *Cellulose Chemistry and Technology*, *46*(9–10), 611–623.

Eom, T. J., & Ow, S. S. K. (1990). Process for removing printing ink from wastepaper. German Patent GB 3,934,772.

Eriksson, K. -E. L., & Adolphson, R. B. (1997). Pulp bleaching and deinking pilot plants use chlorine-free process. *TAPPI Journal*, *80*(6), 80.

Floccia, L. (1988). Fractionation and separate bleaching of wastepaper. *TAPPI international pulp bleaching conference*, Orlando, FL.

Franks, N. E., & Munk, N. (1995). Alkaline cellulases and the enzymatic deinking of mixed office waste. *TAPPI pulping conference*, Chicago, IL.

Gallo, I. Mosele, G., & Elegir, G. (2004). Pilot scale evaluation of alkaline cellulases for enzymatic deinking of mixed office paper. *Ninth international conference on biotechnology in the pulp and paper industry*, book of abstracts, pp. 69–70. Durban, 10–14 October.

Gill, R., Hillerbrand, M., Keijsers, E., Kessel, L. V., Loosvelt, I., Lund, H., et al. (2007). Enzymatic deinking of recycled paper: From laboratory to mill scale. *Tenth international congress on biotechnology in the pulp and paper industry*, Madison, WI, 10–15 June, IndusAPP 2.1, p. 52.

Gu, Q. P., You, J. X., Yong, Q., & Yu, S. Y. (2004). Enzymatic deinking of ONP with lipase/cellulase/xylanase. *China Pulp Paper*, *23*(2), 7.

Haynes, R. D. (2000). The impact of the summer effect on ink detachment and removal. *TAPPI Journal*, *83*(3), 56–65.

Heise, O. U., Unwin, J. P., Klungness, J. H., Fineran, W. G., Jr, Sykes, M., & Abubakr, S. (1996). Industrial scale-up of enzyme-enhanced deinking of non-impact printed toners. *TAPPI Journal*, *79*(3), 207.

Heitmann, J. A., Joyce, T. W., & Prasad, D. Y. (1992). Enzyme deinking of newsprint waste: *International conference on biotechnology in the pulp and paper industry*. Kyoto, Japan: OZEPA.

Ibarra, D., Concepción Monte, M., Blanco, A., Martínez, A. T., & Martínez, M. J. (2012). Enzymatic deinking of secondary fibers: Cellulases/hemicellulases versus laccase-mediator system. *Journal of Industrial Microbiology and Biotechnology*, *39*(1), 1–9. doi:10.1007/s10295-011-0991.

Jeffries, T. W., Klungness, J. H., Sykes, M. S., & Rutledge-Cropsey, K. R. (1994). Comparison of enzyme-enhanced with conventional deinking of xerographic and laser-printed paper. *TAPPI Journal*, *77*(4), 173.

Jeffries, T. W., Sykes, M. S., Cropsey, K. R., Klungness, H., & Abubakr, S. (1995). Enhanced removal of toners from office waste papers by microbial cellulases: *Sixth international conference on biotechnology in the pulp and paper industry*. Vienna: OZEPA.

Kim, T. J., Ow, S. S. K., & Eom, T. J. (1991). Enzymatic deinking method of waste paper. *TAPPI pulping conference*, Orlando, FL.

Knudsen, O., Young, J. D., Yang, J. L. (1998). Mill experience of a new technology at stora dalum deinking plant. *PTS-CTP Deinking symposium*, Munich, Germany, 5–7 May, 17pp.

Leenen, M., & Tausche, J. (2004). Principles of enzymatic deinking (EDT) and practical implementation in a paper mill. *Eighth Pira international conference on paper recycling technology*, Prague.

Ma, J. H., & Jian, C. (2002). Enzyme applications in the pulp and paper industry. *Progress in Paper Recycling*, *11*(3), 36–46.

Magnin, L., Lantto, R., & Delpech, P. (2001). Potential of enzymatic deinking: *Eighth international conference on biotechnology in the pulp and paper industry*. Helsinki: VTT Biotechnology.

Mohammed, S. H. (2010). Enzymatic deinking: A bright solution with a bright future. *IPPTA*, *22*(3), 137–138.

Ow, S. K., Park, J. M., Han, S. H., Srebotnik, E., & Messner, K. (1996). Effects of enzyme on ink size and distribution during the enzymatic deinking process of old newsprint: *Sixth international conference on biotechnology in the pulp and paper industry*. Vienna: OZEPA.

Paik, K. H., & Park, J. Y. (1993). Enzyme deinking of newsprint waste. I: Effect of cellulase and xylanase on brightness, yield and physical properties of deinked pulps. *Journal Korea TAPPI*, *25*(3), 42.

Prasad, D. Y. (1993). Enzymatic deinking of laser and xerographic office wastes. *Appita Journal*, *46*(4), 289.

Prasad, D. Y., Heitman, J. A., & Joyce, T. W. (1992a). Enzymatic deinking of flexographic printed newsprint: Black and colored inks. *Papiripar, Papir-Es Nyomdaipari Mueszaki Egyesuelet*, *36*(4), 122.

Prasad, D. Y., Heitman, J. A., & Joyce, T. W. (1992b). Enzyme deinking of black and white letterpress printed newspaper waste. *Progress in Paper Recycling*, *1*(3), 21.

Puneet, P., Bhardwaj, N. K., & Singh, A. K. (2010). Enzymatic deinking of office waste paper: An overview. *IPPTA*, *22*(2), 83–88.

Putz, H. J., Renne, K., Gottsching, L., & Jokinen, O. (1994). Enzymatic deinking in comparison with conventional deinking of offset news. *TAPPI pulping conference*, San Diego, CA.

Rushing, W., Joyce T. W., & Heitmann, J. A. (1993). Hydrogen peroxide bleaching of enzyme deinked old newsprint. *Seventh international symposium on wood and pulping chemistry*, Beijing.

Rutledge-Cropsey, K., Jeffries, T., Klungness, J. H., & Sykes, M. (1994). Preliminary results of effect of sizings on enzyme-enhanced deinking. *TAPPI recycling symposium*, Boston, MA.

Spiridon, I., & Belgacem, M. N. (2004). Enzymatic deinking of laser printed papers. *Progress in Paper Recycling*, *13*(4), 12.

Spiridon, I., & de Andrade, A. M. (2005). Enzymatic deinking of old newspaper (ONP). *Progress in Paper Recycling*, *14*(3), 14.

Sykes, M., Klungness, J., Abubakr, S., & Rutledge-Cropsey, K. (1995). Enzymatic deinking of sorted mixed office waste: Recommendations for scale-up. *TAPPI recycling symposium*, New Orleans, LA.

Tausche, J. (2002). Mill-scale benefits in enzymatic deinking. *Seventh Pira international recycling technology conference*, Brussels.

Tausche, J. (2005a). Furnishing better deinking: Tailoring enzyme to suit your recycling needs. *Pulp and Paper International*, *47*(7), 20.

Tausche, J. (2005b). Deinking mills dodge financial crunch with customized enzymes. *Pulp Paper*, *79*(10), 49–51.

Tausche (2007). More tonnes, less energy, and reduced total costs via enzymatic deinking. *Tissueworld nice.*

Wang, S., & Kim, M. (2005). Study on old newsprint deinking with cellulases and xylanase: *59th Appita annual conference and exhibition*. Auckland: ISWFPC.

Welt, T., & Dinus, R. J. (1995). Enzymatic deinking – a review. *Progress in Paper Recycling*, *4*(2), 36.

Woodward, J., Stephan, L. M., Koran, L. J., Jr, Wong, K. K. Y., & Saddler, J. N. (1994). Enzymatic separation of high-quality uninked pulp fibers from recycled newspaper. *Biotechnology*, *12*(9), 905.

Xu, Q. H., Qin, M. H., Shi, S. L., Zhang, A. P., & Xu, Q. (2004). Synergistic deinking of ONP by cellulase/hemicellulase combined with laccase-mediator system. *China Pulp Paper*, *23*(8), 6.

Yang, J. L., Ma, J., Eriksson, K. E. L., Srebotnik, E., & Messner, K. (1995). Enzymatic deinking of recycled fibers – development of the Enzynk process. *Sixth international conference on biotechnology in the pulp and paper industry*, Vienna.

Yehia, A., & Reheem, H. A. R. (2012). Some physico-chemical aspects related to the deinking process using xylanase as a surface modifier. *Elixir Applied Chemistry*, *48*, 9498–9502. Available online at <www.elixirpublishers.com> (Elixir International Journal).

Zeyer, C., Joyce, T. W., Heitmann, J. A., & Rucker, J. W. (1994). Factors influencing enzyme deinking of recycled fiber. *TAPPI Journal*, *77*(10), 169.

Zeyer, C., Heitmann, J. A., Joyce, T. W., & Rucker, J. W. (1995). Performance study of enzymatic deinking using cellulase/hemicellulase blends. *Sixth international conference on biotechnology in the pulp and paper industry*, Vienna.

Zhang, S. F., & Hu, X. G. (2004). Enzymatic deinking of postconsumer printing paper. *China Pulp Paper*, *23*(2), 10.

Zhang, X., Renaud, S., & Paice, M. (2008). Cellulase deinking of fresh and aged ONP/OMG. *Enzyme and Microbial Technology*, *42*(2), 103–108.

Zuo, Y., & Saville, B. A. (2005). Efficacy of immobilized cellulase for deinking of mixed office waste. *Journal of Pulp and Paper Science*, *31*(1), 3.

9 Bleaching of Secondary Fibres*

9.1 Introduction

The recycled waste paper component in consumer paper products has grown appreciably, motivated by the encouragement to conserve fibre supply and by mounting pressures on landfill sites. The quality and brightness of the deinked waste fibre furnish will dictate the type of paper product it can be targeted for; the higher the brightness and fibre quality, the more extensive are its end uses. Although the deinking process is designed to optimise brightness gains, frequently this process must be supplemented by an additional bleaching stage or sequence to recapture some of the original brightness of the fibre. To meet a requirement for high brightness, the combination of efficient removal of the ink and chemical treatment is necessary.

Many bleaching chemicals have been used for bleaching deinked pulp (DIP). Because of its negative effects on the environment, chlorine bleaching has very little importance in recycled pulp processing. Chlorine-free bleaching of secondary fibres is more common. It uses hydrogen peroxide, oxygen, ozone, sodium dithionite and formamidine sulphinic acid (FAS). Hydrogen peroxide, oxygen and ozone are oxidative bleaching chemicals; sodium dithionite and FAS are reductive bleaching chemicals (Ackermann, 2000; Fluet, 1995; Gangolli, 1982; Karmakar, 2004; Kronis, 1992, 1997; Magnin, Angelier, & Gailand, 2000; Matzke & Kappel, 1994; Muguet & Sundar, 1996; Patt, Gehr, & Kordsachia, 1993; Renders, 1992, 1995; Renders & Hoyos, 1994; Renders, Chauveheid, & Pottier 1995; Süss, 1995). Hydrogen peroxide bleaching is theoretically efficient for bleaching wood-containing pulp but is also often used to improve the brightness of wood-free DIP (Gehr, 1997; Helmling et al., 1986; Jiang et al., 1999). Ozone and oxygen have also been used for bleaching or decolourising secondary fibres. The use of oxygen has become more common in commercial applications. Ozone is the only chlorine-free bleaching chemical that can almost completely destroy optical brighteners in DIPs. However, its use has not played any significant role in the bleaching of secondary fibres.

No industrial applications of peracids have been reported. Sodium dithionite and FAS are used industrially in some cases to bleach DIPs containing mechanical pulp. They are also used for decolourising dyed brokes or DIP resulting from mixtures containing dyed papers.

Lachenal (1994) has classified these chemicals into two categories: non-degrading and degrading agents. Hydrogen peroxide, sodium hydrosulphite and FAS

*Some excerpts taken from Bajpai (2006) with kind permission from Pira International, UK and Bajpai (2012) with kind permission from Elsevier B.V.

Recycling and Deinking of Recovered Paper. DOI: http://dx.doi.org/10.1016/B978-0-12-416998-2.00009-X

belong to the category of non-degrading agent. Their action is limited to the destruction of carbonyl groups (and azo groups for the reductive agents). Coloured organic molecules are modified into colourless molecules. Pulp yield is not considerably affected. Oxygen and ozone belong to the category of degrading agents. They can destroy phenolic groups, carbon–carbon double bonds and conjugated aromatic structures. Chromophore groups are dissolved away from the fibre structure and then removed. Pulp yield is significantly affected. When high brightness is required, combinations of oxidative and reductive bleaching are sometimes used.

9.2 Chlorine Bleaching

Wood-free DIP is usually bleached with chlorine-containing chemicals in the USA. Recycled fibre with a mechanical fibre content up to 5% is not only very effectively bleached but its colour is stripped. When larger proportions of mechanical fibres are present, yellowing reactions on the mechanical pulp components accompany the treatment, so bleaching is not possible by the usual single-stage process. The residual lignin has to be removed first.

9.2.1 *Bleaching with Hypochlorite*

To achieve high brightness gains, the lignin content should be very low. If higher amounts of lignin are present in the fibre stock, hypochlorite bleaching causes discolouring, especially at low doses. The pulp turns pink and requires large quantities of chemicals to brighten it. Therefore, in conventional chemical pulp bleaching, a common practice was chlorination with subsequent alkali extraction of the chlorinated lignin compounds before the hypochlorite stage. Depending on the requirements of the DIP, the bleaching conditions can vary within certain limits. The quantity of active chlorine is 1–6% on oven-dry DIP; 2–3% is used in most cases. The temperature is 50–55°C for a bleaching time of 2–3h and the pH is 9–10. Towers or chests are used as reaction vessels in which the DIP is stored with consistencies of 3–15%. Treatment with hypochlorite usually uses a single-stage process. Depending on the composition of the recovered paper, brightness levels of 70–80% ISO are possible. Sometimes hypochlorite bleaching also uses two stages. A washing stage is then added between the two reaction vessels to remove the oxidised lignin. This gives better use of the chemicals. Higher brightness levels at the end of the bleaching process reflect this (Bajpai, 2006).

If the proportion of mechanical fibres in the DIP exceeds 10%, chlorination occurs before the hypochlorite stage to degrade the residual lignin content. This ensures the effectiveness of the subsequent hypochlorite stage even when using DIP with a low mechanical fibre content. The treatment with elemental chlorine uses low consistencies of about 3% and temperatures of 30–40°C. Following this, the DIP is washed to remove the chlorinated, degraded lignin products. After that the pulp is bleached with hypochlorite in the medium-consistency range of 12–15%. Small amounts of sodium hydroxide are occasionally added during washing to eliminate the oxidised lignin from the DIP more effectively.

Earlier, a complete extraction stage with alkali such as in chemical pulp bleaching was applied, especially when bleaching recycled fibre pulp with an even greater lignin content. Optimum results are possible with this bleaching sequence by adding greater amounts of bleaching agent at the hypochlorite stage. This improves the brightness and its stability against colour reversion. Because of its negative effects on the environment, this bleaching sequence has very little importance in future recycled pulp processing. Instead, bleaching combinations with alternative chemicals such as FAS or with sodium dithionite will be increasingly used for colour stripping of wood-free recycled pulp.

9.3 Chlorine-Free Bleaching

9.3.1 Bleaching with Hydrogen Peroxide

Strong environmental concerns make hydrogen peroxide more attractive as a bleaching agent for secondary fibres. The chemistry of hydrogen peroxide bleaching has been reviewed by Renders (1995) and Ackermann (2000). The bleaching effect uses the dissociation of hydrogen peroxide in water to form hydronium (H_3O^+) and perhydroxyl ions (HO_2^-). The perhydroxyl anion acts as a nucleophilic bleaching agent. Increasing its concentration is necessary to achieve a high bleaching effect, which is possible by increasing the hydrogen peroxide concentration and by adding sodium hydroxide.

Helmling, Süss, and Berndt (1985) reported that, at low doses of sodium hydroxide, the hydrogen peroxide bleaching liquor does not have sufficient activation. In the highly alkaline range, a loss of brightness occurs with additional sodium hydroxide. This is due to the increased yellowing reactions of excessive hydroxide ions on the lignin structures of mechanical fibres. With hydrogen peroxide/sodium hydroxide proportions at which optimum bleaching occur, the ratio between the rates of bleaching and yellowing reactions is most favourable. Residual hydrogen peroxide content at the end of the bleaching process is a sign of an optimum hydrogen peroxide activation. If the residual hydrogen peroxide is less than 10% of the original hydrogen peroxide dose, the yellowing reactions dominate over the bleaching reactions. Higher brightness is obtained with increasing dose of hydrogen peroxide. The required amount of sodium hydroxide to obtain the maximum bleaching effect also increases. Bleaching should be conducted under optimum hydrogen peroxide/sodium hydroxide conditions for economic reasons. The chemical oxygen demand (COD) of the process water increases with increasing alkalinity, so it is important to keep the alkali amount to a minimum (Berndt, 1982). Table 9.1 shows the brightness development with increasing dose of hydrogen peroxide (Helmling et al., 1985). The maximum increase occurs with hydrogen peroxide additions of less than 1.5%.

Bleaching results are affected by the process parameters such as stock consistency, temperature, reaction time, and the chemicals used (Ackermann, 2000; Renders, 1992, 1995). Brightness is higher at higher pulp consistency (Helmling et al., 1985). Higher consistency gives a higher relative concentration of bleaching

Table 9.1 Effect of H_2O_2/NaOH Dose on Brightness

H_2O_2 (%)	NaOH (%)	Brightness (% ISO)
0.25	0.7	57.6
0.50	0.9	58.9
1.0	1.2	60.4
1.5	1.4	61.8
3.0	1.6	63.6

Source: Based on data from Helmling et al. (1985)

chemicals that come into direct contact with the fibres, and the displacement of the aqueous phase eliminates dissolved matter that can unfavourably affect the bleaching reaction. Mixing problems are encountered when the pulp consistency is above 30%. So the consistency is mainly limited to 25% in separate bleaching stages. The temperature should be in the range 40–70°C for economic reasons. Typical reaction times are in the range 1–3 h when bleaching occurs in a separate process stage. A lower dose of sodium hydroxide is required at higher temperatures for optimum hydrogen peroxide activation.

Although bleaching with hydrogen peroxide is less efficient for secondary fibre than for virgin fibres, careful control of some process parameters can help to maximise the brightness gain. The easiest way to detect poor hydrogen peroxide efficiency is to determine, in parallel with the brightness, the hydrogen peroxide consumption. Generally, a total hydrogen peroxide consumption means a significant loss of hydrogen peroxide efficiency. Some of the important parameters causing this hydrogen peroxide decomposition are discussed below.

Trace metals catalyse the decomposition of hydrogen peroxide (Galland, Bernard, & Vernac, 1989; Kutney & Evans, 1985a). An acid wash at pH 2 is found to be effective in removing transition metal ions for chemical pulp bleaching. Such a pretreatment before a hydrogen peroxide stage has been applied successfully to waste paper containing 40% newsprint and 60% magazines; a temperature beyond 70°C simultaneously destroys catalase (Patt et al., 1993). However, according to Renders (1995a,b), this approach does not appear to be suitable for secondary fibres because filler material such as calcium carbonate can greatly increase the acid demand. Chelating agents and sodium silicate are commonly used in secondary fibres to deactivate the heavy metal ions that contribute to the wasteful decomposition of hydrogen peroxide. Sodium silicate chelates metal ions so it is able to reduce the hydrogen peroxide decomposition. Silicate is also a source of alkalinity; its addition in the pulper helps ink removal by flotation (Kutney & Evans, 1985b; Renders,1992, 1995a,b). Table 9.2 shows the increase in brightness and the reduction of consumed hydrogen peroxide with increasing dose of silicate for a 50/50 mixture of newsprint and magazines (Renders & Hoyos, 1994). Table 9.3 compares the chelating agents diethylenetriaminepentaacetic acid (DTPA) and the sodium salt of diethylenetriami nepenta(methylenephosphonic) acid (DTPMP) plus sodium silicate in post-deinking

Table 9.2 Increase in Brightness and Reduction in Consumed Hydrogen Peroxide with Increasing Dose of Silicate for a 50/50 Mixture of Newsprint and Magazines

Sodium Silicate (%)	Brightness (% ISO)		H_2O_2 Consumption (%)
	After Disperger	After Post-Flotation	
0	56.0	58.8	100.0
1.6	59.5	65.0	97.5
3.4	62.0	67.5	65.0

Source: Based on data from Renders and Hoyos (1994)
Pilot plant optimisation: 50/50 newsprint/magazine mixture, 70°C, 1% H_2O_2, 1% total alkalinity.

Table 9.3 Hydrogen Peroxide Stabilisation for ONP/OMG and MOW Bleaching

Stabilisation	ONP/OMG		Mixed Office Waste	
	Brightness (% ISO)	H_2O_2 Consumption (%)	Brightness (% ISO)	H_2O_2 Consumption (%)
No stabilisation	58.9	100	70.8	100
1.0% DTPA	63.6	92	72.6	97
1.0% DTPMP	63.9	90	–	–
1.6% DTPMP	–	–	72.6	98
2.5% Silicate	66.0	80	73.7	75
5.0% Silicate	67.2	67	–	–
2.5% Silicate + 1.0% DTPA	66.7	78	73.7	73
2.5% Silicate + 1.0% DTPMP	67.8	67	–	–
2.5% Silicate + 1.6% DTPMP	–	–	73.1	68

Source: Based on data from Robberechts et al. (2000)

bleaching on ONP/OMG mixture and mixed office waste (MOW) (Robberechts, Pyke, & Penders, 2000).

An uncontrolled decomposition of hydrogen peroxide is initiated by catalase, which is an enzyme. The main known sources of catalase-forming micro-organisms are waste papers and slime in the wastewater from the paper machine which is sent back to the deinking plant. The presence of catalase can easily be detected (Galland, Bernard, & Vernac, 1989). The destruction of catalase by a thermal treatment (>70°C) has been described. Catalase-forming micro-organisms are killed by the application of peracetic acid, and catalase formation is also inhibited.

Sodium dithionite is sometimes used in combination with hydrogen peroxide in a two-stage bleaching of DIP for colour removal. For use of hydrogen peroxide in a deinking line, several addition points can be considered: in the pulper, in the dispersion unit and as post-bleaching at the end of stock processing line. The use of hydrogen peroxide in the pulping stage is not a perfect solution because ink and contraries present in the pulper reduce the bleaching efficiency of hydrogen peroxide. Its main advantage is that no additional equipment is required.

Hydrogen peroxide is very effective for improving the brightness of DIP. The bleaching result is sufficient for many applications using DIP to satisfy the optical requirements of the paper produced. A prerequisite for successful bleaching is stabilisation of the hydrogen peroxide to the reactions that cause its decomposition. The effects on chromophores of the lignin that use oxidative reactions are limited. If higher optical quality demands are made on the DIP, additional bleaching with reductive bleaching chemicals will be necessary.

Ackermann (2000) has summarised diverse experiences on hydrogen peroxide bleaching of secondary fibres. Bovin (1984) and Blechschmidt and Ackermann (1991) have recommended separate bleaching with hydrogen peroxide after a deinking process (flotation or washing). The increase in brightness in this case is higher compared with those obtained by hydrogen peroxide bleaching before a deinking stage (Bovin, 1984; Putz, 1987). Most deinking plants take advantage of the synergistic effects of flotation and bleaching by adding hydrogen peroxide during pulping and during dispersion or post-bleaching. Ackermann, Putz, and Göttsching (1992) and Helmling, Süss, and Eul (1986) have reported that a prerequisite for maximum brightness gain at the post-bleaching stage is the optimum removal of ink particles at the deinking stages. A higher content of ink particles actually gives an inferior bleaching effect.

Zou, Hsieh, Agrawal, and Matthews (2007) studied tetraacetylethylenediamine (TAED) activator for peroxide bleaching of recycled pulp. Preliminary results indicated that TAED-activated peroxide could provide better brightness, brightness stability, bulk, strength and lower some chemical requirements. In addition, data demonstrated that the use of excess TAED in bleaching was capable of providing brightness equal to or better than that of peroxide-only bleaching. Reaction time, pH and TAED charge were found to have significant effects on the brightness, particularly reaction time at low pH levels and low chemical charges. The addition of DTPA increased tensile strength and improved brightness stability.

Table 9.4 presents the conditions for hydrogen peroxide bleaching.

9.3.2 Bleaching with Dithionite

Sodium dithionite is also known as sodium hydrosulphite. It is a reductive bleaching chemical. It is used not only for bleaching but also for removing colour from coloured recovered paper and carbonless paper (Dumont, Fluet, Giasson, & Shepperd, 1994; Fluet & Shepperd, 1997; Fluet et al., 1994; Hache, Brungardt, Munroe, & Teodorescu, 1994). Most of the acidic and direct dyes are permanently decolourised by sodium dithionite because it breaks the azo groups. Some of the basic dyes are temporarily decolourised. Dithionite is sometimes used in combination with an oxidising agent, because some dyes that are not reactive with oxidising chemicals can react with some reducing agents.

Sodium dithionite decomposes rapidly when exposed to air. When exposed to water, the solid form liberates sulphur gases, which are corrosive to equipment and

Table 9.4 Bleaching Conditions for Hydrogen Peroxide

Parameters	Values
Hydrogen peroxide dose (%)	0.5–3
Sodium hydroxide dose (%)	0.5–2.0
pH	9–11
Time (minutes)	30–90
Pulp consistency (%)	10–15 (medium consistency) 25–30 (high consistency)
Temperature (°C)	60–90 100°C for wood-free papers
Stabilisers	Silicate 1–3% DTPA 0.1–0.3%

Source: Based on Renders (1995) and Ackermann (2000)

buildings. The aqueous solution is typically stored in closed tanks with a nitrogen pad. Sodium dithionite is generally supplied as dry powder. These commercial products may contain stabilisers, buffers (phosphates, carbonates) and chelates.

Dithionite was produced for the first time in 1906 by BASF in Germany in powder form. Initially, the product was obtained by the zinc dust process. Later on, a process was developed using the sodium amalgam from the mercury cell electrolysis of sodium chloride solution, converting this directly to sodium dithionite with sulphur dioxide. This process produces a material that is free of heavy metals, so it is stable. BASF later developed the formiate process, in which sodium formiate is converted with bisulphite to sodium dithionite. Sodium formiate is produced from carbon monoxide and sodium hydroxide. Dithionite can also be produced on site using Borol by the Ventron process.

Sodium dithionite is present in the bleaching solution with about 85% active substance. The advantage of in situ production is that the preparation of the bleaching solution can be controlled. So there is no loss of active substance owing to the transport and storage of the naturally unstable dithionite solution (Hache, Fetterly, & Crowley, 2001; Meyers, Wang, & Hache, 1999).

Dithionite was used for the first time for bleaching of mechanical pulp in the 1930s. Initially sodium dithionite was used in small amounts as a powder. The bleaching effect was limited because the bleaching time and the temperatures were too low. Good results were obtained when atmospheric oxygen was excluded. With the introduction of continuous bleaching processes such as tower bleaching, the use of solutions of dithionite proprietary blends became necessary to ensure their homogeneous mixing with stock. Another option is the use of cooled, alkalised solutions of sodium dithionite. At temperatures below 10°C, these solutions are so stable that they can be stored for long periods. This eliminates the complicated process of dissolving the powder. Dithionite is a stronger reducing agent in alkaline media.

Heavy metal ions have a harmful effect on bleaching. Treatment of the pulp with chelating agents has a favourable effect on bleaching efficiency (Melzer & Auhorn, 1981). Chelating agents are often premixed dry with the sodium dithionite by the

Table 9.5 Effect of Dithionite Dose at Different Temperatures on Brightness

$Na_2S_2O_4$ (%)	Brightness (% ISO) at Different Temperatures				
	25°C	40°C	60°C	80°C	95°C
0.5	59.4	58.6	59.4	60.4	60.3
1.0	60.3	62.9	63.6	64.0	63.6
1.5	60.7	63.9	64.6	65.9	64.7
2.0	61.6	64.5	–	67.1	66.3
2.5	–	–	64.9	66.9	–
3.0	62.6	64.4	65.6	67.5	66.4

Source: Based on data from Putz (1987)
Reaction time 60 min, pH 8.4 and consistency 3%.

supplier. Reducing dithionite has a higher kinetic reaction rate compared with hydrogen peroxide. Reaction time is significantly shorter in dithionite bleaching. Studies have shown that the bleaching reaction takes a few minutes. Bleaching is good at higher temperatures that favour diffusion of the dithionite ions into the fibre cell wall (Melzer, 1985). Owing to the sensitivity of dithionite to oxygen, a separate bleaching stage is essential. So, a combination with the pulping stage of the stock processing operation is not possible as in case of hydrogen peroxide bleaching. Dithionite bleaching is usually done at low pulp consistencies, because at low consistency air is low. The optimum consistency in case of dithionite bleaching is in the range 3–5%. The bleaching chemicals are mixed and the pulp is directed into an upflow bleach tower so that the fibres reacting with the dithionite do not come into contact with atmospheric air. At the top of the tower the ambient air can react with unused dithionite and convert it into bisulphate and bisulphite. The homogeneous mixing and distribution of the bleaching chemicals in the pulp is very important because of the short reaction time. Good mixing reduces decomposition of dithionite through exposure to air. Low-consistency mixers and fluidising medium-consistency pumps that empty the air from the pulp slurry and simultaneously mix the bleaching chemicals are standard components in the dithionite bleaching process. By shifting the stock consistency into the more economical medium-consistency range, dithionite bleaching became a common practice in secondary fibre pulp processing. Higher brightness is obtained at higher temperature. Brightness gain is greater at temperatures beyond 60°C (Table 9.5) (Putz, 1987). Maximum increase in brightness is obtained at about pH 7.0 and above a temperature of 60°C. A higher dose of dithionite and higher reaction temperatures require higher initial pH (Table 9.6) (Putz, 1987).

Fluet (1995) reported that in MOW, brightening and colour stripping efficiencies are better at a temperature of 80–100°C, a pH of 7–8 and medium or high consistency. When temperature and consistency are both in the higher range, the retention time is short. There is some brightness reversion if the retention time is too long and the temperature is high. Kaichang, Collins, and Eriksson (2000) have reported that some detergents can enhance the efficiency of dithionite. Brightness reversion is more in the case of dithionite than FAS (Putz, 1987).

Table 9.6 Effect of pH and Dithionite Dose on Brightness

pH	Brightness (% ISO) at Different $Na_2S_2O_4$ Doses		
	0.5%	1.0%	2.0%
4.0	59.9	–	–
5.0	–	63.1	61.6
6.0	62.6	64.0	63.9
7.0	–	–	64.6
8.0	61.7	63.9	64.8
10.0	59.5	63.2	64.1

Source: Based on data from Putz (1987)
Temperature 80°C, reaction time 60 min and consistency 3%.

Table 9.7 Bleaching Conditions for Sodium Dithionite

Parameters	Values (Range)	
	Low Consistency	Medium Consistency (or High Consistency, Disperger)
Dose (%)	0.2 – 2.0	0.2 – 2.0
Time (min)	60 – 120	30
Consistency (%)	3 – 5	10 – 15(30)
pH	6 – 8	7 – 8
Temperature (°C)	60 – 80	80 – 100

Source: Based on Ackermann (2000)

The optimisation of hydrosulphite bleaching in recycled pulps is necessary to achieve target brightness at a minimised cost for mills applying the neutral paper-making process. The level of brightness in pulps brightened by sodium hydrosulphite under neutral conditions is enhanced by increasing bleaching temperature (Mozaffari, Krishnan, Burch, & Rangamannar, 2006). This can result in higher costs because more energy is needed. During hydrosulphite brightening, increased paper brightness and decreased chemical consumption are offered by chelant treatment. Hydrosulphite bleach consumption can be maintained at an optimum level through proper monitoring and controlling of residual ink concentration in DIP. To reduce cost, optical brighteners should only be used after the hydrosulphite bleaching process has obtained its maximum brightness. A total system approach should be implemented using feedback process controllers to determine brightness and control chemicals usage. The advance quality control programme is implemented at the mill to minimise consumption of bleaching chemicals and paper brightness variability at lower operating costs.

Conditions for dithionite bleaching are summarised in Table 9.7.

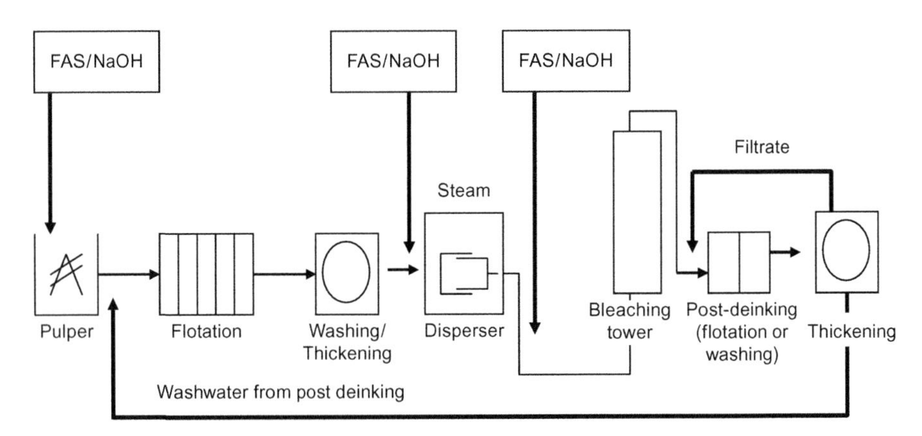

Figure 9.1 Opportunities for application of FAS in a deinking process.
Source: Kronis (1994); reproduced with permission from RISI.

9.3.3 Bleaching with FAS

FAS is excellent for brightening and colour stripping secondary fibres. Its flexibility allows different recycling mills to achieve high brightness pulps for fine paper applications. Its use as bleaching agent was first proposed in the textile industry. The ability of FAS to colour strip effectively results in a substantial reduction in the filtrate colour as well. Different types of application such as in pulper, bleaching tower or disperser have been described (Figure 9.1) (Kronis, 1997). FAS is a low-odour, crystalline reducing agent and can be used on all types of waste paper. Its use is particularly recommended when using a furnish containing dyed paper. A process patented by Süss and Krüger (1983) proposes bleaching conditions for mechanical pulp and secondary fibre pulp that can apply in a single-stage and a two-stage process with other bleaching chemicals. FAS contains thiocarbamide sulphur dioxide according to X-ray structure analysis. The material forms at low temperatures under acidic to neutral pH conditions in a reaction involving two hydrogen peroxide molecules and one thiocarbamide molecule. FAS is a white to slightly yellow powder, odourless and non-flammable. It is also known as thiourea dioxide. FAS is oxidised by atmospheric oxygen like all reducing bleaching chemicals, but compared with dithionite it is considerably less prone. This allows a wider range of application possibilities for bleaching, not only as a separate bleaching stage but also in combination with other stages of the stock processing operation. FAS also has a lower sulphur content than dithionite. This has a positive effect on the sulphate load of the white water loop. The sulphate level in the bleaching effluent can be reduced by as much as 75%. As a result, the corrosion susceptibility of the equipment and instruments is lower. Also, the production of unpleasant odours due to hydrogen sulphide is lower.

FAS is only slightly soluble in water; under alkaline conditions the solubility increases. Only about 27 g/L enter the solution as sulphinate; the solubility under alkaline conditions increases up to 100 g/L. Alkaline solutions of FAS have a higher solubility, but its aqueous solution decomposes very rapidly. For this reason, alkaline bleaching solutions are prepared only shortly before their addition in a continuous process. Consumption must then occur as soon as possible. FAS hydrolyses to the sulphinate anion and urea under heat and alkaline conditions, as follows.

The sulphinate anion has a high negative redox potential and is responsible for reductive bleaching (Fallon, 1994). Compared with dithionite, FAS has a slightly higher reductive potential when both chemicals are under alkaline conditions. Important parameters for FAS bleaching are temperature, time, alkalinity and consistency. The furnish type and the degree of brightening or colour stripping required are also important in determining the feasibility of using FAS. Sodium hydroxide is commonly used as the alkaline source. The most important process parameter of FAS bleaching is the reaction temperature. Kronis (1992) has reported that increasing the reaction temperature from 40°C to 90°C almost doubled the bleaching result of a wood-free DIP after a reaction time of 30 min. At lower temperatures, increasing the reaction time is necessary to develop brightness. Increasing the reaction time alone is insufficient to compensate fully for the lower brightness level that occurs with lower temperatures. Only small quantities of FAS are necessary to perform effective bleaching if the temperature is set at an optimum value (Kronis, 1992). A brightness of about 72% ISO is obtained at a temperature of 80°C, with only 0.2% FAS, whereas at a temperature of 50°C three times more FAS is required to attain the similar brightness.

Kang, Van Heiningen, and Ni (1999) and Taylor and Morrison (1999) have explored the possibility of using sodium silicate instead of sodium hydroxide. They have also examined the effect of chelating agents on FAS bleaching. Sodium silicate improves FAS bleaching by efficiently deactivating transition metals. Addition of chelating agents to the FAS stage in a FAS Peroxide (FP) sequence improved the bleaching of the FAS stage and the subsequent hydrogen peroxide stage (Table 9.8) (Kang et al., 1999). An additive effect is observed when both chelating agent and sodium silicate are added to the FAS treatment in a FP sequence. Significantly higher brightness can be achieved at higher temperatures. At lower temperatures, longer times are necessary to develop brightness. FAS performs very well in a hot disperser application because it operates at high temperature. A hot dispersing unit combines the benefits of high intensity mixing with high temperature (90–120°C).

Work by Fluet (1995) and Kronis (1992) has shown that results are comparable with sodium dithionite and FAS. Kronis (1992) reported that with coloured paper, FAS and Dithionite (Y) provide comparable results, the economics slightly favouring FAS. Fluet (1995) reported that Y performs similarly to FAS on the same dose basis on MOW but at a lower cost. Compared with dithionite, the sulphate levels in the bleaching effluent are lower with FAS (Kronis, 1992). The formation of FAS in situ in the presence of pulp by consuming the residual hydrogen peroxide of a peroxide stage (P stage) is presented as a potentially cost-effective process (Fallon, 1994); however, the safe handling of thiourea, which is a potential carcinogen, must be seriously considered. In this process, the residual hydrogen peroxide from a tower stage is acidified and thiourea is added. Then alkaline conditions allow the FAS bleaching to take place.

When both hydrogen peroxide and FAS are used in a deinking line, it is necessary to recognise that a residual of hydrogen peroxide will consume FAS, and vice versa

Table 9.8 Effect of Silicate and DTPA on FAS and FAS + P Bleaching

Stage or Sequence	Brightness (% ISO)	B* Value	Residual H_2O_2 of P Stage (% on Pulp)
F	61.4	10.0	–
F_Q	62.3	10.8	–
F_{Q+Si}	62.7	10.6	–
FP	76.0	6.0	0.12
$F_{Si}P$	77.5	5.7	0.18
F_QP	78.1	5.3	0.32
$F_{Q+Si}P$	78.6	5.3	0.37
$F^*_{Q+Si}P$	79.4	5.1	0.42

Source: Based on data from Kang et al. (1999)
F: 0.5% FAS, 0.25% NaOH, 90°C, 1 h.
F_Q: 0.5% FAS, 0.2% DTPA, 90°C, 1 h.
F_{Q+Si}: 10% pulp cons, 0.2% DTPA, 1% Na_2SiO_3, 0.5% FAS, initial pH 9.6, 90°C, 1 h.
F^*_{Q+Si}: 10% pulp cons, 0.2% DTPA, 1% Na_2SiO_3, 0.5% FAS, initial pH 9.6, 90°C, 1 h, pulp slurry was acidified to pH 5 before washing.

(Kronis, 1997). This is especially important if FAS treatment immediately follows a hydrogen peroxide bleaching stage, because of the significant hydrogen peroxide residual that normally results and the more expensive FAS that could be consumed. Conversely, a residual of FAS before hydrogen peroxide is normally not of concern because of the low levels of FAS applied and the negligible amounts remaining at the end of an FAS stage. It is important to destroy or reduce the residual hydrogen peroxide before the FAS treatment. Thought should be given to having a post-flotation stage between the hydrogen peroxide and FAS. Dilution that occurs after the hydrogen peroxide treatment before flotation, followed by thickening of the deinked stock before the FAS stage, would ensure a negligible hydrogen peroxide residual. When hydrogen peroxide is used in pulper and FAS is used downstream, there is little chance that hydrogen peroxide will exist until the FAS stage. If a residual of hydrogen peroxide needs to be destroyed immediately before FAS treatment, bisulphite can be used.

FAS can be used for brightening and colour stripping secondary fibres (Bajpai, 2012). Mill trials show it to be an acceptable alternative to conventional bleaching agents, although in some applications it necessitates the use of additional biocide. Bleaching with FAS has been shown to be 12% cheaper than sodium hypochlorite bleaching, excluding the cost of the biocide (Faltas, Bradford, & Conti, 1997). Despite concerns over biological control, full implementation of FAS bleaching can be achieved with the application of proper controls.

Kang et al. (1999) studied the addition of sodium silicate and chelant to the FAS stage to bleach recycled fibres. They found that the addition of a chelant to the FAS stage in a FP sequence for recycled fibres improved the bleaching performance of not only the FAS stage, but also the subsequent peroxide stage. The replacement of sodium hydroxide with sodium silicate as the alkaline source improved the FAS

Table 9.9 Bleaching Conditions for FAS

Parameters	Values
Dose (%)	0.3–1.0
pH	8–11
	pH is adjusted with 1 part FAS:0.5 part NaOH
Pulp consistency (%)	12–15 (medium consistency)
	25–30 (high consistency)
Temperature (°C)	80–120
Time (minutes)	15–90

bleaching of recycled fibres. This is due to reduced FAS decomposition induced by transition metal ions, because sodium silicate can efficiently deactivate the transition metal ions. When both chelant and sodium silicate are added to the FAS treatment in a FP sequence, an additive effect was observed.

Zhao et al. (2004) studied the factors affecting FAS bleaching of DIP. The brightness of DIP increases with a dose of FAS between 0.25% and 0.5%, with the optimum dose of FAS being 0.4–0.5%. The optimum ratio of FAS and sodium hydroxide is 1:1. It appears that there is no significance of the length of bleaching time on the resultant pulp. Acidifying after bleaching is necessary but this can reduce the brightness of the pulp. This brightness reduction does not occur with alkaline peroxide mechanical pulp.

Deinked pulp made from ONP was bleached with either sodium hydrosulphite or FAS (Qian et al., 2001). For both of these, the brightness improved and there was less reversion of it, and the strength properties improved, compared with the results when using hydrogen peroxide. These chemicals also cost less than hydrogen peroxide. It was suggested that they could be used instead of hydrogen peroxide to bleach DIP when making newsprint.

Conditions for FAS bleaching are summarised in Table 9.9.

9.3.4 Bleaching with Oxygen

Improved pulp cleanliness can be achieved with oxygen bleaching of secondary fibres. Oxygen treatment is used for the production of writing and printing papers with high standards of optical quality or in processing wood-free DIP as a chemical pulp substitute. The brightness gains depend on the grade of secondary fibre. Oxygen bleaching represents an effective process step within secondary fibre processing, especially where unbleached fibres negatively influence the optical cleanliness of wood-free DIP. With increasing contents of unbleached fibres in recycled pulp, the advantage of the hydrogen-peroxide-supported oxygen stage compared with hydrogen peroxide bleaching grows in importance.

Putz, Ackermann, and Gottsching (1995) reported an alkaline yellowing of the pulp if there is a high groundwood content, and the application of oxygen alone on a wood-containing pulp is not appropriate. In most cases, MOW largely consisting of wood-free, bleached fibres is used. Use of the oxygen stage for high-quality paper depends to a considerable degree on economic criteria. In North America particularly, the operation of the appropriate plants shows that their profitability links with the costs of cleaning, screening, bleaching and the treatment of the resulting residues. Industrially, the efficiency is in doubt. The market situation of virgin hardwood pulp also has a main influence because both types of fibre compete as the raw material for the production of wood-free paper grades. The success of the process is therefore linked to some degree to the political pressure for using secondary fibres. Oxygen treatment is conducted at medium consistency (12–15%), high pressure 100 psi (7 bars), alkaline pH (9.5–10.5) and high temperature (100–110°C). The reactions of the oxygen bleaching process are very fast radical chain reactions. These reactions are called auto-oxidation and are responsible particularly for degrading lignin and for forming hydrogen peroxide with the participation of the oxygen and its radical reaction products. The brightness increase that accompanies bleaching is due to the subsequent, relatively slow reactions of the hydrogen peroxide with the chromophores. Several electrophilic (radical) and nucleophilic (ionic) reactions occur (Ackermann, 2000; Gratzl, 1992). These reactions result in formation of many heterolytic and homolytic fragmentations and hydroperoxides. The hydroperoxides finally decompose. Gratzl (1992) has reported that homolytic reactions for other oxygen chemical bleaching stages such as hydrogen peroxide are unusable because they result in the decomposition of the bleaching agent, but processes involving radicals are absolutely essential at the oxygen stage. The electrophilic character of oxygen is very weak because of its very low reduction and oxidation potential. Increasing the temperature or activating the system using a base such as sodium hydroxide is necessary to initiate a reaction. The activation involves an ionisation of functional groups, especially phenolic structures and structures with acidic hydrogen such as quinone methide in the lignin. Temperatures higher than 80°C are necessary to achieve feasible reaction times. Reactions on the contaminants, introduced with the secondary fibres in the oxygen treatment of DIP besides those on the chromophores of the lignin, occur. This is in contrast with chemical pulp bleaching. This results in improved pulp cleanliness. Markham and Courchene (1988), Magnotta and Elton (1983) and Patt, Gehr, and Matzke (1996) have reported that the following effects are important in this background: colour stripping of water soluble dye; delignification of wood-derived dirt particles; detackification of stickies; wet strength resin removal; improvement in laser and xerographic ink removal; fragmentation of adhesives such as hot melts and waxes; and solubilisation of binders that hold together flakes of coating wet strength paper and inks.

Several researchers have studied the bleaching of secondary fibres with oxygen (Darlington et al., 1992; Duxbury, Thomas, Hristofas, Yee, & Magnotta, 1995; Economou, Economides, & Vlyssides, 1996; Kulikowski, Naddeo, & Magnotta, 1991; Magnin et al., 2000; Marlin, Magnin, Chirat, & Lachenal, 2001, 2002; Marlin, Magnin, Lachenal, & Chirat, 2002; Patt et al., 1996; Putz et al., 1995; Strasbourg & Kerr, 1998; Süss, 1995). The optical effect improves as the oxygen content increases (Ackermann, 2000). Higher increase in brightness and improved pulp cleanliness are obtained. Higher oxygen concentrations produce a stronger optical homogeneity, particularly for pulp mixtures having greatly fluctuating proportions of contaminants. Large gas bubbles are formed with large oxygen quantities, particularly in medium consistencies. Some pulp is dragged along by its uncontrolled movement so the retention time is difficult to control. The pressure is often increased to keep the gas bubbles small. The most effective method is to ensure an optimum distribution of the oxygen in the pulp with suitable mixing units. Oxygen levels of 5 kg/tonne pulp will then be sufficient to provide the desired bleaching effects.

Oxygen bleaching of secondary fibres with low mechanical pulp content (16%) has been examined by Economou et al. (1996). The furnish contained 40% kraft bags, 40% old corrugated containers, 10% coloured ledger and 10% newspapers. It was found that bleaching improved when the pulp consistency was increased to 25%. The dose of the sodium hydroxide was found to be the most important parameter. Increasing the dose of sodium hydroxide reduced the kappa number by up to 56% and increased the brightness of the oxygen bleached pulp by up to 12.7 ISO points. When the amount of sodium hydroxide was increased beyond 6% w/w, a noticeable drop in selectivity was observed while there was no change in the breaking lengths of the bleached pulp. The increase in oxygen partial pressure reduced the kappa number and increased the brightness. The breaking length of the handsheets improved slightly. Increase of oxygen pressure beyond 6 bars reduced the selectivity of delignification and the tear properties.

Ackermann et al. (1996) observed that oxygen treatment can cause a dramatic loss in brightness in the bleaching of wood-containing DIP owing to yellowing of the mechanical fibre component. Only the addition of hydrogen peroxide can prevent this brightness loss. Air Products has developed high-efficiency process under the trade name OXYPRO (O_R) for oxygen bleaching of secondary fibres (Duxbury et al., 1995). This process is found to be very efficient for bleaching and very effective in removing the colour (Darlington et al., 1992). The colour stripping is through the formation of various different oxidant species, which may include some or all of the superoxide radical ions, hydrogen peroxide ions, hydroperoxide ions, hydroperoxyl and hydroxyl radicals, and small amounts of hydrogen peroxide (Darlington et al., 1992). Depending on the furnish composition, the O_R stage may also contain hydrogen peroxide. This treatment is found to be equal or exceed the colour stripping and brightening ability of sequences containing H, P, Z, Y and FAS. The O_R stage is reported to improve optical homogeneity by removing the colour from dyed and brown fibres. Kulikowski et al., (1991) has found that the treatment is effective for removing contaminants such as stickies, speck dirt, laser printed inks and wet strength resins.

Duxbury et al. (1995) has reported the optimisation of the O_R system for brightening, colour stripping and cleaning of MOW. Sodium silicate is found to have the most beneficial effect on the brightness. The incorporation of the O_R stage in a mill deinking line (Weir, Scotland) has been successful. The O_R stage can deal with high quantities of groundwood. Hydrogen peroxide was found to be very effective in increasing the brightness of high groundwood (20–30%) pulp. Patt et al. (1996) and Magnin et al. (2000) have reported that OP treatments are more effective than conventional P stages. The optical homogeneity is considerably better although there is little increase in brightness. The mottled appearance of mixed recovered papers containing brown fibres is found to be more effectively reduced. Duxbury et al. (1995) have found that the difference in brightness between P and OP stages becomes more distinct with increasing amounts of unbleached kraft fibres in a wood-free pulp (50% white ledger and 50% coloured ledger). Hydrogen peroxide bleaching under oxygen pressure OP in the case of pulp containing bleached chemical pulp and 10% brown fibres from kraft envelopes was studied by Marlin et al. (2001). The difference in brightness between OP and P was always found to be positive (up to 10 points). It increased with temperature and with an increase in sodium hydroxide dose, particularly at high temperature. The reason could be attributed to higher reactivity of oxygen with lignin at high temperature and high alkalinity. It was also found that the superiority of OP versus P is governed by the amount of oxygen and not by the pressure. A study on the bleaching kinetics of P and OP stages and the O + P sequence revealed that at similar hydrogen peroxide consumption, the OP stage reached higher brightness values than the P stage, indicating that oxygen had an additional action on the lignin when introduced in a P stage. Oxygen and hydrogen peroxide had an additive effect because the O + P sequence reached the same brightness as that of the OP stage but the kinetics of OP were much faster than that of O + P.

Strasbourg and Kerr (1998) have reported the use of oxygen in a deinking line that uses MOW in a North American mill. This technology was found to be better for handling unbleached and semi-bleached fibre commonly present in MOW furnishes. Without the oxygen, even a relatively small amount of unbleached fibre was evident in the final stock, whereas with oxygen, it was possible to have up to 3% unbleached fibres in the furnish. Other studies (Magnin et al., 2000; Marlin, 2001; Putz et al., 1995; Süss, 1995) did not show any improvement by using oxygen in a hydrogen peroxide stage. According to Putz et al. (1995), the bleaching results of an OP stage correspond to a P stage on a wood-containing pulp. An additional detachment of residual ink from the fibres is affected by mechanical forces in the high-shear mixer but not by oxygen treatment. Hydrogen peroxide bleaching of a pulp mixture composed of 70% refined bleached chemical pulp and 30% refined unbleached mechanical pulp at various temperatures and various doses of sodium hydroxide was investigated under oxygen pressure (Marlin et al., 2001). The additional brightness gain was 1 point only at lower doses of sodium hydroxide (Table 9.10). The addition of a reductive stage after the P or OP stages did not increase the brightness difference between OP and P stages. The inferior effect of oxygen in the case of mechanical pulp could be attributed to the creation of coloured quinone groups on TMP lignin during the oxygen treatment and to the poor delignifying action of oxygen because of the higher molecular size of TMP lignin and its lower amount of free phenolic groups (Marlin, 2002). The efficiency of the OP stage seems to depend on the composition of the waste papers, and on the process conditions.

Table 9.11 presents the effects of oxygen treatment on DIPs (Gehr & Borschke, 1996). The best results are obtained with a low dose of alkali and the simultaneous application of hydrogen peroxide at a temperature of 80°C. COD load is also reduced. The increase in brightness is usually 4–8% ISO points and even 12% ISO points in some cases.

Table 9.10 Brightness After P and (OP) Stages on Fully Bleached Chemical Pulp Contaminated with 30% Unbleached Mechanical Pulp

Temperature (°C)	Brightness (% ISO)					
	1.5% NaOH		1.0% NaOH		0.5% NaOH	
	P	(OP)	P	(OP)	P	(OP)
70	71.7	70.2	74.7	74.5	75.1 (68.7)[a]	76.2 (59.3)[a]
80	70.6	69.7	73.4	74.7	76.0 (76.4)[a]	77.1 (78.4)[a]
90	70.6	68.4	74.6	74.0	76.2 (69.8)[a]	77.0 (66.1)[a]

Source: Based on data from Marlin et al. (2001)
[a]Consumption of H_2O_2/initial H_2O_2 (%).

Table 9.11 Effects of Oxygen Bleaching on Deinked Pulps

	High alkali application High temperature	Low alkali application Low temperature Hydrogen peroxide addition
Wood-containing DIP	High delignification High COD load Loss of brightness	Low delignification Relatively low COD load Increase of brightness
Wood-free DIP	Delignification High COD load Low brightness increase	Low delignification Relatively low COD load High brightness increase

Source: Based on data from Gehr and Borschke (1996)

9.3.5 Bleaching with Ozone

Ozone is a powerful oxidiser, second only to fluorine. It oxidises organic compounds containing mainly >C=C< linkages which are oxidised to carbonyl groups >CO. Ozone also attacks the aromatic groups. The heterolytic decomposition of ozone in which different polyoxide stages occur results in the splitting of CC bonds. The carboxyl and carbonyl groups are formed simultaneously. Oxygen and hydrogen peroxide are formed in situ as part of this process. Aromatic, especially phenolic, structures are also attacked by ozone and decomposed into simple aliphatic acids. These reactions are not selective and last only a few seconds. Ozone has a limited stability in aqueous systems, which leads to homolytic decomposition. This is one of its drawbacks. Decomposition of ozone is catalysed by the presence of even negligible amounts of transition metal compounds, heavy metals or bases (pH > 3). The radical reactions also result in degradation of carbohydrate structures, which is not desirable, besides forming hydrogen peroxide and oxygen.

Ozone is the only oxygen-based bleaching chemical able to destroy optical brighteners that enter the secondary fibre pulp primarily through wood-free writing and printing papers. It is a powerful oxidiser. Ozone transforms dyes into colourless compounds as a result of its high reactivity towards conjugated bonds (Karp

& Trozenski, 1996). Ozone bleaching of secondary fibres has been studied by several researchers (Abadie-Maubert & Soteland, 1985; Castillo, Vidal, & Colm, 1995; Gangolli, 1982; Karp & Trozenski, 1996; Kogan, Perkins, & Muguet, 1994, 1995; Magnin et al., 2000; Muguet & Sundar, 1996; Patt et al., 1996; Roy, 1994; van Lierop & Liebergott, 1994).

Ozone is usually generated by corona discharge in pulp and paper applications. It is a relatively unstable, reactive gas, and must be generated at site. Ozone bleaching is commonly done at high consistency. Medium consistency can be an alternative. Castillo et al. (1995) has studied the effect of several process variables on bleaching. The effects of ozone dose (0.2–1%), temperature (25–45°C) and pulp consistency (30–45%) on brightness development were studied. Brightness depended linearly on ozone charge and pulp consistency. Low consistency decreased the final brightness because of the reaction of ozone with water. Temperature had no effect. Usually, acidic conditions are used for ozone bleaching of virgin fibres because metals ions accelerate ozone decomposition and a pH of 2.5 helps to solubilise and remove these ions. Chelating agents can also be used to inactivate the detrimental ions. Some reports in the literature show that ozone can efficiently remove the dyes under neutral or alkaline pH also (Kogan et al., 1994). Kogan et al. (1994) showed that the radical reactions of the ozone caused by the increased concentration of OH ions can also lead to colour stripping of dyes, although less effectively than at a low pH. Acidic conditions actually have many disadvantages like the need for an acid, pH shock inducing stickies problems, risk of $CaSO_4$ deposits, loss of $CaCO_3$, etc. (Muguet & Sundar, 1996). However, in the case of bleaching pulp for tissues, acidic conditions can be justified because low levels of ash are required for tissues. The highest increase in brightness was observed in the case of fibre furnish containing either unbleached or semi-bleached chemical fibres. Very little or no bleaching effect was observed in secondary fibres containing high amounts of mechanical fibres. However, improvement in mechanical properties was noted (Gangolli, 1982). Treatment of mechanical pulp with ozone affects the optical properties both through light adsorption and light scattering. Ozone bleaching is greatly influenced by the composition of the DIP. Abadie-Maubert and Soteland (1985) and van Lierop and Liebergott (1994) reported that improvement in mechanical or optical properties depends on the composition of waste papers. Higher content of mechanical fibres exceeding 20% prevents any optical gain (Muguet & Sundar, 1996). Higher brightness is obtained when ozone is used in combination with hydrogen peroxide. A larger proportion of mechanical fibres is beneficial for the development of strength characteristics. As a result of the formation of new carboxyl groups, an increase in fibre bonding accompanies ozone treatment. This results in a significant increase in strengths such as tensile index and burst index. This effect does not occur with wood-free DIPs (van Lierop & Liebergott, 1994).

Several reports in the literature have shown that ozone is able to destroy fluorescence (Kogan et al., 1994, 1995; Magnin et al., 2000; Patt et al., 1996). Brightness increases with increasing ozone dose; however, fluorescence decreases, 80% of which is found to be removed at 1% ozone on a mixed white/coloured paper (Muguet & Sundar, 1996) (Table 9.12. Magnin et al. (2000) reported that on DIP

Table 9.12 Effect of Ozone Charge on the Brightness Development of Mixed Paper (White/Coloured Ledger)

Ozone Consumption (%)	Brightness (% ISO)	
	Without UV	With UV
0.00	60.8	70.9
0.20	65.8	71.1
0.45	69.9	73.6
0.50	72.4	75.9
0.85	74.0	76.5
1.20	76.5	78.6

Source: Based on data from Muguet and Sundar (1996)

produced from 100% white office papers free of coloured papers, nearly 70% of the fluorescence could be destroyed with ozone. Owing to the great number of the multiple bonds in dye structures and the high reactivity of ozone on these bonds, ozone is considered an efficient chemical for colour stripping of dyed paper. Karp and Trozenski (1996) found that ozone exhibits similar bleaching characteristics to that of sodium hypochlorite when applied to a wide spectrum of colourants used to dye paper.

Air Liquid has developed the Redoxal process, which is an ozone-based reductive/oxidative process combining Ozone (Z), P and Y or FAS. ZP is found to be effective in bleaching a large range of furnishes. Excellent optical properties, high colour removal and acceptable yield are obtained (Kogan et al., 1995; Muguet & Sundar, 1996). Hydrogen peroxide is used after ozone without washing to stabilise the previously achieved brightness, protecting the pulp from brightness reversion. Other researchers have also obtained good results with a ZPY combination in bleaching wood-free secondary fibres containing coloured papers (Magnin et al., 2000).

Ozone treatment allows almost complete destruction of optical brighteners. This is particularly important with DIPs in papers used for food applications. For such applications, no chemical substances can transfer to the foodstuffs regardless of whether they represent a potential health risk. Considering the considerable costs involved, this effect may not offer sufficient reason to introduce an ozone stage within secondary fibre processing. Roy (1994) reported that by using chemical quenchers, achieving a comparable effect is possible, in which the fluorescent materials are masked and therefore made ineffective. Kappel and Matzke (1994) reported that the higher ozone content can even cause a small loss of brightness in DIPs produced from old newspapers and magazines (Table 9.13).

Roncero, Colom, Pastor, Blanco, and Vidal (1998) determined the effectiveness of ozone and FAS as bleaching agents for different coloured papers (green, red and blue), enhancing colour stripping in secondary fibres by enzymes (cellulase A). The action of bleaching agents depends on the colour of paper used: ozone is a good

Table 9.13 Effect of Ozone on Brightness of Wood-Containing Grades Based on 50% Old Newspapers and 50% Magazines

Ozone Charge (%)	Brightness (% ISO)	
	Pulp A	Pulp B
0.0	58.8	58.2
0.5	57.9	57.3
1.0	57.4	57.0
1.5	57.0	56.9
2.0	56.9	56.6

Source: Based on data from Kappel and Matzke (1994)

reagent, whereas FAS treatment appears to be effective enough for blue and red. In the conditions used, it is possible to destroy the colour with bleaching sequence ZF, except for the red colour paper. The novel enzyme cellulase A was found to be effective and to enhance colour stripping.

Cardona et al. (2001) studied the effects of cellulases on the bleaching kinetics of coloured secondary fibres by ozone. Two laboratory cellulases from *Bacillus* sp. BP-23 (CelA and CelB), as well as one commercial cellulose, were used. Results revealed that the cellulases made bleaching by ozone easier, although the results were dependent on the enzymes and the paper used. The most effective cellulase was the laboratory CelB, with the rate of chromophore removal being approximately twice that without cellulases. There was also reflectance improvement across the spectrum, and improvements in dye removal, colour stripping and bleaching indexes.

Fernandes and Floccia (2000) investigated an innovative bleaching sequence using ozone. Industrial-scale pilot studies were also conducted at Matussiere and Forest Turckheim mill, France. Bleaching stages, comprising ozone alone or in combination with other oxidative agents, were examined for pH, temperature and concentration of other bleaching agents. Results demonstrated the efficiency of ozone in colour stripping, partial sterilisation of circuits and deinking. Colour removal is explained by the combined action of ozone and peroxide and attack on carbon–carbon double bonds and phenolic groups. There was no significant effect on bleaching by the use of an ozone/peroxide stage.

Cooperation between Ponderosa Fibres and Air Liquide resulted in the first commercial application of ozone for bleaching secondary fibres (Muguet & Sundar, 1996). The newly patented process, Redoxal, benefits from the combined bleaching power of ozone and hydrogen peroxide, reinforced by the colour stripping ability of a reducing chemical. The process is a highly cost-effective sequence.

9.3.6 Bleaching with Peroxyacids

Peroxyacids have been studied only on a laboratory scale for bleaching secondary fibres (Dubreuil, 1995; Kapadia et al., 1992; Magnin et al., 2000; Szegda, 1994;

Thorp, Tieckelmann, Millar, & West, 1995; Tschirner & Segelstrom, 2000; Walsh, Hill, & Dutton, 1993). The most commonly reported peroxyacids are peracetic acid and peroxymonosulphuric acid (Caro's acid). Dubreuil (1995) reported that at a temperature of 60–80°C, peracetic acid and Caro's acid react as electrophiles. They hydroxylate aromatic agents to facilitate the nucleophilic oxidative degrading in subsequent bleaching stages. They are able to destroy optical brighteners like ozone or chlorine and chlorine dioxide. In addition, peracids oxidise alcohols and aldehydes to carbonic acids and can hydroxylate double bonds. This destroys the conjugation in chromophores. The disadvantages are the acidic reaction conditions that cause gypsum to form and reduce the pulp yield when carbonate is present. Potassium permonosulphate, which is the salt of permonosulphuric acid, has been proposed for the removal of colour (Kapadia et al., 1992). It was found to work well at alkaline conditions (pH 9–11) and a temperature of 70°C. This chemical has been proposed to replace sodium hypochlorite for re-pulping wet strength papers (Kapadia et al., 1992). It has been found that activated alkali metal persulphates re-pulp neutral/alkaline wet strength broke and decolourise certain dyes and optical brighteners more effectively than non-activated persulphates (Thorp et al., 1995).

A mixture of peracetic acid and permonosulphuric acid has been found to be very effective for bleaching secondary fibres (Szegda, 1994). This mixture is obtained by addition of acetic acid to Caro's acid. Permonosulphuric acid increases both brightening and colour removal of reductive bleaching agents used in the same sequence beyond the ability of hydrogen peroxide. A combination of Caro's acid and sodium chloride was used by Walsh et al. (1993) to remove colour in wood-free papers. Caro's acid oxidises the halide to the halogen, which in turns removes the colour. A significant effect on lignin removal, ink removal and brightness improvement for low grades of secondary fibres having high lignin content was observed by peracid pretreatment (Tschirner & Segelstrom, 2000). Better results are obtained when the mixture of peracetic and peroxymonosulphuric acids is used. Bleaching response in the subsequent alkaline hydrogen peroxide stage is significantly improved. The effects of peracids on deinking MOW are found to be less pronounced. Dubreuil (1995) and Magnin et al. (2000) found that fluorescence can be destroyed by peracetic acid, Caro's acid and potassium permonosulphate under acidic conditions. The considerable costs of the bleaching agent and the unsuitable pH conditions for DIP are the most important reasons negating industrial application of peracids.

9.3.7 Direct Borohydride Injection Bleaching

This process comprises direct borohydride injection (DBI) using Borol solution. It is based on the sequential addition of bisulphite solution and sodium borohydride solution to the recycled fibre mass in medium- to high-consistency mixing equipment (Meyers et al., 1999). This process offers advantages in terms of colour stripping efficiency, brightness gains and costs compared with FAS, hydrosulphite and peroxide. The process is also simple to implement, requires no retention time and results in less brightness variation in the end product. A trial at Ponderosa Fibres, USA, compared the performance and economics of DBI against sodium hydrosulphite

(Hache et al., 2001). The DBI process outperformed sodium hydrosulphite from both performance and economic viewpoints, providing a higher brightness gain, improving the amount of "b" value reduction and reducing the reductive bleaching chemical application costs by 52%. DBI has allowed the mill to reduce oxidative bleaching chemical rates and to realise further savings.

The use of DBI is growing in recycled mills throughout the world (Crowley, Rangamannar, & Reynoso, 2002). Process conditions to ensure maximum pulp response include high temperature, good mixing and medium/high pulp consistency. The best mixing conditions are found in pulp lines that use a disperser or medium consistency mixing equipment.

DBI technology is found to offer a better performance and more economic advantages in the reductive bleaching and colour stripping of recycled fibres, compared with FAS and other conventional bleaching chemicals, such as sodium hydrosulphite and hydrogen peroxide. These results were obtained from two mill-scale trials for towel and napkin-grade pulps (Rangamannar, Bettano, & Hebert, 2005). In both trials, DBI achieved mill brightness targets and accommodated more coloured paper and mechanical pulps than the FAS method. The process involves injecting sodium bisulphite with an aqueous mixture of sodium borohydride and sodium hydroxide to produce reductive bleaching liquor. This liquor synergistically promotes powerful and cost-effective colour stripping when directly added to a pulp stream in medium- or high-consistency mixing equipment at process temperatures. Specifically, the cost analysis shows that implementing the DBI process can reduce reductive bleaching costs by 20% for napkin-grade pulps and by 28% for towel grade.

A study conducted on the basis of results from a tissue-producing paper mill in Malaysia during a 4 month period in 2005 showed that the patented technology of colour stripping using the DBI process enabled the manufacturers producing tissue from waste paper to reduce costs (Abu Hasan, 2005). Despite the fact that lower-grade waste paper with a larger share of inks and coloured materials was being used, the required quality for brightness (maintained within the range of 76–80% ISO) and cleanliness was nevertheless accomplished. As a result, the overall cost per tonne of tissue paper was significantly lowered. The method contributed to the reduction of costs for colour stripping/bleaching and other chemicals, changed the waste paper mix in favour of lower-quality material and eliminated the need for a retention tower for bleaching. The overall approximate saving of 9% translated into an annual cost reduction of US$0.53 million for a plant with a production capacity of 100 tonnes/day.

References

Abadie-Maubert, F. A., & Soteland, N. (1985). Utilization of ozone for the treatment of recyclable papers. *Ozone Science Engineering*, *7*(3), 229.

Abu Hasan, M. J. (2005). DBI: Cost reduction for wastepaper based tissue manufacturers. *Paper Asia*, *21*(10), 25–28.

Ackermann, C. (2000). Bleaching of deinked pulp. In L. D. Gottsching & H. Pekarinen (Eds.), *Recycled fiber and deinking – Papermaking science and technology* (7, pp. 307). Helsinki: Fapet Oy.

Ackermann, C., Putz, H. -J., & Göttsching, L. (1992). Waste paper treatment for the production of high quality graphic papers. *Wochenblatt Fur Papierfabrikation, 120*(11/12), 433.

Ackermann, C., Putz, H. -J., & Gottsching, L. (1996). Do alternative chlorine-free bleaching agents revolutionise the bleaching of wood containing DIP? *Das Papier, 50*(6), 320.

Bajpai, P. (2006). *Advances in recycling and deinking* (180 pp.). U.K: Pira International.

Bajpai, P. (2012). *Environmentally benign approaches for pulp bleaching* (2nd ed.). Amsterdam: Elsevier Science B.V.

Berndt, W. (1982). The chemicals of the deinking process. *Wochenblatt Fur Papierfabrikation, 110*(15), 539.

Blechschmidt, J., & Ackermann, C. (1991). Chemo-physical basis of the deinking process. *Wochenblatt Fur Papierfabrikation, 119*(17), 659.

Bovin, A. (1984). Improved flotation deinking by development of the air mixing chamber: *TAPPI pulping conference proceedings* (p. 37). Atlanta, GA: TAPPI Press.

Cardona, C., Roncero, B., Colom, J. F., Pastor, F. I. J., Diaz, P., & Vidal, T. (2001). Upgrading of coloured recycled paper by ozone and cellulases. P. Vahala, & R. Lantto (Eds.), *Eighth international conference on biotechnology in the pulp and paper industry* (pp. 267–268). Helsinki, 4–8 June.

Castillo, I., Vidal, T., & Colm, J. F. (1995). Upgrading of recycled paper by ozone. *Eighth international symposium on wood and pulping chemistry* (Vol. 1, p. 353). Helsinki.

Crowley, T. R., Rangamannar, G., & Reynoso, A. (2002). Case studies of applied new technologies: Borohydride-based bleaching in US and Latin American recycle mills. *2002 Fall technical conference and trade fair* (7 pp.). San Diego, CA, 8–11 September.

Darlington, B., Jezerc, J., Magnotta, V., Naddeo, R., Walier, F., & White-Gaebe, K. (1992). Secondary fiber color stripping: Evaluation of alternatives. *TAPPI pulping conference proceedings* (p. 67). Boston, MA.

Dubreuil, M. (1995). Introduction to fluorescence in fibre recycling. *Progress in Paper Recycling, 4*(4), 98.

Dumont, I., Fluet, A., Giasson, J., & Shepperd, P. (1994). Two applications of hydrosulphite dye-stripping: Yellow directory and colored ledger. *Pulp and Paper Canada, 95*(2), 136.

Duxbury, P., Thomas, C. D., Hristofas, K., Yee, T. F., & Magnotta, V. L. (1995). Laboratory and mill-scale optimisation of an oxygen stage for bleaching of mixed office waste. *Fourth international wastepaper technology conference*. London Gatwick.

Economou, A. M., Economides, D. G., & Vlyssides, H. (1996). Oxygen bleaching of secondary fibers with low mechanical pulp content. *Progress in Paper Recycling, 5*(3), 53.

Fallon, K. C. (1994). In situ formation of formamidine sulfinic acid: An oxidative/reductive bleaching process for recycled fiber. *TAPPI pulping conference proceedings* (p. 263). San Diego, CA.

Faltas, R. W., Bradford, R., & Conti, N. P. (1997). FAS reductive bleaching of recycled fibre: A case study. *51st Appita annual general conference* (Vol. 2, pp. 427–436). Melbourne, 28 April–2 May.

Fernandes, J. C., & Floccia, L. (2000). Use of ozone in deinking and bleaching secondary fibres. *2000 recycling symposium* (Vol. 1, pp. 191–198). Washington, DC, 5–8 March.

Fluet, A. (1995). Sodium hydrosulfite brightening and colour-stripping of mixed office waste furnishes. *TAPPI pulping conference*. Chicago, IL.

Fluet, A., Dumont, I., & Beliveau, D. (1994). Sodium hydrosulphite brightening: Laboratory versus mill results. *Pulp and Paper Canada, 95*(8), 37.

Fluet, A., & Shepperd, P. (1997). Color stripping of mixed office papers with hydrosulphite-based bleaching products. *Progress in Paper Recycling, 6*(2), 74.

Galland, G., Bernard, E., & Vernac, Y. (1989). Recent progress in deinked pulp bleaching. *Paper Technology*, *30*(12), 28.

Gangolli, J. (1982). The use of ozone in pulp and paper industry. *Paper Technology and Industry*, *26*(5), 152.

Gehr, V. (1997). Bleaching of secondary fibre stocks – What can the white magic achieve? *Papier*, *51*(11), 580–585.

Gehr, V., & Borschke, D. (1996). Bleaching as an integrated step of modern deinking plants. *Wochenblatt Fur Papierfabrikation*, *124*(21), 929.

Gratzl, J. S. (1992). The chemical principles of pulp bleaching with oxygen, hydrogen peroxide and ozone – A short review. *Das Papier*, *46*(10A), V1.

Hache, M., Fetterly, N., & Crowley, T. (2001).North American Mill experience with DBI. *87th annual meeting, PAPTAC*. Montreal.

Hache, M. J. A., Brungardt, J. R., Munroe, D. C., & Teodorescu, G. (1994). The color stripping of office wastepaper with sodium hydrosulphite. *Pulp and Paper Canada*, *95*(2), 120.

Helmling, O., Süss, U., & Berndt, W. (1985). Steigerung der Effektivität der Chemikalien für Flotations-Deinking-Verfahren. *Wochenblatt Fur Papierfabrikation*, *113*(17), 657.

Helmling, O., Süss, U., & Eul, W. (1986). Upgrading of waste paper with hydrogen peroxide: *TAPPI pulping conference proceedings*. Atlanta, GA: TAPPI Press. (p. 407).

Jiang, Y., Qian, Y., Yu, T., & Zhan, H. (1999). Hydrogen peroxide-sodium dithionite – Two stages bleaching of mixed office waste paper deinked pulp. *China Pulp Paper*, *18*(6), 1–5.

Kaichang, L., Collins, R., & Eriksson, K. E. L. (2000). Removal of dyes from recycled paper. *Progress in Paper Recycling*, *10*(1), 37.

Kang, G. J., Van Heiningen, A. R. P., & Ni,Y. (1999). Addition of sodium silicate and chelant to the FAS stage to bleach recycled fibers. *TAPPI pulping conference* (Vol. 2, p. 571). Orlando, FL.

Kapadia, P. C., Tessier, H. G., & Langlois, S. (1992). 'Agent de Retrituration Sans Chlore', *Conference de Technologie Estivale, Pointe au Pic.*

Kappel, J., & Matzke, W. (1994). Chlorine-free bleaching chemicals for recycled fibers: *TAPPI recycling symposium proceedings*. Atlanta, GA: TAPPI Press. (paper no. 8-1).

Karmakar, G. R. (2004). Oxidative and reductive bleaching of secondary fibres. *IPPTA*, *16*(3), 115–118.

Karp, B. E., & Trozenski, R. M. (1996). Non-chlorine bleaching alternatives: A comparison between ozone and sodium hypochlorite bleaching of colored paper: *International pulp bleaching conference proceedings*. Atlanta, GA: TAPPI Press. (p. 425).

Kogan, J., Perkins, A., & Muguet, M. (1994). Ozone bleaching of deinked pulp. *TAPPI recycling symposium proceedings* (p. 237). Boston, MA.

Kogan, J., Perkins, A., & Muguet, M. (1995). Bleaching deinked pulp with ozone-based reductive-oxidative sequences. *TAPPI recycling symposium proceedings* (p.139). New Orleans, LA.

Kronis, J. D. (1992). Adding some color to your waste paper furnish. *TAPPI pulping conference proceedings* (p. 223). Boston, MA.

Kronis, J. D. (1994). Practical aspects of applying FAS in secondary fiber processing-opportunity for chlorine-free bleaching, *Proceedings of the International Pulp Bleaching Conference* (p. 77). Portland, USA: Oregon.

Kronis, J. D. (1997). Optimum conditions play major role in recycled fiber bleaching with FAS. In K. L. Patrick (Ed.), *Advances in bleaching technology*. California: Miller Freeman Books. (p. 104).

Kulikowski, T., Naddeo, R., & Magnotta, V. (1991). Oxidative cleaning and bleaching. *TAPPI pulping conference* (p. 3). Orlando, FL.

Kutney, G. W., & Evans, T. D. (1985a). Hydrogen peroxide stabilization of bleaching liquors. *Svensk Papperstidn*, *88*(9), R84.
Kutney, G. W., & Evans, T. D. (1985b). Peroxide bleaching of mechanical pulps, part 2. Alkali deinking hydrogen peroxide decomposition. *Svensk Papperstidning*, *88*(9), R84.
Lachenal, D. (1994). Bleaching of secondary fibers – Basic principles. *Progress in Paper Recycling*, *4*(1), 37.
Magnin, L., Angelier, M. C., & Gailand, G. (2000). Comparison of various oxidising and reducing agents to bleach wood-free recycled fibres. *Ninth PTS CTP deinking symposium* (p. 9). Munich, May.
Magnotta, V. L., & Elton, E. F. (1983). Process for recovering fiber from wet-strength resin-coated paper. US Patent No. 4,416,727.
Markham, L. D., & Courchene, C. E. (1988). Oxygen bleaching of secondary fiber grade. *TAPPI Journal*, *71*(12), 168.
Marlin, N. (2002). *Comportement de Mélanges de Pates Papetiereschimiques et Mecaniques Lors de Traitements Par Leperoxyde D'hydrogene en Presence D'oxygene – Application au Blanchiment de Fibres Recyclees*, Thesis for the grade of INPG PhD, Ecole Francaise de Papeterie deGrenoble.
Marlin, N., Magnin, L., Chirat, C., & Lachenal, D. (2001). Effect of oxygen on peroxide bleaching of recycled fibres, part 1. Case of fully bleached chemical pulp contaminated with kraft brown fibres of mechanical pulp. *Progress in Paper Recycling*, *10*(3), 11.
Marlin, N., Magnin, L., Lachenal, D., & Chirat, C. (2002). Effect of oxygen on the bleaching of pulp contaminated with unbleached kraft or mechanical fibres – Application to recycled fibres. *Seventh European workshop on lignocellulosics and pulp*. Turku/Abo.
Matzke, W., & Kappel, J. (1994). Present and future bleaching of secondary fibers: *TAPPI recycling symposium proceedings*. Atlanta, GA: TAPPI Press. (p. 325).
Melzer, J. (1985). Kinetics of bleaching mechanical pulps by sodium dithionite. *International pulp bleaching conference proceedings* (p. 69). CPPA. Montreal.
Melzer, J., & Auhorn, W. (1981). Optimisation du Processus de Blanchiment de la Pate Mecaniquepar l'emploi de Sequestrants. *BASF international symposium* (19 p.). Ludwigshafen.
Meyers, P., Wang, D., & Hache, M. (1999) DBI, a novel bleaching process for recycled fibers. *TAPPI 99 – Preparing for the next millennium* (Vol. 2, pp. 373–385). Atlanta, GA, 1–4 March.
Mozaffari, S., Krishnan, S., Burch, D., & Rangamannar, G. (2006). Hydrosulfite bleaching of recycled pulps at Abitibi Thorold: Impact of process variables on bleach performance. *92nd annual meeting of the Pulp and Paper Technical Association of Canada* (pp. C307–C311), Book C. Montreal, 7–9 February.
Muguet, M., & Sundar, M. (1996). Ozone bleaching of secondary fibers. *Wastepaper VII*. Chicago, IL, 4–7 June.
Patt, R., Gehr, V., & Kordsachia, O. (1993). Bleaching and upgrading of secondary fibers by chlorine free chemicals. *Eucepa conference book of paper* (p. 305). *Vienna.*
Patt, R., Gehr, V., & Matzke, W. (1996). New approaches in bleaching of recycled fibers. *TAPPI Journal*, *79*(12), 143.
Putz, H.-J. (1987). *Upcycling von Altpapier für den Einsatz in Höherwertigen Graphischen Papieren Durch Chemisch-Mechanische Aufbereitung* (p. 202), PhD Thesis, Darmstadt University of Technology Darmstadt, Germany.
Putz, H.-J., Ackermann, C., & Gottsching, L. (1995). Bleaching of wood-containing DIP with alternative chemicals-possibilities and limitations. *Fourth international waste paper technology conference*. London Gatwick.
Qian, Y., Tan, B., Jiang, Y., Liang, Y., Li, X., & Zhan, H. (2001). Reductive bleaching of deinked pulp for newsprint production. *China Pulp Paper*, *20*(6), 8–11.

Rangamannar, G.; Bettano, J., & Hebert, R. 2005. "DBI" bleaching of recycled fibers for the production of towel and napkin grades. *Engineering, pulping and environmental conference* (14 pp.). Philadelphia, PA, 28–31 August, session 43.

Renders, A. (1992). Hydrogen peroxide and related chemical additives in deinking processes. *TAPPI pulping conference proceedings* (p. 233). Tappi Press, Atlanta.

Renders, A. (1995a). Recycled fiber bleaching. In R. W. J. McKinney (Ed.), *Technology of paper recycling*. Glasgow: Blackie Academic and Professional. (p. 157). Tappi Press, Atlanta.

Renders, A., Chauveheid, E., & Pottier, G. (1995b). Bleaching mixed office waste with hydrogen peroxide. *TAPPI pulping conference proceedings* (p. 709). Tappi Press, Atlanta.

Renders, A., & Hoyos, M. (1994). Disperser bleaching with hydrogen peroxide: A tool for brightening recycled fiber. *PTS deinking symposium*. Munich.

Robberechts, M., Pyke, D., & Penders, A. (2000). The use of hydrogen peroxide and related chemicals in waste paper recycling. *Sixth international recycling technology conference budapest* (p. 16). Hungary.

Roncero, M. B., Colom, J. F., Pastor, F. I. J., Blanco, A., & Vidal, T. (1998). Upgrading of coloured waste paper by ozone, formamidine sulfinic acid and enzymes. *International pulp bleaching conference* (pp. 343–347). Book 2, poster presentations. Helsinki, 1–5 June.

Roy, B. P. (1994). How do you remove/destroy/extinguish fluorescent material in deinked pulp? *Progress in Paper Recycling*, *4*(1), 74.

Strasbourg, R., & Kerr, J. C. (1998). Deink market pulp mill – An operations perspective on the design and construction aspects. *Paper in South Africa*, *18*(2), 9.

Szegda, S. J. (1994). Use of high conversion peroxyacids in non-chlorine bleaching sequences for recycled fibers. *TAPPI pulping conference proceedings* (p. 272). San Diego, CA, November.

Süss, H.-U. (1995). Bleaching of waste paper pulp – Chances and limitations. *Fourth international waste paper technology conference* (p. 13). London Gatwick.

Süss, H.-U., & Krüger, H.. (1983, March). German Patent DE 3,309,956 C1.

Taylor, R. C., & Morrison, C. R. (1999). The effects of various chelating agents on FAS bleaching efficiency, for brightening of secondary fiber in a flotation deinking process. *TAPPI 99 – Preparing of the next millennium* (Vol. 2, p. 357). Atlanta, GA.

Thorp, D. S., Tieckelmann, R. H., Millar, D. J., & West, G. E. (1995). Chlorine free wet strength paper repulping and decolorizing with activated persulfates. *Papermakers conference*. Chicago, IL, April.

Tschirner, (2000)., & Segelstrom, R. (2000). Peracid pre-treatment for improved ink removal and enhanced bleachability of recycled. *Fiber Progress in Paper Recycling*, *9*(2), 15.

van Lierop, B., & Liebergott, N. J. (1994). Bleaching of secondary pulps. *Pulp and Paper Science*, *20*(7), J206.

Walsh, P. B., Hill, R. T., & Dutton, D. B. (1993). Secondary fiber processing: Color destruction in wood-free furnishes. *Progress in Paper Recycling*, *3*(1), 9.

Zhao, N. -Z., Wang, H. -Y., Tan, X. -Y., Zhang, D. -J., Shen, Z. -B., & Fang, G. -G. (2004). Formamidine sulphinic acid bleaching of DIP and acidifying treatment of bleached pulp. *China Pulp Paper*, *23*(12), 29–31.

Zou, Y., Hsieh, J. S., Agrawal, C., & Matthews, J. (2007). TAED activator for peroxide bleaching of recycled pulp. *2007 TAPPI eighth research forum on recycling* (22 pp.). Niagara Falls, Ontario, 23–26 September.

10 Refining of Recycled Fibres*

10.1 Introduction

Economics, conservation policies and mandated recycled-content regulations have combined to make the recycling of recovered paper a rapidly growing segment of the pulp and paper industry. Newly built recycling mills competing for quality recovered paper have dramatically affected the business of paper. The use of recycled fibre (RCF) has seen a significant development in printing paper grades, especially magazine papers in recent years. RCF shows a high potential in supercalendered (SC) grades because of the contribution of fillers and chemical pulp providing competitive gloss, oil absorption, brightness and strength. On the other hand, some limitations have been reported in light scattering, compressibility, bulk and surface properties, as well as a certain sensitivity in rotogravure printability. Compared with a standard RCF process for newsprint, an RCF line for magazine grades usually includes some of the following stages: post-bleaching, second dispersion, refining and washing. One of the expected benefits of post-refining is to enhance fibre structure so that paper surface and absorption properties together with tensile strength are improved with limited effect on drainage, bulk and tear strength. For that purpose, different technologies are available, among them high-consistency refining, low-consistency refining and dispersing (Bajpai, 2006).

Refining has the potential to 'reactivate' the bonding potential of RCFs. Baker (1999, 2000) has reported that the papermaking properties of recycled furnishes can be improved by refining. The effects depend on the furnish. RCFs are unable to absorb as much energy during refining as virgin equivalents, hence they require gentler refining conditions. The swelling and bonding abilities of fibres are reduced when they pass through the papermaking process. In papermaking, pressing and drying are more intensive than in pulp drying. Also, the slushing and cleaning of already-used fibres decreases the amount of fines and fibrils that are needed for good fibre bonding.

Various RCF treatments regenerate the swelling and bonding abilities a little, but not enough. So, a more intensive treatment, such as refining, is needed to redevelop the fibres. Correct refining can improve the papermaking potential of RCFs and reduce paper manufacturing cost (Lumiainen, 1992a,b, 1994a,b 1995a,b, 1997).

Refining creates fibrils, which improve the bonding ability. However, every reuse weakens the fibres and causes irreversible changes. These make the RCFs more

*Some excerpts taken from Bajpai (2006) with kind permission from Pira International, UK and Bajpai (2012) with kind permission from Springer Science & Business Media.

Recycling and Deinking of Recovered Paper. DOI: http://dx.doi.org/10.1016/B978-0-12-416998-2.00010-6

sensitive to refining errors than virgin fibres. If not refined correctly, the result can be disastrous. Negative effects include high increases in the drainage resistance, a heavy fibre length and a reduction in tear strength values. These can be avoided by choosing the equipment and conditions correctly (Guest, 1991; Levlin, 1976). Finally, RCFs often contain shives from the mechanical pulp components. Because these can be harmful, refining is needed to remove them when producing fine or coated paper grades.

In most recovered papers, high-consistency re-pulping and refining are advantageous as these develop the strength properties without reducing the drainage rates. High consistencies greatly reduce the amount of net energy that can be applied per unit of throughput in any pass through the system. Although this reduced refining intensity may cost more in energy, it is more than compensated for by the better drainage properties achieved. RCFs tend to magnify any papermaking problems as the variability in the raw material can lead to large quality changes.

Refining applications vary widely as the objectives of refining RCFs depend on the final product and the recovered paper grade. To restore fibre strength, at least partly, the fibre surface is reactivated and the fibre fines lost during processing are regenerated. Strength enhancement is often a refining goal in packaging paper production. Old corrugated container (OCC) raw material used in North America can therefore be made sufficiently strong to eliminate the need for a sizing press. The double-sorted old corrugated carton grades used in Europe and elsewhere do require surface sizing. In either case, refining can still reduce the sizing outlay necessary to meet strength demands.

Refining is also useful for reducing unacceptably high R14 or R30 fractions in furnishes for SC and lightweight coated (LWC) papers (Holik, 2000). The stock can be suitably prepared under appropriate refining conditions by eliminating the R14 component and reducing the R30 fraction to the desired level. Tables 10.1 and 10.2 list the main refining applications in RCF processing for white and brown grades (Holik, 2000). Wood-free printing and writing papers and tissue are not normally refined or use only a brushing refiner. With brown grades, fluting and test liner definitely require refining.

Certain strength properties are improved by refining. Since fibre properties change because of refining, other paper characteristics are also affected. For instance, decreasing freeness value decreases bulk and drainage. Optical characteristics such as brightness, opacity, light scattering coefficient and tear strength are negatively influenced to a greater or lesser extent. An optimum compromise is therefore necessary between not only cost-effectiveness and quality but also between the various quality parameters. This compromise can occur by varying the refiner fillings, the energy input, the refiner type and the processing stages before refining. RCF pulp is mainly refined at low consistencies of about 3–6%. Sometimes high consistencies up to 30% and above have been used, such as in high-consistency refining.

RCF pulp usually comprises a mixture of different fibre types, which produces significant differences in refining resistance or potential between the stock components. In some cases, fractionation can reduce this problem so that the different fibre components produced can be treated separately. This kind of selective refining

Table 10.1 Refining of RCF – White Grade

	Standard Newsprint	Improved Newsprint SC	LWC Paper
Application	Partly required	Required	Required
Target	Strength improvement	Strength improvement Fibre coarseness reduction	Strength improvement Fibre coarseness reduction
Net refining energy	<50	50–100	~100

Source: Based on Holik (2000)

Table 10.2 Refining of RCF – Brown Grade

	Fluting or Test Liner	Board
Application	Required	Required
Target	Strength improvement	Strength improvement
Net refining energy	~100	~150 (200)

Source: Based on Holik (2000)

instead of full-stream refining further consolidates the coarse R14 fibres in deinked pulp (DIP), for example. The separate components can be mixed again or allocated to the different layers in multilayer products.

10.2 Refining Effects

Investigations into the effects of recycling have been numerous and varied. These have shown a dramatic loss of tensile and bursting strengths on repeated remaking of low-yield chemical pulp into paper (Figure 10.1) (Brancato, 2008; Nazhad & Paszner, 1994) The review of the literature on recycling of chemical pulps clearly reveals that the losses in paper strength are brought about by loss of bonding or, in other words, a reduction in the resistance of bonds to an applied force. On the other hand, the loss of bonding could be a function of two parameters: fibre flexibility and surface condition. Flexibility in this context refers to the wet-plasticity of pulp fibres. It is defined as the inverse of the product of the moment of inertia of the body and the modulus of elasticity of the material from which the body is made. It is not yet certain why loss of surface condition or flexibility of fibres occurs during recycling. Furthermore, it has not been established which of the above factors in recycling plays the major role in the strength loss of recycled paper.

Lundberg, De Ruvo, Fellers, and Kolman (1976) looked at how high fibres and low-consistency refining affected pulps. They found that a mixture of high- and

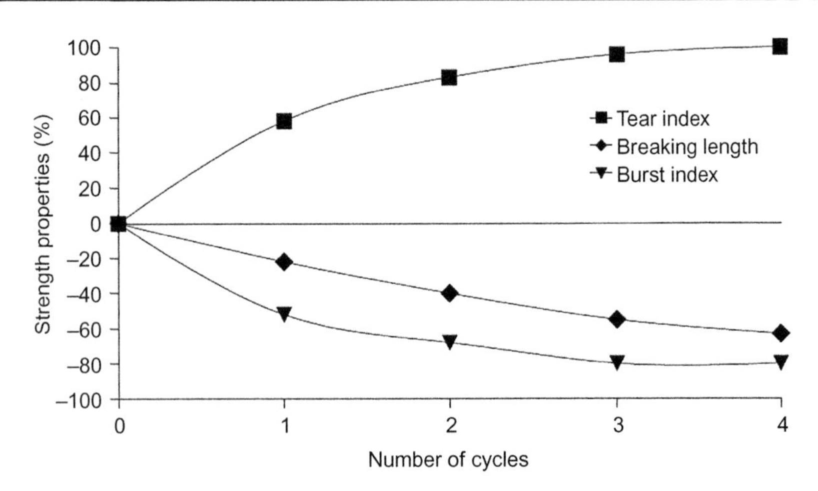

Figure 10.1 Effect of recycling on bleached kraft sheet properties.
Source: Based on Nazhad and Paszner (1994) and Brancato (2008).

low-consistency refining was needed to obtain adequate strength at reasonable drainability. Studies on a commercial carton-board machine demonstrated that high-consistency refining produced adequate levels of strength while reducing steam consumption. In another study, Levlin (1976) investigated the refining of newspapers, magazines, corrugated board, folding boxboard and a mixture of other grades. The amount of refining and the way in which the stocks were refined were varied. He found that the papermaking properties could be improved by refining, but that the method of refining was critical. The optimum refining conditions were furnish dependent: low-intensity refining should be used to develop the properties of furnishes with a high mechanical-fibre content.

The use of OCC to produce linerboard and corrugating medium has increased dramatically, hence the quality has deteriorated. So there has been greater interest in refining to enhance quality. Many mills now use 100% recycled paper and some use mixed paper. Nazhad and Awadel-Karim (2001) investigated the possibilities for upgrading OCC pulp. The role of specific energy and intensity on the strength development of OCC pulps was studied to achieve optimum strength quality. Soaked samples of OCC were disintegrated and refined at a specific edge load (SEL) of 0.5, 1 and 3 W s/m at refining energies of 0–400 kW h/tonne. Pulp comparable to virgin pulp was achieved by refining OCC pulp at specific energies of 80–100 kW h/tonne, but papermaking quality deteriorated beyond this range. Refining at a SEL of 0.5 W s/m produced higher tensile or burst strength, and an SEL of 3 W s/m was detrimental. Tear strength slightly increased with low, gentle refining at 10 kW h/tonne but it decreased with continued refining. Optimum tensile strength and burst strength were achieved using a specific energy range of 70–90 kW h/tonne, irrespective of SEL. The freeness for this range is 250–350 Canadian standard freeness (CSF).

Rihs (1992) examined refining using OCC, deinking ledger and old newsprint (ONP) to see if it enhanced these pulps. He found that, with the correct refining

Table 10.3 Effect of Refining on Strength Properties

Property	Change in Properties (%)		
	0.5 W s/m	1 W s/m	2 W s/m
Tensile index	+36	+22	+7
Burst index	+35	+14	+7
Tear index	−24	−36	−39
Scott ply bond	+100	+157	+114
Kenley stiffness	+20	+17	+6

Refining up to a maximum energy input of 150 kW h/tonne.
Source: Based on Baker (1999)

conditions, secondary fibre can respond like virgin fibre. In another experiment, Fisher (1980) refined mixed secondary fibres. The slushed stock had a wetness value of 42° Schopper–Riegler (SR) value, a breaking length of 3.95 km and a tear index of 10 mN/mg. On refining to a wetness of 60° SR, the breaking length increased by 35% and the tear strength decreased by 25%.

In an investigation by Peixoto Silva and Chaves de Oliveira (2003), an elemental chlorine-free pulp from *Eucalyptus urograndis* and an unbleached pulp from pinus were submitted to four cycles of handsheet forming and recycling. After each cycle the recycled pulp was refined in a PFI-refiner at three freeness levels to recover the original properties. The results showed that the strongest reduction in mechanical properties was achieved during the first cycle but that the second one also had a significant influence. The energy consumption required to regain the initial properties was 89 W h for the pinus pulp and 38 W h for the eucalyptus. Recycling was less detrimental for eucalyptus than for pinus pulp and the properties in both were positively affected by refining.

Moore, Cathie, Crow, and Smith (1995) have reported that upcycling methods can be used successfully to produce printing and writing grades from mixed office papers and the strength properties can be improved by refining. The mixed furnish was obtained from a board mill using RCF so that the stock had the same chemical and mechanical treatment as mill stock. This allowed the fibres to react in a similar way on refining, making the results more likely to reflect the outcomes obtained at the mill, under the same refining conditions. The stock was thickened to 3.5% consistency before refining. As expected, refining had beneficial effects on some fibre properties and negative effects on others (Table 10.3). These results clearly show that if the right refining conditions are chosen, strength properties of a mixed paper furnish can be improved without necessarily showing an increased resistance to drainage and reduction in fibre length.

Iyengar (1996) investigated system designs that provide a basis for using mixed papers in a containerboard facility. The approach of treating mixed papers and OCC in two separate systems provided the best quality product and allowed a mixed paper content as high as required. The only difference from conventional OCC systems is

Table 10.4 Effect of Fibre Type on Strength Properties

Property	Office Paper	Bleached Hardwood	Bleached Softwood
Breaking length (km)	4.3	2.5	3.9
Burst index (kPa m^2/g)	2.8	1.05	2.5
Tear index (mN m^2/g) 1.7	1.7	2.1	2.9

Source: Based on Baker (1999)

Table 10.5 Burst Improvements at 100 kW h/tonne

Paper Grade	Burst Improvement (kPa m^2/g)
OCC	1.2–1.6
Newsprint	0.25–0.30
Office paper	1.1
Hardwood pulp	1.4–1.8
Softwood pulp	3.0–3.5
CTMP	0.7

Source: Based on Baker (1999)

medium-consistency pulping and washing. However, the capital costs can exceed those expected for an OCC processing facility.

The effects of refining conditions such as SEL and cutting edge length tend to be property specific, therefore mills could tailor refining conditions to the properties of greatest importance to them. For example, stiffness development is far greater at an SEL of 2.0 W s/m, but tensile and burst development are much lower. Guest (1991) emphasises that RCFs require a different refining strategy to retain strength properties than the strategy used for virgin pulps (Table 10.4). Further comparisons of the refining of mixed office papers with virgin fibres also showed the potential to substitute hardwood and softwood virgin pulps (Tables 10.5). In the untreated state, the recovered office paper has better burst strength and tensile strength than the virgin fibre, but it has a higher wetness or lower freeness and lower tear index. When refined to the levels used for printing and writing, the recovered office paper has a higher tear tensile relationship than hardwood, but is not as strong as the softwood pulp. However, it is still probable that the mixed paper would equate to a standard furnish of 30% softwood and 70% hardwood.

The influence of refining recycled kraft pulps at high and low consistency was studied by Lundberg et al. (1976). They observed that a mixture of high- and low-consistency refining was needed to obtain adequate strength at reasonable drainability. Studies on a commercial carton-board machine demonstrated that high-consistency refining, as opposed to low-consistency refining, produced adequate strength while reducing steam consumption.

The OptiFiner concept from Metso Paper is aimed at developing key fibre properties (Kankaanpaa & Soini, 2001). It focuses on deflaking, refining and dispersion subprocesses of stock preparation. A study has examined results from the treatment of OCC and ONP and old magazines (OMG). The OptiFiner concept incorporates conical dispersion and low-consistency refining in the same line. The investment costs of the dispersion process are considerably higher than the investment costs of the refining process, but significantly improved RCF qualities are obtained by combining conical dispersion and low-consistency refining stages in the same line. Conical dispersion offers a large processing area and low energy intensity followed by gentle fibre treatment. Large amounts of water in low-consistency refining act as lubricants in processing and the entire treatment is conducted in the fluidised phase. Freeness reduction with low-consistency refiner is higher than with the high-consistency disperser. The tensile index of OCC pulp also developed better with the low-consistency refiner. Combining dispersion and refining stages in the same line means all fibre potential can be used to give the best quality end product.

The effect of refining on recycled chemical-bleached bagasse and wheat straw pulps was studied by Garg and Singh (2004). The pulps were beaten to a freeness of 350 mL CSF. Standard handsheets were prepared and a proportion of the backwater recycled. Pads were formed from the remainder of the pulp and reslushed and used for the preparation of further handsheets and pads. It was largely possible to recover the tensile strength of recycled pulps through refining in a PFI mill, but there was a sharp decrease in freeness. This reduction in freeness was more severe for wheat straw pulp than in bagasse pulp after refining. The considerable content of very fine particles in the wheat straw pulp significantly contributed to its slowness. The decrease in freeness reduced when the pulps were given an alkali treatment before refining. A 50:50 mixture of refined coarse fraction and unrefined fines fraction in the recycled pulps gave much higher freeness than found in the refined whole pulp for equivalent strength values.

Lumiainen (1992a) studied the refining behaviour of the Conflo refiner on RCFs over a 2-year period. Refining OCC pulp improves properties such as tensile strength, burst strength, tensile energy absorption, internal bonding and stiffness. Because RCFs have been irreversibly weakened by earlier refining and other stresses, subsequent refining must be done with care. Mill installations using Conflo refiners have improved bonding ability with a minimum increase in drainage resistance and a minimum decrease in fibre length. Efficient pretreatment improves the initial fibre properties before refining, but these properties can be improved even more by refining (Lumiainen, 1992a, 1994a). The better the pretreatment, the better the properties of the RCF. The fibre responds well when refining at low consistency, consuming a moderate amount of energy for using Conflo refiners. A typical energy consumption is 30–60 kW h/tonne. The conditions for refining RCFs must be selected correctly to avoid excessive refining. Refining improves the natural bonding capacity of RCFs, which reduces the need for chemical bonding agents. By improving the bonding capacity, more RCFs can be used in making paper and board.

Metso Paper offers a range of solutions for stock preparation of fibres based on RCF (Kremsner, 2003). The continuous vat pulper concept is generally used for slushing OCC and mixed waste paper. The main pulper features a perforated screen plate of 10–14 mm and a slushing rotor in the bottom part, with a combined lightweight and heavy impurity connection in a higher position. A continuous slushing drum is generally used for slushing ONP, OMG and sorted mixed waste paper. The drum pulper concept is becoming increasingly important for OCC-based fibres, combining the advantages of gentle slushing, efficient reject removal, lower energy consumption and a simple process. Screen baskets and screen plates are used for coarse screening and deflaking of OCC and mixed waste paper. A three-stage screening system is used for ONP, OMG and sorted waste. Fractionation is used in test liner, fluting and multilayer cardboard production. Multistage fractionation is necessary to achieve an optimum split of short-fibre and long-fibre fractions. Fine screening is only performed on the long-fibre fraction. Refining of RCFs is essential; high- or low-consistency refining is generally used, and sometimes a combination of the two.

Amcor Research and Technology has commissioned a pilot facility comprising a 16 in. (406 mm) double-disc refiner and an 8 in. (203 mm) multipurpose screening system for projects relating to fibre quality. PFI mill data do not adequately establish optimal refining conditions for commercial refiners. Low-intensity refining results in higher strength properties plus substantial savings in net energy. A study on the upgrading of clarifier reclaimed fibre demonstrated that multipurpose screening using a suitable basket and rotor combination may be successfully used to upgrade low-quality fibre (Ghosh & Vanderhoek, 2001). In a case study of a mill producing virgin pulp, 60–80% of the total mass rejected by the secondary screening system could be recovered by the installation of a small screen of the type used in the pilot equipment. In further case studies, a multipurpose screening system with an appropriate basket could be used to improve the quality of reclaimed fibre from the clarifier of an integrated mill, and the multipurpose screen could also be used to improve the dewaxing of fibre from saturated waxed boxes.

The use of mechanical pulp and RCF is increasing in newsprint furnishes, but the quality of RCF is decreasing. The papermaker's aim is to extract maximum strength and performance from the available fibre, while maximising paper machine performance, minimising the use of expensive, low-yield fibres and maintaining the quality of the end product. It is critical to use refining strategies to accomplish this. For relatively weak fibres, low-intensity refining is a proven approach to an optimum result (Demler & Silveri, 1995). This practice is well established for virgin and recycled mechanical pulps such as thermomechanical pulp, groundwood and deinked newsprint. Increases in burst strength and tensile strength have been demonstrated by machine trials and enabled a reduction in softwood addition of 7.0%.

Demler and Silveri (1995) performed pilot and mill trials that examined low-consistency refining of an ONP/OMG DIP to define optimal intensity and energy requirements for maximum pulp property development. Results showed that low-intensity refining is required to maximise strength properties. A 20% improvement in strength properties was obtained by fractionating the pulp then refining the long-fibre fraction. Newsprint contains a large amount of mechanical fibres, so it has a poor response to medium-intensity refining. The short, weak fibres found in newsprint require low-intensity impacts and energy inputs. In optimisation trials at Pira

Table 10.6 Change in Strength Properties of Newsprint at Different Specific Edge Loads

Property	Change in Properties (%)	
	0.25 W s/m	0.5 W s/m
Tensile index	+13%	+17%
Burst index	+6%	+8%
Tear index	−30%	−23%
Scott ply bond	+83%	+36%
Kenley stiffness	−14%	+6%

Source: Based on Baker (1999)

International, low SEL values of 0.25 and 0.5 W s/m were chosen to give gentler refining of the mechanical fibres. They used net refining energies of up to 100 kW h/tonne. The ONP and OMG furnish was obtained from a newsprint deinking mill, so the stock had the same chemical and mechanical treatment as stock refined in the mill. The stock was taken from the decker chest before the paper machine at 4% consistency. Most of the results were similar for the newsprint and magazine stock refined at the two different SEL values, but it was believed that an SEL of 0.5 W s/m generally exhibited better results (Table 10.6). Strength development was marginally higher at 0.5 W s/m for most properties and the CSF was higher. The SR value also showed a more gradual decrease at 0.5 W s/m. The main effect was the greatly increased ply bond found at the lower SEL load of 0.5 W s/m. The overall impression is that it is possible to over-refine some types of recycled pulps and care must be taken as the type of refining will be influenced by the content of the furnish and the required properties.

Research by Sampson and Wilde (2003) has shown the suitability of the pre-refining strategy for strength development in recycled furnishes. This strategy involves a preliminary fractionation stage. The long-fibre fraction was separately refined and blended with the short-fibre fraction and the whole pulp was refined as an equivalent process to co-refining the two fractions. For pre-refining, the short-fibre fraction was refined then blended with the long-fibre fraction before a co-refining stage. The pre-refining strategy on RCFs gives improved tensile strength without increasing the stock's net energy and without reducing its density.

Lumiainen (1992b) performed trials with OCC scrap at the typical low-refining consistency. Results indicated that refining improves the natural bonding ability of secondary fibres, which in turn reduces the need for chemical bonding agents. Refining naturally lowers the tear strength, fibre length and bulk of DIP. Improved binding ability allows papermakers to use increasing amounts of secondary fibres in the furnish.

The optimal conditions for refining OCC have been determined through pilot-plant trials (DeFoe, 1991; Rihs, 1992). The work involved studying two pulps, one produced commercially and one produced in a pilot plant by blending rolls of liner and corrugating medium. The liner and medium were produced from 30% OCC and

70% virgin pulp. Pulp properties and energy requirements of the pulps produced using three-plate patterns run at 3.5% consistency were compared with those produced by a high-speed single-disc refiner operated at 30% consistency. The results indicated that low-consistency refining was the better operation for enhancing OCC properties.

Many researchers have examined what might be considered fundamental problems in recycling: how fibres are affected by recycling processes and what effect their response has on the properties of paper made from these fibres. The effectiveness of two regeneration methods, namely improvement of the papermaking potential for five different grades of RCFs, was investigated by Olejnik, Stanislawska, Wysocka-Robak, and Przybysz (2012). In the first method, only the refining operation was used; in the second, removal of fines and subsequent refining were applied. The dewatering ability of every regenerated pulp (measured by the SR test) and basic strength properties of papers made from these pulps were tested. The results showed that it was possible to increase the papermaking potential of all pulps tested; however, higher effectiveness was achieved when the removal of fines was followed by the refining operation. Further analysis of the results also indicates that recycled pulp grades based on kraft paper have sufficiently high papermaking potential, and in many cases they do not need additional regeneration. Pulp regeneration is strongly recommended in the case of pulps based on LWC, SC and newsprint papers. Experiments have also proved that the selection of an optimal method for improved papermaking potential always has an individual character, and the method always has to be adjusted to a specific pulp grade and the paper grade made from it.

10.3 Use of Enzymes in Upgrading Secondary Fibre

For a more efficient utilisation of RCF products, their physical strength properties must be improved to a competitive level. The strength loss of secondary fibres is caused by hornification (decrease of swellability and conformability of fibres) (Nazhad & Paszner, 1994) and loss of relative bonded area (Garg & Singh, 2006). The strength of fibres can be recovered to a certain degree by beating. However, generation of fines during the beating process decreases the runnability of the paper machine. Decrease of drainage properties of pulp caused by the fines becomes an important limiting factor for recovering paper strength by beating. Mechanical pulp content in waste paper grades used for corrugated medium is an additional source for generating more fines during the beating.

Table 10.7 presents some of the literature results of the degree of hornification observed on various pulp samples (Tze & Gardner, 2001; Wistara & Young, 1999).

The use of enzymes in upgrading secondary fibre has been extensively studied by several researchers (Bhardwaj, Bajpai, & Bajpai, 1995, 1997; Pommier et al., 1990; Putz, Wu, & Gottsching, 1990; Sarkar, Cosper, & Hartig, 1995). In most of the studies, commercial mixtures of different cellulases and hemicellulases were used to treat recycled pulps, and the treatments resulted in improved drainage properties. Stork et al. (1995) used isolated cellobiohydrolases and endoglucanases of

Table 10.7 Degree of Hornification Observed in Different Pulp Samples

Pulp	Recycling Level	Water Retention Value	Degree of Hornification (%)
Black spruce, bleached, kraft	Unrecycled	2.19	–
	1 × Recycled	1.9	13.24
	2 × Recycled	1.81	17.35
Hardwood, bleached, kraft	Unrecycled	1.92	–
	1 × Recycled	1.10	42.71

Source: Based on Tze and Gardner (2001) and Wistara and Young (1999)

Penicillium pinophilum to treat recycled pulps and measured the effects on the water retention value. They concluded that endoglucanases are necessary for improving the dewatering of pulp.

Pulp fibrillation by cellulases for enhancing strength properties was already patented in 1959 (Bolaski, Gallatin, & Gallatin, 1959). The effect of different hydrolases on the improvement of the refining process by reducing energy consumption has been a matter of intensive research in the past two decades (Bajpai, 2012). Different cellulases and hemicellulases have been proved to have a beneficial effect on refining and on surface fibre morphology, giving rise to better bonding properties and a closer structure of paper. Furthermore, cellulase-free xylanase application has improved pulp fibrillation and water retention, shortened the time of refining in virgin pulps, enhanced the restoration of bonding and increased both the freeness in RCFs and the selective removal of xylan from dissolving pulps.

Some basic effects of the enzymatic treatment were studied by Jackson, Heitmann, and Joyce (1993) with a preparation made of xylanases and two different mixtures of cellulases and xylanases on a softwood kraft pulp. With cellulases, pulp drainability improved and it was verified that fibre hydrolysis occurs as a consequence of the treatment. Low enzyme doses produced reductions in fines content, which were related to a possible effect of flocculation produced by the enzyme, similar to that of polymeric drainage aids. The highest enzyme doses led to increased fines contents, which were attributed to the fibre disintegration produced by cellulases.

The treatment of RCF with an enzyme (0.2 wt% Pergalase A40) after refining significantly improved freeness (Eriksson, Heitmann, & Venditti, 1997, 1998). Wash treatments also significantly improved freeness levels, but the yield loss was considerably greater in tests. Enzyme pretreatment was coupled with reduced refining to maintain freeness at the same level as the control. The researchers focused on blended colour ledger in manufacturing board from 100% RCF. The enzyme dosage was 0.2 wt% Pergalase A40 to achieve optimum freeness improvement at the lowest cost and with the least detrimental impact on the physical properties of the pulp, although a lower enzyme dose could be economically more favourable while providing reasonable freeness improvement. They concluded that enzyme-assisted drainage improvement of RCF is economically viable but it depends on production rate improvements, process optimisation and enzyme recycling.

Pala, Mota, and Gama (1998) investigated refining, refining plus enzymatic treatment, enzymatic treatment plus refining and enzymatic treatment of RCFs. Physical and mechanical properties of pulp and paper were measured to evaluate the effectiveness of the methods. They found that the most suitable method for upgrading recycled pulps was refining plus enzymatic treatment. Refining increases burst and tensile resistance, whereas enzymatic treatment produces better drainage under certain conditions. Fibre that has been recycled more than once has lower papermaking qualities than virgin or once-recycled fibre. By using an enzyme blend with RCF, some of the lost freeness can be restored. Pergalase, a trademark of Ciba Specialty Chemicals, is a blend of enzymes that improves the freeness of the fibre but does not reduce the fibre strength. The enzyme is effective at an optimum pH of 5.5–6 but remains active at pH 4.5–7. The optimum temperature is 50–60°C. Enzymes need time to be effective; a 15 min retention time has been adequate, providing there is good mixing.

Moran (1996) reported that addition of enzyme before refining has a very different effect compared with post-refining treatment. Pre-refining applications result in improved refining efficiency, whereas post-refining treatment results in increased furnish freeness. A combination of the two can provide optimised strength and drainage benefits. Pre-refining treatment involves the addition of the enzyme to the stock that feeds an unrefined stock chest. In this application, the enzymes begin to hydrolyse the outer walls of cellulose fibres. This surface treatment prepares the fibre surface for improved refining efficiency. Strength properties can be developed with fewer refining energy requirements. This application has the following benefits:

- lower refining energy to meet strength specifications;
- improved strength properties at equivalent refining energy.

The interaction of the enzyme with the furnish in a well-agitated chest after the refiner(s) results in a freeness recovery. The enzyme in this application selectively hydrolyses the amorphous cellulose on the fibre surface, especially the fines. This makes the fibres and fines less hydrophilic, increasing the freeness of the stock. For post-refining in drainage improvement applications, mill trials have shown that an increase in freeness of up to 100 mL CSF can be obtained through an enzyme treatment of refined OCC-based pulp. Results vary with mixing, pH, temperature, pulp consistency and retention time. The two methods for adding enzyme can be combined for strength and drainage benefits. Keeping in mind that the enzyme is a catalyst, its function should continue from a pre-refining application into a post-refining effect. In other words, the enzyme should weaken the fibre walls before refining, then continue to work on the fibres, improving drainage after refining. Field results have shown this to be true. However, by splitting the addition, the new fibre surfaces created by refining can be treated by a fresh dose. This has been found to be most effective in providing strength and drainage benefits. The ratio of pre-refining to post-refining addition, as well as the dose, will depend on enzyme–fibre interaction time and furnish characteristics. The benefits of an enzyme-enhanced drainage

Table 10.8 Laccase Treatment Providing Increased Fibre Surface Carboxylic Acid Group Content of RCF from a Coloured Office Paper

U Laccase/odg Pulp	mM Acid Grps/100 g od Pulp (%)
10,600	4.67
21,000	8.86
31,700	30.27

Source: Based on enhancing pulp properties of RCFs with oxidative enzymes, www.ipst.gatech.edu/faculty/ragauskas_art/research_opps/bleaching%20recycled%20paper.pdf
od = oven dried

programme have been demonstrated on grades such as tubestock, gypsum linerboard and corrugated medium. Other applications include linerboard, coated boxboard, saturating kraft and bag grades. Outside the recycled area, enzymes are also improving papermaking operations for glassine, greaseproof and release papers.

The development of methods to improve paper strength and optical qualities enables mills utilising RCF to shift to lower-cost, lower-quality grades while maintaining product quality. Chemical oxidation of cellulose fibres to increase the number of fibre surface carboxylic acid groups has been shown to improve fibre strength properties. Enzymatic fibre oxidation with laccase has been reported to result in improvements in paper strength, thought to be caused by the increase in fibre-surface carboxylic acid groups (Barzyk, Page, & Ragauskas, 1997; Le, Daneault, & Chabot, 2006; Mansfield, 2002; Pei, Shi, Wei, & Zhao, 2005). Preliminary work in Ragauskas' group has shown that treatment of RCFs with an oxidative enzyme such as laccase increases the surface carboxylic acid group content of recycled mixed office waste (Table 10.8) and results in improved paper strength properties (Chandra & Ragauskas, 2005).

Biricik and Atik (2012) investigated the positive effect of using commercial cellulase enzymes on the physical strength properties of corrugated medium. The long-fibre fraction of pulp is the most appropriate application point of a cellulase enzyme. The pulp used in this research was obtained at the fractionation screen levels of Modern Karton's paper mill in Corlu, Turkey. The long-fibre fraction of the pulp for corrugated medium was pre-treated with four different commercial enzymes: Roglyr Bio 1537, Maximyze 2520, Benstone and Bensoft Plus. The handsheets for physical testing were prepared after blending beaten long-fibre and unbeaten short-fibre fractions. The obtained test results indicated that the application of enzyme on appropriate fibre fractions has positive effects on the strength properties of the corrugated medium. The short-span compression test is one of most important strength properties of corrugated board; in this case, results showed an increase from 8.41% to 20.5% with enzymatic pretreatment and beating of the long-fibre fraction.

Effects of recycling on properties of elemental chlorine-free bleached softwood kraft pulp were evaluated in the laboratory (Oksanen, Pere, Paavilainen, Buchert, & Viikari, 2000). The tensile strength, fibre flexibility and water retention value lost during drying of the pulp were recovered by refining between the cycles, which, however, resulted in deteriorated drainage properties. The recycled pulps were treated with purified *Trichoderma reesei* cellulases and hemicellulases, and the changes in fibre properties due to enzymatic treatments were characterised. The endoglucanases (EG I and EG II) significantly improved pulp drainage already at low dosage levels, and EG II was found to be more effective at a given level of carbohydrate solubilisation. Combining hemicellulases with the endoglucanase treatments increased the positive effects of the endoglucanases on pulp drainage. However, as a result of the endoglucanase treatments, a slight loss in strength was observed. Combining mannanase with endoglucanase treatments appeared to increase this negative effect, whereas the impact of xylanase was not significant. Although the drainage properties of the pulps could be improved by selected enzymes, the water retention capacity of the dried hornified fibres could not be recovered by any of the enzymes tested.

Maximino et al. (2011) analysed the effects of the enzymatic treatment with Pergalase A40 (a blend of cellulases and hemicellulases) on an industrially recycled pulp made up of OCC fibres, kraft liners and a low percentage of white office paper. The enzymatic treatment was done through a 2^3 experimental design, by varying pulp concentration, enzyme dosage and treatment time. Furthermore, a combined treatment (enzyme plus PFI mechanical refining) was also evaluated, to obtain a greater improvement in drainability, while maintaining or improving the properties of secondary fibres. Enzymatic pretreatment of RCFs without refining increases the initial freeness degree of pulp, almost without any loss in tensile strength, for most of the conditions analysed. Generally, combined treatments (enzyme + refining) show that a higher tensile index level may be attained, with significant drainability improvement and minor specific energy consumption compared with the reference pulp.

Chauhan, Parihar, Dixit, and Kumar (2011) found enzyme treatment of the shredded currency waste of Reserve Bank of India (RBI) very effective in its recycling for making handmade paper in terms of the improvement in brightness and strength properties of the pulp produced. In addition to this, they found that all the properties showed an increasing trend with a rising dose of enzyme. Therefore, to get a brightness equivalent to the control pulp, less pulping chemical might be sufficient, thereby possibly saving the chemicals used. As a result, the quality of effluents also improves and a better quality handmade paper can be produced by incorporating an enzymatic treatment stage in the conventional process of recycling the shredded currency waste of RBI. Thus, utilisation of biotechnology in processing shredded currency waste helps improve the environmental status of the Indian handmade paper industry besides addressing the problem of global warming and solid waste disposal.

References

Bajpai, P. (2006). *Advances in recycling and deinking* (180 pp.). UK: Pira International.

Bajpai, P. (2012). *Biotechnology for pulp and paper processing* (414 pp). NewYork, USA: *Springer Science + Business Media.*

Baker, C. (2000). *Refining technology* (p. 197). UK: Pira International.

Baker, C. F. (1999). *Refining recycled fibres, paper recycling challenge – Process control and mensuration*. Appleton, WI, USA: Doshi & Associates. (Chapter 9).

Barzyk, D., Page, D.H., & Ragauskas, A. (1997). Carboxylic acid groups and fiber bonding. *Fundamentals of papermaking materials, transactions of the fundamental research 11th symposium* (Vol. 2, pp. 893–907), Cambridge, September.

Bhardwaj, N. K., Bajpai, P., & Bajpai, P. K. (1995). Use of enzymes to improve drainability of secondary fibres. *Appita Journal*, *48*(5), 378–380.

Bhardwaj, N. K., Bajpai, P., & Bajpai, P. K. (1997). Enhancement of strength and drainage of secondary fibers. *Appita Journal*, *50*(3), 230–232.

Biricik, Y., & Atik, C. (2012). Effect of cellulase treatment of long fiber fraction on strength properties of recycled corrugated medium. *African Journal of Biotechnology*, *11*(58), 12199–12205.

Bolaski, W., Gallatin, A., & Gallatin, J.C. (1959). Enzymatic conversion of cellulosic fibres. US Patent No. 3,041,246.

Brancato, A. (2008). *Effect of progressive recycling on cellulose fiber surface properties*, PhD Thesis, Georgia Institute of Technology, December.

Chandra, R. P., & Ragauskas, A. J. (2005). Modification of high-lignin kraft pulps with laccase. Part 2. Xylanase-enhanced strength benefits. *Biotechnology Progress*, *21*(4), 1302–1306.

Chauhan, S., Parihar, S., Dixit, P., & Kumar, S. R. (2011). Effect of enzyme treatment on recycling of shredded currency waste of RBI for making handmade paper. *Current World Environment*, *6*(1), 77–85.

DeFoe, R.J. (1991). Refining variables relating to OCC property development. *TAPPI pulping conference*. Orlando, FL.

Demler, C., & Silveri, L. (1995). Strength developments of mechanical and de-inked newsprint pulps through low intensity refining. *49th Appita annual general conference*. Hobart.

Eriksson, L.A., Heitmann, J.A., & Venditti, R.A. (1997). Drainage and strength properties of OCC and ONP using enzymes with refining. *Recycling symposium* (p. 423). Chicago, IL: TAPPI Press, 14–16 April.

Eriksson, L.A., Heitmann, J.A. Jr., & Venditti, R.A. (1998). Freeness improvement of recycled fibres using enzymes with refining, enzymes applications in fiber processing. In K. E. L. Eriksson (Ed.), *ACS Symposium Series: Vol. 687* (p. 340). Washington, DC: American Chemical Society.

Fisher, H. S. (1980). Report on a high-consistency de-inking process for mixed secondary fibres. *Paper Trade Journal*, *164*(17), 54.

Garg, M., & Singh, S. P. (2004). Response of bagasse and wheat straw recycled pulps to refining. *TAPPI Journal*, *3*(10), 11.

Garg, M., & Singh, S. P. (2006). Reasons of strength loss in recycled pulp. *Appita Journal*, *59*(4), 274–279.

Ghosh, A.K., & Vanderhoek, N. (2001). Improving stock quality through refining and screening. *55th Appita annual conference*. Hobart.

Guest, D.A. (1991). Refining wastepaper – Theory and practices. *Pira international conference on current and future technologies of refining*. Birmingham.

Holik, H. (2000). In L. Gottsching & H. Pakarinen (Eds.), *Unit operations and equipment in recycled fibre processing, recycled fibre and de-inking*. Helsinki, Finland: Fapet Oy. (Chapter 5).

Iyengar, S. (1996). Mixed waste provides supplement for traditional OCC-fibre supply. *Pulp and Paper*, *70*(3), 143.

Jackson, L. S., Heitmann, J. A., & Joyce, T. W. (1993). Enzymatic modifications of secondary fiber. *TAPPI Journal*, *76*(3), 147–154.

Kankaanpaa, V., & Soini, P. (2001). OptiFiner – Approach to the conical dispersion and refining of recycled fibre. *Fibre Paper*, *3*(3), 42.

Kremsner, M. (2003). Solutions in slushing, screening and refining of recycled fibres. *30th DITP international annual symposium*. Slovenia.

Le, R. J., Daneault, C., & Chabot, B. (2006). Acidic groups in TMP oxidized fibres by TEMPO to improve paper strength properties. *Pulp and Paper Canada*, *107*(4), 39–41.

Levlin, J.E. (1976). On the beating of recycled pulps. *EUCEPA symposium on secondary fibres and their utilisation in the paper industry*. Bratislava.

Lumiainen, J. (1992a). Do recycled fibres need refining? *Paperi ja Puu*, *74*(4), 319.

Lumiainen, J. (1992b). Refining recycled fibre – Advantages and disadvantages. *TAPPI Journal*, *75*(8), 92.

Lumiainen, J. (1994a). Do recycled fibres need refining? *Papel*, *2*, 36.

Lumiainen, J. (1994b). Is the lowest refining intensity the best in low-consistency refining of hardwood pulps. *TAPPI papermakers conference*. San Francisco, CA.

Lumiainen, J. (1995a). Refining of secondary fibres. *Third international refining conference*. Atlanta, GA: Pira International.

Lumiainen, J. (1995b). The specific surface edge load theory. *Third Pira international refining conference*. Atlanta, GA.

Lumiainen, J. (1997). Refining of ECF and TCF bleached Scandinavian softwood kraft pulps under the same conditions. *Paperi ja Puu*, *79*(2), 109.

Lundberg, R., De Ruvo, A., Fellers, C., & Kolman, M. (1976). Recycling of waste paper – Influence of process conditions and machinery. *EUCEPA symposium on secondary fibres and their utilisation in the paper industry*. Bratislava.

Mansfield, S. D. (2002). Laccase impregnation during mechanical pulp processing – Improved refining efficiency and sheet strength. *Appita Journal*, *55*(1), 49–53.

Maximino, M. G., Taleb, M. C., Adell, A. M., & Formento, J. C. (2011). Application of hydrolytic enzymes and refining on recycled fibers. *Cellulose Chemical Technology*, *45*(5–6), 397–403.

Moore, G.K., Cathie, K.,Crow, H., & Smith, J. (1995). Developing mixed office waste as a furnish component for printings and writings. *Third Pira international refining conference*. Atlanta, GA.

Moran, B. R. (1996). Enzyme treatment improves refining efficiency, recycled fibre freeness. *Pulp and Paper*, *70*(9), 119.

Nazhad, M. M., & Paszner, L. (1994). Fundamentals of strength loss in recycled paper. *Tappi Journal*, *77*(9), 171.

Nazhad, M.M., & Awadel-Karim, S. (2001). Possibilities for upgrading OCC pulp. *55th Appita annual conference*. Hobart.

Oksanen, T., Pere, J., Paavilainen, L., Buchert, J., & Viikari, L. (2000). Treatment of recycled kraft pulps with *Trichoderma reesei* hemicellulases and cellulases. *Journal of Biotechnology*, *78*, 39–48.

Olejnik, K., Stanislawska, A., Wysocka-Robak, A., & Przybysz, P. (2012). Evaluation of the possibilities of upgrading the papermaking potential for different recycled pulp grades. *Fibres & Textiles in Eastern Europe*, *20*(2(91)), 102–106.

Pala, H., Mota, M., & Gama, F. M. (1998). Effects of enzymatic treatment and refining on the properties of recycled pulp. *Associacao Portuguesa dos Tecnicos das Industrias de Celulose e Papel*, 478.

Pei, J. C., Shi, S. L., Wei, H. L., & Zhao, W. B. (2005). Li–Na Strength properties enhancement of unbleached kraft pulp through laccase catalyzed oxidation. *Chung-kuo Tsao Chih/China Pulp and Paper*, *24*(6), 1–4.

Peixoto Silva, R., & Chaves de Oliveira, R. (2003). Effect of refining on recovery of physico-mechanical properties of recycled fibres from pinus. *Papel*, *64*(8), 87.

Pommier, J. C., Goma, G., Fuentes, J. L., Rousset, C., & Jokinen, O. (1990). Using enzymes to improve the process and the product quality in the recycled paper industry. Part 2: Industrial applications. *TAPPI Journal*, *73*(12), 197–202.

Putz, H. J., Wu, S., & Gottsching, L. (1990). Enzymatic treatment of waste paper. *Das Papier*, *40*(10A), V42–V48.

Rihs, J.D. (1992). Refining of recycled fibres from brown and white grades. *TAPPI papermakers conference*. Nashville, TN.

Sampson, W.W., & Wilde, R. (2003). Application of a novel refining strategy to strength development of recycled furnishes. *Progress in pulp refining seminar*. Vancouver.

Sarkar, J.M., Cosper, D.R., & Hartig, E.J. (1995). Applying enzymes and polymers to enhance the freeness of recycled fiber. *TAPPI Journal*, *78*(2), 89.

Stork, G., Pereira, H., Wood, T. M., Dusterhoft, E. M., Toft, A., & Puls, J. (1995). Upgrading recycled pulps using enzymatic treatment. *TAPPI Journal*, *78*(2), 79.

Tze, W. T., & Gardner, D. J. (2001). Swelling of recycled wood pulp fibers: Effect on hydroxyl availability and surface chemistry. *Wood and Fiber Science*, *33*(3), 364–376.

Wistara, N., & Young, R. A. (1999). Properties and treatments of pulps from recycled paper. Part I. Physical and chemical properties of pulps. *Cellulose*, *6*(1), 291–324.

11 Improving Drainability of Recycled Fibres*

11.1 Introduction

Paper consumption has been increasing rapidly all over the world for decades. Because of concerns about environmental issues and economic aspects, the importance of recycling and utilisation of secondary fibres has been recognised (Payne, 1997). Nowadays, more than 50% of all paper produced is recovered (Bajpai, 2006).

Generally, pulp properties of recycled fibres decline during reprocessing. Thus there is a limitation to paper recycling. Deterioration of recycled fibre properties is mainly due to the irreversible changes occurring in fibre structure caused by repeated chemical and mechanical treatments and drying. Recycled fibres can be upgraded by refining and adding chemicals and primer fibres (Bajpai, 2006). Refining forms more fibrils on the surface of fibres, therefore improving the bonding characteristics. In this way, the mechanical strengths of the paper can be mended. However, in this step the originally high content of fines (small cellulose fibres) in the recycled pulp increases further. By definition, fines have been identified as the fraction of the furnish solids that passes through a 200 mesh screen. The main problem with fines is that because of their high relative surface area, the dewatering rate is lower than primary pulp. Thus, the productivity of the papermaking process is considerably decreased compared with an operation using virgin fibre.

The currently known methods of improving the drainage or restoring the strength of secondary fibres may be grouped into seven generalised categories: mechanical treatment, chemical additives, fractionation, blending, chemical treatments, enzymatic treatments and paper machine modifications. The enzymatic treatments of secondary fibres with cellulase and mixtures of cellulase and hemicellulase have been reported to increase drainability by the hydrolysis of high specific area materials, such as fines and fibrils (Bajpai, 2012; Bajpai & Bajpai, 2001). Enzymes in combination with polymeric flocculants have also been tried to enhance the freeness of recycled fibres, which translates into improved drainage (Bhardwaj, Bajpai, & Bajpai, 1997; Sarkar, Cosper, & Hartig, 1995).

Decreasing the amount of fines could be the key solution for improving drainage as the fines are considered responsible for the deteriorated drainability. When cellulolytic enzymes are used for partial hydrolysis of cellulose chains, and thus to form a better recycled fibre structure, it is important to find the balance between two opposite requirements. On the one hand, by hydrolysing the fines, improved dewatering rates are obtained. On the other, enough fines have to be left in the pulp to

*Some excerpts taken from Bajpai (2006, 2012).

Recycling and Deinking of Recovered Paper. DOI: http://dx.doi.org/10.1016/B978-0-12-416998-2.00011-8

obtain optimal interfibre bonding, which is required for good strength properties of the end product. Moreover, enzyme action should not result in excessive hydrolysis, as this means loss of weight and thus production.

Cellulolytic enzymes are produced by fungi and cellulolytic bacteria. The cellulase system contains three types of enzymes:

- cellobiohydrolases,
- Endoglucanases,
- β-D-glucosidases.

The cellulase enzymes are classified based on their mode of action.

- Endoglucanases randomly attack the amorphous regions of cellulose substrate, yielding high degrees of polymerisation oligomers.
- Cellobiohydrolases are exoenzymes that hydrolyse crystalline cellulose, releasing cellobiose. Both types of enzyme hydrolyse β-1,4-glycosidic bonds.
- β-D-glucosidase or cellobiase converts cello-oligosaccharides and cellobiose to glucose.

Using mixtures of cellulases can be disadvantageous for certain pulp properties. Applying purified enzymes, specific regions of the cellulose fibres can be attacked, so the desired part of the pulp can be modified. In secondary fibres, the fines and fibrils, which cause low rates of drainage, consist of amorphous cellulose. Because the amorphous cellulose is more accessible than crystalline cellulose, it is not necessary to use the whole cellulase complex for hydrolysis. Thus, applying endoglucanases may be effective enough.

Most cellulases both from bacterial and fungal sources have a general structure. According to Linder and Teeri (1997) three well-distinguished domains can be identified. These are as follows:

- catalytic domain,
- cellulose-binding domain (CBD),
- linker region, which establishes the linkage between the first two domains.

For the efficient hydrolysis of cellulose, the catalytic domain has to be adsorbed on the substrate. The significance of the CBD in the hydrolysis of cellulose has not been clearly described. However, it has been shown that the adsorption of the cellulase enzymes and therefore their hydrolytic activity are reduced if the CBD is removed from the protein.

Researchers have proposed that the enzymatic attack may involve a peeling mechanism, which removes fibrils and fibre bundles that naturally have a high affinity for water, and leaves the fibres less hydrophilic and easier to drain. Also, there are reports that enzymes act preferentially on fines, which have a propensity to block up interstices in the fibre network. The increase in drainage could be due to the cleaving of amorphous cellulose on the surface of fines. Owing to the high specific surface area of the fines, the attack of cellulases is specific towards this fraction (i.e. the fines) of the pulp. The fibre surface is stripped through the enzymatic hydrolysis of subsequent layers or fibrils. Enzymes can either flocculate or hydrolyse fines and remove fibrils from the surface of large fines. Enzyme-aided flocculation occurs when a low enzyme dose is used. In this case, fines and small fibre particles aggregate with each other or with the larger fibres, decreasing the amount of small particles in the pulp and consequently improving pulp drainage. On the other hand, at higher enzyme concentrations, flocculation becomes less significant, and hydrolysis of fines begins to predominate. Enzyme specificity also plays a very important role (Jackson, Heitmann, & Joyce, 1993; Kantelinen & Jokinen, 1997; Mansfield, Jong, Stephens, & Saddler, 1997; Mansfield & Wong, 1999; Pere, Siika-aho, Buchert, & Viikari, 1995).

The paper properties are affected differently by endo- and exoglucanases. These enzymes have different modes of action.

- Endoglucanases are more active on amorphous cellulose and randomly attack the inner part of the cellulosic chain. Endoglucanase action is probably the main determinant of drainage improvement.
- Exoglucanases hydrolyse both crystalline and amorphous cellulose by removing cellobiose from the terminal part of cellulose chains.

11.2 Effect of Enzymes and Chemical Additives on Drainage

The improvement in drainage rates of recycled fibres by cellulase mixtures was first discovered by Fuentes and Robert (1986). Researchers from La Cellulose du Pin showed that a mixture of cellulase and hemicellulase increases the freeness of pulp. Improved drainage and faster machine speeds, resulting from increased freeness, yields significant savings in energy and thus in overall cost. The endoglucanase activity is a prerequisite for drainage improvement of recycled pulps. Treatment with a mixture of cellulase and hemicellulase at low concentration increases the drainage of pulp. The increase in drainage can enhance the capacity of a secondary fibre preparation plant, increase machine speed or pulp dilution in the head box and ultimately produce paper of better quality. In addition to an increase in drainage, regular use of enzymes under optimum conditions may produce beneficial secondary effects such as greater reliability of the paper machine. A less substandard paper is found to be produced, partly because of a lower frequency of breaks. In some cases, enzymes can be also used with normal retention/drainage agents to give a significant increase in the freeness to the pulp. The key to a successful enzyme application is the careful selection of the right enzymes for a mill's specific furnish, process conditions and water chemistry.

Several commercial enzymes are available which improve the drainage of secondary fibres. Novozymes has developed FibreCare® D, which is a cellulase enzyme with endoglucanase activity. Novo researchers studied the efficacy of this enzyme in the laboratory and paper mills for improving the drainability of different types of recycled pulp (Shaikh & Luo, 2009). On treating the refined pulps with enzyme, improvements in Canadian standard freeness (CSF) were observed of 40.5% in mixed waste (MW), 13.1% in old newsprint (ONP) and 19.3% in old corrugated container (OCC) (Table 11.1). Using OCC and addition of enzyme at different levels, the CSF value increased with the increase in enzyme dose. But there was no significant effect on tensile and compression strengths of the enzyme pulps (Tables 11.2 and 11.3). The endoglucanase activity of cellulase partly hydrolyses the amorphous and low molecular mass components of cellulose present in the form of fine, fibrils and colloids and thus helps in dewatering the pulp. These components have very high specific surface area and hold maximum water (Bajpai, 2006).

The drainage improvement caused by enzymes was demonstrated in a North American mill to study the change in machine runnability (Shaikh & Luo, 2009). An increase in machine speed was observed for the production of 200 g/m^2 liner on enzyme dosing. A 6–7% reduction in steam consumption within in the dryer section was also observed owing to increase in dewaterability. The ring crush increased by

Table 11.1 Increase in CSF of Different Pulps by Enzyme

Pulp Type	Increase in CSF (%)
MW	40.5
ONP	13.1
OCC	19.3

Source: Based on Shaikh & Luo, 2009

Table 11.2 Effect of Enzyme Dose on OCC

Enzyme Dose (g/TP)	CSF (mL)
Nil	300
50	353
100	368
200	388
500	422

Source: Based on Shaikh & Luo, 2009

Table 11.3 Effect of Enzyme Dose on Tensile Strength and Compression Strength of OCC

Enzyme Dose (g/TP)	Tensile Strength (Nm/g)	Compression Strength (Nm/g)
Nil	45.0	22.0
50	45.5	22.5
100	45.5	23.0
200	43.0	21.5
500	43.0	22.0

Source: Based on Shaikh & Luo, 2009

two points. A similar trial with enzyme treatment 100 g per Tonne of pulp (100 g/TP) before refining in a European mill, producing towel and tissue, showed an increase in machine speed from 1650 to 1750 m/min for tissue and from 1600 to 1750 m/min for towel productions. The specific refining energy also was reduced by 12.5%, probably because of the presence of some cellobiohydrolase activity in the enzyme product, which helps in pulp refining (Bajpai, 2006). In Asian mill trials producing various basis weight papers from OCC, steam savings on enzyme treatment were observed.

Pergalase A-40 which is a commercial cellulase enzyme has been used in several mills to improve drainage (Eriksson, Heitmann, & Venditti, 1997, 1998; Pommier, Goma, Fuentes, Rousset, & Jokinen, 1990). It is applied after refining/beating the pulp, mainly to improve dewatering. The effects of enzymatic treatment on the pulp and paper properties of recycled paper were investigated by Dienes, Egyházi, and Réczey (2004). They used two commercial *Trichoderma* cellulase enzyme preparations to improve pulp properties of recycled paper. Pergalase A-40 was compared with endoglucanase III (IndiAge Super L) based on their effects on drainage, water retention value, fibre length distribution and paper properties (air permeability, tensile index, burst index, tear index). Enzyme characteristics were investigated to establish the industrial applicability. IndiAge Super L containing only EG III proved to be more stable than Pergalase A-40 against fluctuations in temperature and pH. It also proved to be more efficient in improving drainage. The effects of enzymatic treatments on pulp and paper properties showed that the type of enzyme component is probably very important (Tables 11.4 and 11.5). Moreover, existence or absence of a CBD might play a determining role.

It has been suggested that CBDs disrupt the structure of cellulose fibre surfaces but have no detectable hydrolytic activity (Din et al., 1991; Xiao et al., 2001). The binding of CBDs to cellulose, under a wide range of environmental conditions and without the need for chemicals, makes them attractive molecules for designing new paper-modifying agents that are environmental friendly (Levy, Nussinovitch,

Table 11.4 Effect of Enzymatic Treatment on Pulp Properties

Treatment	**Pulp Properties**					
	Freeness, SR (CSF)	**Water Retention Value %**	**Air Permeability (mL/s)**	**Burst Index (kPa m^2/g)**	**Tear Index (Nm^2/kg)**	**Tensile Index (Nm/g)**
Untreated	37 ± 0.5 (340)	144 ± 6	2.99 ± 0.23	3.06 ± 0.07	10.81 ± 0.49	51.4 ± 1.5
IndiAge Super L	31 ± 0.7 (414)	168 ± 7	3.47 ± 0.15	3.23 ± 0.08	8.06 ± 0.27	52.2 ± 2.2
Pergalase A-40	35 ± 0.8 (364)	135 ± 6	3.70 ± 0.31	3.21 ± 0.04	9.49 ± 0.90	53.9 ± 2.1

Source: Based on Dienes et al., 2004

Table 11.5 Effect of Enzyme Treatment on Physical Characteristics of Pulps

Treatment	Pulp Properties			
	Mean Fibre Length (mm)		Fines (<0.2 mm) (%)	
	Length Weighted	Arithmetic	Length Weighted	Arithmetic
Untreated	1.68	0.87	2.04	20.18
IndiAge Super L	1.78	0.91	1.94	21.79
Pergalase A-40	1.65	0.85	2.14	22.46

Source: Based on Dienes et al., 2004

Shpigel, & Shoseyov, 2002; Shoseyov, Levy, Shani, & Mansfield, 2003). Treatment of secondary paper fibres with CBDs allows improvements in pulp drainability and mechanical properties of paper (Pala, Lemos, Mota, & Gama, 2001; Pala, Pinto, Mota, Duarte, & Gama, 2003). The interface system fibre–water–fibre, and after drying, fibre–air–fibre, may be affected by CBD treatment, influencing the technical properties of pulp and paper (Pala et al., 2003). Similarly, Levy et al. (2002) showed that a CBD from *Clostridium cellulovorans* improves the mechanical properties of Whatman paper sheets, as well as transforming it into a more water-repellent paper, effects that were even more significant when a double CBD was used (Levy et al., 2002). Pinto, Amaral, Costa, Gama, and Duarte (2004) and Pinto, Moreira, Mota, and Gama, 2004 studied the influence of CBDs on cellulosic fibre properties in different pulp samples, namely recycled (from kraft and office waste papers) and chemical (eucalypt and pine) pulps. They observed that the fibres treated with CBDs exhibited improved pulp drainability with reduced Schopper–Riegler (SR) value by up to 15%, and increased air permeability of pulp handsheets, while the strength properties were not significantly affected. The surface charge of the fibres decreased with this treatment, but the effect of the addition of a cationic polymer (cationic starch) was not affected by it, at least in the studied pulp sample. They also noted that the behaviour of the different fibres for the tested properties was dependent on the fibre morphology and origin.

Bhardwaj, Bajpai, and Bajpai (1995) studied several commercial carbohydrate-modifying enzymes for improving the drainage of secondary fibre. Drainage improvement compared with control was significant with Pergalase A-40 (Table 11.6). The drainage improvement was 11.7% (with 0.1% enzyme addition) and 21.3% (with 0.2% enzyme addition) at a reaction time of 30 min for low freeness pulp. An increase in the reaction time to 180 min improved the drainage by 25.4% (with 0.1% enzyme) and 31.7% (with 0.2% enzyme). The pulp retained most of the required strength properties when treated with Pergalase A-40 either at 0.1% enzyme and a reaction time of 45 min or at 0.2% enzyme and a reaction time of 30 min. Increase in reaction time beyond 30 min with 0.2% enzyme resulted in deterioration of strength properties. Pergalase A-40 treatment on pulps of different initial freeness

Table 11.6 Effect of Pergalase Enzyme on Drainability of OCC

Enzyme Dose (% o.d. pulp)	Reaction Time (min)	Improvement in Drainage (%)
0	–	–
0.1	30	11.7
0.1	45	15.9
0.1	60	17.1
0.1	120	20.3
0.1	180	25.4
0.2	30	21.3
0.2	60	27.6
0.2	180	31.7

Note: o.d., oven dry.
Source: Based on Bhardwaj et al., 1995

showed that the lower the initial freeness, the higher the gain. When the pulp was treated with enzyme, the freeness increased without any loss of the mechanical properties in the paper; when mechanical refining preceded the enzymatic treatment, better physical properties were obtained at a freeness similar to the control. In other words, better physical properties can be obtained at an identical drainability.

Jeffries (1992) compared the effects of a commercial cellulase, a commercial xylanase containing cellulase activity and a cellulase-free xylanase from *Aureobasidium pullulans*. The commercial cellulase was Celluclast™ and the commercial xylanase containing cellulase activity was Pulpzyme HA™. The effects of the enzymes were significant. In the case of the *A. pullulans* xylanase acting on chemical fibres, the freeness was better than that observed with Celluclast™ when both were used at the same dose.

The effects of purified cellulases and hemicellulases on the properties of recycled kraft pulps have been studied (Oksanen, Buchert, Pere, & Viikari, 1995). Elemental chlorine-free bleached kraft pulp was recycled by subsequent drying, slashing and refining. The recycled pulps were treated with purified *Trichoderma reesei* hemicellulases and cellulases, and their combinations. Changes in fibre properties caused by the enzymatic treatments were characterised by measuring the water retention value, SR value and handsheet strength. The strength properties and water retention values that were lost upon drying could be recovered almost to the initial level by refining between the cycles. The tensile index increased as a function of recycles whereas tear index decreased slightly. However, extensive refining resulted in deteriorated drainage properties, namely increased SR value. Of the single enzymes, endoglucanases were most effective in improving the drainage whereas cellobiohydrolases had practically no effect. The pulp strength was, however, negatively affected even with rather low endoglucanase doses. Xylanase and mannanase treatments improved the change only slightly. Although, the drainage could be improved by enzymatic treatments, none of the enzymes could improve the swelling of the recycled fibres. In another study, Oksanen et al. (2000) found combining hemicellulases with the endoglucanase treatments increased the positive effects of the endoglucanases on pulp drainage. However, as a result of the endoglucanase treatments (high dose) a slight loss in strength was observed. Endoglucanase alone has been found to be more detrimental to strength properties than endoglucanase plus xylanase at a given level of cellulose hydrolysis.

Stork et al. (1995) found that presence of endoglucanase activity is a prerequisite for improving drainage of recycled fibres by enzymatic means. When cellobiohydrolase and xylanase activity were present, they acted synergistically with the endoglucanase to improve its effects. Mannanases were not helpful. The effect of cellulases on chemical pulp fibres and fibres containing lignin was quite different. Chemical pulp fibres were severely damaged and disintegrated completely on prolonged incubation whereas the strength properties of recycled mechanical pulp fibres were affected only to a small extent. Strength properties of different classes of primary and secondary pulps were not improved by an endoglucanase treatment. However, this treatment prevented further improvements in the breaking length by beating. Results for tear strength were found to be more complex. Endoglucanase-treated and beaten kraft pulps were reduced in tear strength whereas thermomechanical pulp and groundwood were not. Treated chemithermomechanical pulp followed by beating gave superior pulp strength than control pulps.

The most suitable method for upgrading recycled pulps is refining with an enzymatic treatment. Refining increases the burst and tensile resistance, and the enzymatic treatment produces better drainage results under certain conditions. Fibre that has been recycled more than once has lower papermaking qualities than virgin or once-recycled fibre (Pala, Mota, & Gama, 1998). By using an enzyme blend with recycled fibre, some lost freeness can be restored.

Pergalase is a blend of enzymes that improves the freeness of the fibre but does not reduce the fibre strength. The enzyme is effective at an optimum pH of 5.5–6 but remains active at pH 4.5–7. The optimum temperature is between 50 and 60°C. Enzymes need time to be effective. A 15-min retention time is adequate, providing there is good mixing. Trial results from three mills showed that machine speeds were increased when using Pergalase. The benefits of such an enzyme-enhanced drainage programme have been shown on grades including tube stock, gypsum linerboard and corrugating medium (Moran, 1996).

Pala et al. (2001) showed the effect of several glycanases on a secondary fibre pulp. Most enzymes improve drainability, which at the same time leads to a substantial decrease in strength properties. To minimise these losses, many factors have to be accounted for, the pulp composition and fibre structure probably being very important, as they greatly affect enzymatic performance. However, as evidenced by the fact that CBDs alone may modify the fibres' properties, it seems likely that protein adsorption may modify the interfacial properties in such a way that pulp drainability and paper strength become modified. Irrespective of this effect, the hydrolytic activity may strikingly affect the fibres, as demonstrated by the fact that different enzymes lead to different effects. Cellulases have quite detrimental effects on paper strength (Ecostone, Pergalase, P. ocitanis L/S). Xylanases, while improving drainability, cause no damage to the strength (Xylanase C482). Each pulp represents a new challenge, requiring a careful study to define the more effective enzyme and process. The data shows that no upgrade is possible without establishing a drainage/strength compromise. Low enzyme doses and short reaction periods seem to be advisable with the advantage that sugar solubilisation is lower. Actually, by treating old paperboard containers pulp with Celluclast 1.5L and Primalco G, it was possible to increase drainage with limited strength loss. Whatever the mechanism of fibre modification by the enzymes, it is very interesting to note that little time is required to introduce such modifications. It has also been shown that the response of fibres to enzymes depends on their length. Generally, the shorter fibres are more susceptible to enzymatic attack than longer ones, as they present a wider specific surface area to the enzymes. The use of CBDs may be a very attractive method for treating secondary paper pulps. Pala et al. (2001) reported for the first time the possibility of achieving a simultaneous increase in the resistance and drainage rate. The action of the CBDs on the fibres seems to be due to their adsorption to the fibre surface, which alters the fibre surface characteristics.

Stork and Puls (1995) and Stork et al. (1995) used isolated cellobiohydrolases and endoglucanases of *Penicillium pinophilum* to treat recycled pulps. They found that the action of endoglucanases was necessary to improve the drainage of recovered paper. The effect did not appear to be due to a selective hydrolysis of the fines fraction but was a consequence of the hydrolysis of amorphous cellulose on the surface of the fibres. Depending on the origin and history of primary and secondary fibres, the endoglucanase treatment decreased the strength properties to differing degrees. It has been reported by Pere et al. (1995) that endoglucanases dramatically decreased pulp viscosity. These enzymes appear to attack cellulose at sites where even a low level of hydrolysis reduces pulp viscosity, resulting in a marked deterioration of strength properties.

To improve the drainability of secondary fibre, treatment with cellulase alone generally leads to a loss in pulp brightness. Furthermore, when used in combination with a drainage aid polymer, the loss in brightness is even more pronounced. A method using a mixture of cellulase and pectinase has been described to accomplish the goal of simultaneously increasing freeness without a loss of brightness and physical properties (Olsen, Zhu, & Hubbe, 2000). As seen in Table 11.7, treatment with cellulase alone gave a very high improvement in CSF value (190 mL) but not the brightness, whereas treatment with pectinase alone resulted in greater

Table 11.7 Effect of Cellulase and Pectinase[a] on Drainage of Recycled Pulp

	Treatment			
Pulp Properties	**No Enzyme**	**Cellulase**	**Pectinase**	**Cellulase + Pectinase**
CSF (mL)	330	520	400	460
Brightness	80.7	81.6	82.3	82.3

[a]Both the enzymes were used at equal dose level of 1 L/tonne.
Source: Based on Olsen et al., 2000

Table 11.8 Cationic Polyacrylamide Requirement for Drainage Control of OCC With and Without Enzyme

Treatment	10s Filtrate Weight (g)				
	470	**540**	**600**	**650**	**700**
No enzyme treatment	0.0	0.13	0.60	0.84	1.10
After enzyme treatment	–	0.0	0.35	0.58	0.80

Source: Based on Shaikh & Luo, 2009

enhancement of brightness (1.6 points) but very little improvement in CSF (only 70 mL). However, treatment with a combination of both enzymes resulted in appreciable improvement in CSF (130 mL) and 1.6 points enhancement in brightness (the same as that with pectinase).

Hemicellulases have been found to increase the drainability of mechanical pulp (Karsila et al., 1990). Xylanase improves the freeness of deinked recycled pulp while having no detrimental effect on fibre tensile strength. By comparison, the tear indices of recycled pulps treated with cellulases decreased. These findings suggest that xylanases might be much more effective than cellulases or crude xylanase/cellulase mixtures. Xylanases, however, remove hemicelluloses, which promote interfibre bonding. This effect can also lead to poor paper properties

Functional additives are generally introduced to improve retention and drainage on the paper machine and the quality of the final product. The high specific surface area of the colloidal fraction of the furnish readily adsorbs a great amount of the additives without a perceptible benefit. The unproductive consumption of additives often necessitates compensation with higher additive dosing levels. Preferential endoglucanase activity upon the colloidal fraction is expected to modify the overall response of the furnish to functional additives (Cabrera et al., 1996). The impact of enzymatic treatment on the drainage of OCC pulp, conditioned with various levels of cationic polyacrylamide, was evaluated (Shaikh & Luo, 2009). It was found that the requirement of cationic polyacrylamide decreased greatly for the same drainage rate when pulp was treated with enzyme (Table 11.8).

Bhardwaj et al. (1997) studied the use of several chemical additives including modified polyacrylamides, starches and enzymes to improve the drainage and strength of secondary fibres containing corrugated kraft cutting and corrugated boxes. Drainage improved by 39.6% with enzyme over control without any appreciable change in pulp strength. However, the treatment of pulp with various chemical additives resulted in a significant improvement in drainage and strength. The best results were observed with an anionic polyacrylamide.

The interaction of various types of cellulase and polymer for increasing the freeness of a laboratory and mill furnish was examined by Sarkar et al. (1995). Both enzyme and polymer significantly increased the freeness of pulp. There was not much change in tensile and burst strength of handsheets. Although an independent treatment of pulp suspension, either with enzyme or polymer, can improve the freeness of pulp stock, a combination of lower doses of enzyme and polymer substantially increases the freeness. Treatment of recycled fibre with polymer alone can produce large flocs. However, by using enzyme with lower levels of polymer, a potentially more uniform sheet can be produced. Enzyme treatment followed by polymer addition could provide a new biochemical method for increasing the freeness of recycled fibre.

References

Bajpai, P. (2006). *Advances in recycling and deinking* (180pp). Leatherhead, Surrey, UK: Pira International.

Bajpai, P. (2012). *Biotechnology for pulp and paper processing* (414pp). New York: USA: Springer Science + Business Media.

Bajpai, P., & Bajpai, P. K. (2001). Status of biotechnology in pulp and paper industry. *Inpaper International* Indian Agro Papermill Association, 2001, July–December 29.

Bhardwaj, N. K., Bajpai, P., & Bajpai, P. K. (1995). Use of enzymes to improve drainability of secondary fibres. *Appita Journal*, *48*(5), 378–380.

Bhardwaj, N. K., Bajpai, P., & Bajpai, P. K. (1997). Enhancement of strength and drainage of secondary fibers. *Appita Journal*, *50*(3), 230–232.

Cabrera, C. F., Sarkar, J. M., Didwania, H. P., Espinoza, E., & Benavides, J. C. (1996). Paper mill evaluation of a cellulolytic enzyme and polymers for improving the properties of waste paper pulp: *Papermaker conference* (p. 481). Philadelphia: TAPPI Press.

Dienes, D., Egyházi, A., & Réczey, K. (2004). Treatment of recycled fiber with *Trichoderma* cellulases. *Industrial Crops and Products*, *20*, 11–21.

Din, N., Gilkes, N. R., Tekant, B., Miller, R. C., Warren, A. J., & Kilburn, D. G. (1991). Non-hydrolytic disruption of cellulose fibers by the binding domain of a bacterial cellulase. *Bio-Technology*, *9*(11), 1096–1099.

Eriksson, L. A., Heitmann, J. A., & Venditti, R. A. (1997). Drainage and strength properties of OCC and ONP using enzymes with refining: *Recycling symposium* (p. 423). Chicago, IL: TAPPI Press. April 14–16.

Eriksson, L. A., Heitmann, J. A., Jr., & Venditti, R. A. (1998). Freeness improvement of recycled fibres using enzymes with refining, enzymes applications in fiber processing. In K. E. L. Eriksson (Ed.), *ACS symposium series* (vol. 687, pp. 340). Washington, DC: American Chemical Society.

Fuentes, J. L., & Robert, M. (1986). French Patent 2,604,198.

Gill, R. (2008). Advances in use of fibre modification enzymes in paper making. *Conference Aticelca XXXIX congresso annuale*, Fabriano, Italy, 29–30 May.

Jackson, L. S., Heitmann, J. A., & Joyce, T. W. (1993). Enzymatic modifications of secondary fiber. *TAPPI Journal*, *76*(3), 147–154.

Jeffries, T. W. (1992). Emerging technologies for materials and chemicals from biomass. In R. M. Rowell, T. P. Schultz, & R. Narayan (Eds.), *ACS symposium series no. 476*, (p. 313).

Kantelinen, A., & Jokinen, O. (1997). The mechanism of cellulase/hemicellulase treatment for improved drainage. *Biological sciences symposium*, pp. 267–269.

Karsila, S., Kruss, I., & Puuppo, O. (1990). European Patent 351655 A 9001249004.

Levy, I., Nussinovitch, A., Shpigel, E., & Shoseyov, O. (2002). Recombinant cellulose crosslinking protein: A novel paper-modification biomaterial. *Cellulose*, *9*, 91–98.

Linder, M., & Teeri, T. T. (1997). The roles and function of cellulose binding domains. *Journal of Biotechnology*, *57*, 15–28.

Loosvelt, I. (2009). Current applications of fibre modification enzymes in the paper industry and future possibilities. *Fibre engineering* (39pp), Gothenburg, Sweden, 24–26 March.

Mansfield, S. D., & Wong, K. K. Y. (1999). Improving the physical properties of linerboard via cellulolytic treatment of the recycled paper component. *Progress Paper Recycle, 9*, 20–29.

Mansfield, S. D., Jong, E., Stephens, R. S., & Saddler, J. N. (1997). Physical characterization of enzymatically modified kraft pulp fibers. *Journal of Biotechnology, 57*, 205–216.

Moran, B. R. (1996). Enzyme treatment improves refining efficiency, recycled fibre freeness. *Pulp and Paper, 70*(9), 119.

Oksanen, T., Buchert, J., Pere, J., & Viikari, L. (1995). Biotechnology in the pulp and paper industry: Recent advances in applied and fundamental research. *Proceedings of the sixth international conference on biotechnology in the pulp and paper industry*, Vienna, p. 177.

Oksanen, T., Pere, J., Buchert, J., & Viikari, L. (1997). The effect of *Trichoderma reesei* cellulases and hemicellulases on the paper technical properties of never-dried bleached kraft pulp. *Cellulose, 4*(4), 329.

Oksanen, T., Pere, J., Paavilainen, L., Buchert, J., & Viikari, L. (2000). Treatment of recycled kraft pulps with *Trichoderma reesei* hemicellulases and cellulases. *Journal of Biotechnology, 78*, 39–48.

Olsen, W. L., Zhu, H., & Hubbe, M. A. (2000). Method of improving pulp freeness using cellulase and pectinase enzymes. U.S. Patent 6,066,233.

Pala, H., Mota, M., & Gama, F. M. (1998). Effects of enzymatic treatment and refining on the properties of recycled pulp. *Associacao Portuguesa dos Tecnicos das Industrias de Celulose e Papel*, 478.

Pala, H., Lemos, M. A., Mota, M., & Gama, F. M. (2001). Enzymatic upgrading of old paperboard containers. *Enzyme and Microbial Technology, 29*, 353–361.

Pala, H., Mota, M., & Gama, F. M. (2002). Enzymatic modification of paper fibres. *Biocatalysis and Biotransformation, 20*(5), 353–361.

Pala, H., Pinto, R., Mota, M., Duarte, A. P., & Gama, F. M. (2003). Cellulose-binding domains as a tool for paper recycling. In S. D. Mansfield & J. N. Saddler (Eds.), *Applications of enzymes to lignocellulosics, ACS symposium series* (pp. 105–115).

Payne, M. (1997). The increasing use of a strategic resource. *Recovered Paper, 39*(8), 34–38.

Pere, J., Siika-aho, M., Buchert, J., & Viikari, L. (1995). Effects of purified *Trichoderma reesei* cellulases on the fiber properties of kraft pulp. *TAPPI Journal, 78*, 71–78.

Pinto, J. R., Amaral, E., Costa, A. P., Gama, F. M., & Duarte, A. P. C. (2004). Improving papermaking with cellulose-binding domains. *Congresso iberoamericano de investigacion en cellulose y papel*, Vol. 3, Cordoba, pp. 303–305.

Pinto, R., Moreira, S., Mota, M., & Gama, M. (2004). Studies on the cellulose-binding domains adsorption to cellulose. *Langmuir, 20*(4), 1409–1413.

Pommier, J. C., Goma, G., Fuentes, J. L., Rousset, C., & Jokinen, O. (1990). Using enzymes to improve the process and the product quality in the recycled paper industry. Part 2: Industrial applications. *TAPPI Journal, 73*(12), 197–202.

Sarkar, J. M., Cosper, D. R., & Hartig, E. J. (1995). Applying enzymes and polymers to enhance the freeness of recycled fiber. *TAPPI Journal, 78*(2), 89.

Shaikh, H., & Luo, J. (2009). Identification, validation and application of a cellulose specifically to improve the runnability of recycled furnishes. *Proceedings of the ninth international technical conference on pulp, paper and allied industry* (Paperex'09). New Delhi, India, 4–6 December. pp. 277–283.

Shoseyov, O., Levy, I., Shani, Z., & Mansfield, S. D. (2003). Modulation of wood fibers and paper by cellulose binding domains. In S. D. Mansfield & J. N. Saddler (Eds.), *Applications of enzymes to lignocellulosics, ACS symposium series* (pp. 116–131).

Stork, G., & Puls, J. (1995). Recent advances in applied and fundamental research. In *Proceedings of the sixth international conference on biotechnology in the pulp and paper industry*. Vienna, p. 145.

Stork, G., Pereira, H., Wood, T. M., Dusterhoft, E. M., Toft, A., & Puls, J. (1995). Upgrading recycled pulps using enzymatic treatments. *TAPPI Journal*, *78*(2), 79.

Tomme, P., Boraston, A., McLean, B., Kormos, J., Creagh, A. L., Sturch, K., et al. (1998). Characterization and affinity applications of cellulose-binding domains. *Journal of Chromatography B*, *715*, 283–296.

Xiao, Z., Gao, P., Qu, Y., & Wang, T. (2001). Cellulose-binding domain of endoglucanase III from *Trichoderma reesei* disrupting the structure of cellulose. *Biotechnology Letters*, *23*, 7.

12 Effects of Recycled Fibre on Paper Machines*

12.1 Introduction

Papermaking pulps from recycled wastepaper perform differently on the paper machine than similar never-dried pulps. A main factor in the difference between once-dried and never-dried fibre is that dried fibres are homified, that is irreversible fibre bonding within the fibre wall occurs that resists re-swelling (Bouchard & Douek, 1994; Howard, 1990; Howard & Bichard, 1992; Lumiainen, 1994; McKee, 1971; Nazhad & Paszner, 1994; Phipps, 1994). Another factor in the runnability difference is that recycling depolymerises cellulose (Bouchard & Douek, 1994; Howard, 1990), generates fines and shortens the fibre length, so a lower freeness is observed (Howard & Bichard, 1992; McKee, 1971; Nazhad & Paszner, 1994; Phipps, 1994). Specifically, the drying process has been reported to decrease the specific volume and surface areas of pulp (Klungness & Caulfield, 1982). Dried fibres are also more difficult to wet because of their glassy crystalline nature compared with the highly amorphous, hydrophilic virgin fibre (Howard, 1990; Klungness & Caulfield, 1982; Nazhad & Paszner, 1994). In addition, the redistribution of olefinic compounds during drying results in self-sizing (Howard, 1990). The increased crystallinity of dried fibre also renders them more brittle and difficult to bond with other fibres (Howard, 1990). As a result of these changes in the fibre, the papermaker experiences problems with drainage and wet-web strength. An increase in the drainage and wet-web strength allows the papermaker to increase production rates and recycled fibre content levels (Ackermann, Gottsching, & Pakarinen, 2000).

12.2 Effects on Paper Machine Runnability

There can be numerous possible causes behind an individual machine break. Even though the causes fundamentally have physical, chemical or microbial background (or a combination of these), the actual break can be related to pitch deposition, various chemical deposits, stickies and other adhesives in recycled paper processes and microbiological slime (Haapala, Liimatainen, Korkko, Ekman, & Niinimaki, 2010). In addition, paper holes and spots, cuts and sheet breaks have reportedly been caused by shives or flakes, or bubbly gasses of droppings from the paper machine onto the web (Roisum, 1990a,b). Also, breaks due to mechanical problems, condensation

*Some excerpts taken from Bajpai (2006) with kind permission from Pira International, UK.

Recycling and Deinking of Recovered Paper. DOI: http://dx.doi.org/10.1016/B978-0-12-416998-2.00012-X

droplets and many more causes have been reported. Closing the water circuits of paper machines leads to a situation when the amounts of dissolved and colloidal substances increase (Kokko, Niinimaki, Zabihian, & Sundberg, 2004; Wearing, Barbe, & Ouchi, 1985). These materials are derived from wood constituents like lignin, hemicelluloses and extractives. Many additives have effects on water properties as well. Studies have shown that contaminants in whitewater decrease the strength of paper. Tay (2001) suggests that contaminants in whitewater make the fibrous material more hydrophobic and hinder the formation of bonds. This is seen as a reduction in paper strength. Based on the important role of chemical stability, the same phenomena can cause web defects and breaks at a paper machine. Haapala et al. (2010) have shown that elevated conductivity, charge and dissolved calcium levels increased the formation of defects on the paper machine. Chemical stability has widely been acknowledged as a key to undisturbed and clean paper production. Production of chemical deposits has been attributed, for example, to variations in the process pH, temperature and the charge (Haapala et al., 2010; Hubbe et al., 2006; Hubbe & Rojas, 2008; Kallio & Kekkonen, 2005; Sihvonen, Sundberg, Sundberg, & Holmblom 1998; Wathen, 2007). The importance of a stable process is increased when water systems are closed, filler usage is increased and the amounts of different substances in the process and water cycles are increased.

Drainage and pressing taken together are the dewatering steps on a paper machine (Bajpai, 2006). On a commercial paper machine, the fibre, minerals and other chemicals that will become the dry sheet of paper are delivered in a mixture called the furnish that contains about 99% water. The art of making paper is all about how much of, and how fast, this water is removed in various stages. In the forming section, the relative amount of water is reduced from about 99% to about 80%. Fibre strength, bonding potential and drainage rate of the pulp slurry influence paper machine runnability. In the wet end of the paper machine, the fibres are suspended in the whitewater along with a controlled amount of dissolved, colloidal and suspended solids, which are added to promote drainage, the formation of fibre bonds, sizing, opacity, colour and other desired properties. These solids accompany the fibres to the forming fabric. When recovered paper is re-pulped and added to the furnish, many nonfibrous solids are introduced into the papermaking process. These contaminants are removed from the system by physical separation, such as screening, washing and decker action. However, even the best cleaning methods remove only a percentage of these contaminants, because none of the available technologies are 100% efficient. So some of these solids enter the blend chest with the secondary fibre. This creates several problems for paper machine stock systems (Scott, 1993).

A heavy buildup of suspended solids is also observed. The heavy contaminants escape the centrifugal forward-flow cleaners and settle at the bottom of the chests. These contaminants can cause excessive wear on refiner plates, abrade the electropolished finish on the headbox and even damage the paper machine's clothing. However, this problem is relatively rare because the efficiency of stock preparation and of most forward-flow cleaners is quite high. If this type of problem occurs, the stock and water chests are emptied and washed out, the cleaners are inspected for wear and replaced if necessary. When the paper machine is started up again,

Table 12.1 Detrimental Effects Caused by Stickies

Formation of deposits
Web breaks in paper machine
Increased cleaning intervals in paper machine
Paper defaults
Paper breaks in printing press

consistencies, flows and pressure drops are carefully observed to ensure maximum efficiency from the forward-flow cleaners (Scott, 1993).

Experience has shown that proper preparation for the use of recycled fibre may help avoid disastrous results. Recycled fibre brings many positive characteristics to the sheet, ranging from improved formation in tissue and fine papers to improved plybond in paperboard. So the recycled fibre is not inherently bad, only the contaminants the furnish contains. There are several types of contaminant. Potentially tacky substances, called stickies, are now the most feared contaminants in recycled fibre pulps (Delagoutte, 2005). Because of their tacky character, stickies deposit on wires, felts, rolls and other moving parts, especially on the paper machine. Table 12.1 shows a variety of detrimental effects caused by stickies (Ackermann, Putz, & Gottsching, 1996). Faults occur in the paper primarily as thin areas or holes that can cause web breaks in a paper machine and a printing press. The result is longer non-productive time and cleaning time that adversely affect the productivity of the processes. There are several reasons why tacky contaminants have consequences that are more serious: a greater use of pressure-sensitive adhesive-coated papers at home and in the office, the increasingly widespread use of mailing campaigns and the common use of supplements and glued-in inserts in magazines. These ensure an increasing content of adhesives in recovered paper. In addition, increasingly more paper grades are being surface sized, coated or finished in similar ways.

Unless these interfering substances are continuously removed from the papermaking system in a controlled manner, they will accumulate and eventually lead to deposit and runnability problems (Song & Rys, 2004). Technology in place today is based on fixation of the pitch or stickies to the fibre before they have a chance to agglomerate, or to coat the pitch or stickies with a polymer that makes them non-tacky and therefore unable to agglomerate. Mills commonly use one of three chemical methods to control pitch and stickies deposits:

- Detackification
- Stabilisation
- Fixation

These methods are rarely combined because they may conflict with each other. In detackification, a chemical is used to build a boundary layer of water around the pitch and stickies to decrease depositability. Detackification can be achieved by the addition of pitch adsorbents such as talc and bentonite. However, these adsorbents can end up contributing to the pitch deposit problem if talc/pitch particles are unable to be retained in the paper sheet with surfactants and water-soluble polymers.

In stabilisation, surfactants and dispersants are used to enhance colloidal stability chemically and allow pitch and stickies to pass through the process without agglomerating or depositing. Fixation involves the use of polymers to fix pitch and stickies to the fibre and remove them from the whitewater system.

The interfering substances in papermaking are usually anionic in nature. Removal of anionic trash by reducing cationic demand with a cationic polymer is a way of deposit control through fixation. The advantage of using cationic polymeric coagulants for pitch and stickies control is that the pitch and stickies are removed from the system in the form of microscopic particles dispersed among the fibres in the finished paper product. There is still a need, however, for fixatives with increased pitch and stickies fixation power. Alum, starches and low molecular weight (MW) cationic coagulants conventionally used for deposit control can neutralise anionic trash and detrimental substances (pitch and stickies) and form complexes. However, they may not carry sufficient charge and/or MW to fix pitch and stickies complexes to the fibre. If not strongly fixed to the fibres, the complexes will concentrate in the system and lead to deposition problems. To solve these problems, Ciba Specialty Chemicals has developed several diallyldimethylammonium chloride (DADMAC) homopolymers and copolymers with high fixation power for pitch and stickies control. The new polymers include a high MW, structured polyDADMAC with high fixation power, copolymers of DADMAC with a comonomer capable of boosting the polymer's fixation power and terpolymers (polymers of DADMAC copolymerised with two comonomers which can provide synergic enhancement of fixation power to the polymer).

The stickies that are usually present in recycled fibre pulps do not necessarily need to cause problems. Macroscopic contaminants above a certain size (approximately >150 μm) can be effectively removed from a recycled pulp slurry by using suitable screening techniques. Finely dispersed microstickies (size approximately <100 μm) pass through the screens with the accept and cause the aforementioned problems. These stickies are most feared because they form in part during pulp processing by secondary reactions with other components of the pulp or whitewater and are therefore difficult to control. One way to avoid these problems is to wash the recovered pulp and then purify the washing filtrate. Another possible solution is to add talc to deactivate the tacky surface. Because stickies are lighter than water, they tend to collect at the high water lines in stock and whitewater chests. Subsequent surges in chest level can free up masses of stickies, which then enter the stock flow and proceed to a secondary pulper, screen or cleaner, where shear forces break them down even further. The resulting particulate materials move closer to the headbox and eventually enter the forming section with the thin stock flow. Deposits of stickies can collect on internal headbox surfaces, forming fabrics, wet press felts, press rolls, dryer felts, dryer cans and other papermaking equipment. Contamination by stickies adds to existing problems with deposits caused by pitch, barium sulphate, microorganisms and fillers. One solution is to increase the frequency of boilouts to prevent problems associated with sheet contamination and flow the thin stock flow. Deposits of stickies can collect on internal headbox surfaces, forming fabrics, wet press felts, press rolls, dryer felts, dryer cans and other papermaking equipment. Contamination by stickies adds to existing problems with deposits caused by pitch, barium sulphate, micro-organisms and fillers. One solution is to increase the frequency of boilouts to prevent problems associated with sheet contamination and flow disturbances due to deposits in the screens, approach piping and former. Boilout solution chemistry should include high alkalinity, surfactants and sequestrants.

Some mills have found that a solvent action helps to remove latex, pitch and waxy components. Maintaining a high system temperature during the boilout is also important. Proper washing and rinsing should restore the paper machine to an acceptable degree of cleanliness. The deposition of these materials, including heavy stickies, onto forming fabrics creates troublesome sheet holes. Fabrics can be kept free of stickies by applying a low-concentration aqueous solution of a low MW cationic polymer on the face side of the fabric just ahead of the breast roll. The thoroughly mixed polymer can be applied most effectively using a low-pressure, low-volume fan shower designed to give double coverage. The slightly anionic polyester yarn of the fabric attracts the cationic polymer, which in turn reacts with anionic colloids in the system.

Dissolved solids can cause changes in system chemistry on the paper machine. The nature of dissolved solids varies for each source of recycled fibre. Alkaline materials causing pH changes on the paper machine have the greatest impact on operations. Testing the pH and/or anionic or cationic charge of whitewater is the most common way to monitor these materials (Smith, Dorn, & Orr, 1992). Some of the more common dissolved solids are starches, silicates and proteins used in the manufacture of adhesives. Another group comprises the additives used in papermaking, including alum, synthetic polymers, dyes, acids and bases. The wood tissue itself contributes disaccharides, soluble hemicelluloses, soluble organic acids, tannins, etc. This aggregate of chemicals also contains chlorides, sulphates and other anions known to cause corrosion. In some cases the limits to which the whitewater system can be closed are determined by the point where catastrophic corrosion occurs, measured in litres of effluent per tonne of production.

Dissolved solids introduced by secondary fibre can also complicate the wet-end chemical objectives of the papermaker (Linhart, 1988; Tremont, 1995). In ionic form, these solids can provide counterions that stabilise additives and limit their ability to be adsorbed onto the surfaces of fibres and fines. In non-ionic form, they can sterically hinder interactions, such as dyeing, sizing and retention. Dissolved solids must be pumped from the system if their concentration begins to affect machine runnability or paper quality. This means discharging whitewater and replacing it with fresh water. A thorough study is required to determine the best point for adding the fresh water.

Microbial contamination of recycled pulps caused by bacteria and fungi can be the cause of production disturbances and quality problems (Blanco & Gaspar, 1998, chap. 6). Paper mills provide the natural conditions of nutrient supply, temperature and moisture for the breeding of slime-forming micro-organisms, which can cause spoilage of raw materials and additives (pulp, mechanical pulp, recycled paper, starch, filler and pigment dispersions) (Bennett, 1985; Bjorklund, 2000, 2002; Blanco, Negro, Gaspar, & Tijero, 1996; Blanco, Negro, Monte, & Tijero, 2002;

Brown & Gilbert, 1993; Chaudhary, Gupta, Gupta, & Banerjee, 1997; King, 1990; Kulkarni, Mathur, Jain, & Gupta, 2003; Safade, 1988). Intense growth of spore-forming bacteria can cause problems with the hygienic quality of the end products in manufacturing of food-quality packaging, paper and board. Volatile, malodorous metabolic products of microbes may enter the end products. According to Salzburger (1996), the most important economic loss caused by microbes in paper machines arises from the growth of biofilms, namely slime layers on the machine surfaces that are mostly made of stainless steel. Unless microbial growth is controlled, these tiny organisms can bring the large, modern, computer-controlled, hi-tech paper machines – up to 9 m wide and 200 m long – to a standstill (Salzburger, 1996). The reuse of whitewater inoculates previously sterile parts of the system, spreading and magnifying the problem. If the growth of these micro-organisms is unchecked for too long, the deposits they cause may dislodge and be carried into the paper, lowering product quality. The slime may also cause breaks in the paper at the wet end, interfere with the free flow of stock or cause excessive downtime for cleaning (Blankenburg & Schulte, 1997; Jokinen, 1999). Microbial loads are usually due to organic contamination that enters the stock preparation system through the recovered paper or during storage of the recycled pulp. In addition, tightly closed process water loop systems with a high concentration of biologically degradable, dissolved organic substances favour the growth of micro-organisms. These lead to the covering of machine components and pipe systems with slime. Detaching slime clots disturb the sheet formation process and can even lead to losses of production due to web breaks. Because microbial respiration consumes the oxygen contained in the process water, anaerobic conditions occur and organic acids are produced that cause odour emissions and corrosion problems. A microbial contamination of foodstuff by recycled fibre containing packaging paper can be excluded because microbes do not migrate from the paper on dry or moist food (Danielsson, 1998). The measures for containing microbial activity involve the addition of biocides (Yang, Chen, Chen, & Zhao, 2002) and thermal stock treatment. Temperatures of 80°C can achieve considerable decontamination.

The fibre length distribution of secondary fibre is distinctly different from that of virgin pulp. The fibres in secondary fibre furnishes are also limited in the degree to which they can be swollen, because their previous drying causes some irreversible hydrogen bonding in the amorphous structure of the cell wall and near the surfaces of the crystallites (Nazhad & Paszner, 1994). This plays an important role in machine runnability. During sheet formation and in the press section, fines reduce the water permeability of the web by filling voids, which effectively blocks the passage of water through the sheet in the *z*-direction. This slows drainage on the forming fabric and can lead to problems related to sheet consolidation, such as lack of squareness, linting and impaired strength development. In the press section, this loss of water permeability limits the amount of water that can be removed from the sheet in the press nip. This can lead to crushing, sheet holes, press section breaks and a lower consistency sheet delivered to the first dryer. A lower consistency sheet increases the energy cost of drying, hence it significantly affects the direct manufacturing costs. Problems in the dryer section result from the release of fines into dryer fabrics and cans. Here fines often combine with stickies to form a tenacious deposit. Stickies on the dryer fabric or dryer can pull fines or fibre fragments out of the sheet. These fines then dry and attract more stickies. This process continues until a composite is formed. There are several actions that can be taken to solve these problems. Fabric drainage can be improved by the following:

- Changing the design of the fabric
- Increasing the vacuum
- Raising the temperature of the stock in the headbox
- Using polymeric drainage aids.

12.2.1 Performance of Enzymatically Deinked Recovered Paper on Paper Machine Runnability

Fibre strength, bonding potential and drainage rate of the pulp slurry influence paper machine runnability. The application of cellulases on pulp to enhance the freeness has been researched and patented by several research groups (Bhatt, Heitmann, & Joyce,1991; Fuentes & Robert, 1986; Jackson, Heitmann, & Joyce, 1993; Pommier, Goma, Fuentes, Rousset, & Jokinen, 1990; Sarkar, Cosper, & Hartig, 1995). Sarkar et al. (1995) showed that cellulases together with a polymer treatment enhanced the freeness of old corrugated container (OCC) and linerboard/newspaper furnishes without a significant loss in tensile or burst strength. Pommier et al. (1990) reported that cellulase-treated mixed papers containing OCC had better drainage properties while maintaining strength. Bhatt et al. (1991) had similar results with bleached and unbleached recycled softwood pulp. In addition, Pommier et al. (1990) reported that, on an industrial scale, cellulase action increased the freeness to the point where the headbox consistency could be decreased or the machine speed could be increased. Stork et al. (1995) and Stork and Puls (1995) explained that the drainage enhancement by cellulase action was a result of the hydrolysis of amorphous cellulose on the surface of the fibres, and not a selective hydrolysis of fines. However, Jackson et al. (1993) found a selective hydrolysis of the fines fraction that resulted in a larger, longer fibre content of cellulase-treated wastepaper compared with control. Forest Products Laboratory (FPL) researchers (Rutledge-Cropsey, AbuBakr, & Klungness, 1998) reported on the drainage and wet-web strength results of pilot paper machine runs using pulp that was enzymatically deinked on an industrial scale in a cooperative effort between Voith Sulzer, the Forest Products Laboratory and the National Council for Air and Stream Improvement. Three deinking trials were conducted at the Voith Sulzer pilot plant in Appleton, WI, USA. Approximately 3300 kg of sorted high-value office paper consisting of at least 90% laser-printed white paper was used in each of the trials. The recovered paper had a very low coloured-paper content and was approximately 12% ash. The first of the three deinking trials was the control. Heat-killed enzyme was used with surfactant as deinking aids in a high-consistency pulper. The enzyme was a commercial cellulase. In the other two trials, active enzyme and surfactant were used as deinking aids. However, the latter of the two active enzyme trials was the most successful in deinking efficiency. All three deinking trials consisted of the sequence of high-consistency pulping, screening, flotation, cleaning screening, washing and pressing to form wet lap. In general, the enzyme-treated pulp ran better than the control pulp (Tables 12.2–12.4). Mainly, it was the enhanced drainage and wet-web strength that contributed to the improvement in runnability.

Enzymatic treatment of recycled pulp enhances drainage and wet-web strength and decreases vacuum requirements without a loss in dry strength. It has been reported

Table 12.2 Effect of Enzyme on Paper Machine Runnability, Specifically Drainage

	Paper Machine Run		
	Control	**Enzyme 1**	**Enzyme 2**
Machine chest consistency (%)	3.2	3.4	3.0
Machine chest freeness (mL)	440	535	515
Headbox consistency (%)	0.73	0.42	0.38
Headbox flow rate (L/min)	90	128	149
Total drainage rate (L/min)	46	61	107
Total drainage (%)	51	48	72

Source: Based on Rutledge-Cropsey et al. (1998)

Table 12.3 Effect of Enzyme on Pressing Response

Solids (%)	Paper Machine Run		
	Control	**Enzyme 1**	**Enzyme 2**
Couch	26.7	26.9	26.8
First press	38.8	37.5	37.5
Second press	1.91	2.17	2.25
Third press	2.10	2.49	2.45

Source: Based on Rutledge-Cropsey et al. (1998)

Table 12.4 Effect of Enzyme on Wet-Web Tensile Strength

Strength·(m/g)[a]	Paper Machine Run		
	Control	**Enzyme 1**	**Enzyme 2**
Couch	0.93	1.05	1.07
First press	1.78	1.90	1.96
Second press	1.91	2.17	2.25
Third press	2.10	2.49	2.45

[a]Values are normalised to the solids level of the control run.
Source: Based on Rutledge-Cropsey et al. (1998)

that cellulases break down the cellulose that has a high affinity to water but do not contribute to the overall hydrogen bonding potential of the fibres (Pommier et al., 1990). This would explain the increase in drainage without a change in sheet strength (Bajpai, 2012). The increase in wet-web strength with enzyme treatment may be a sign of increased fibre swelling and associated water. These effects will improve paper machine runnability without sacrificing product quality. It is possible that refining will enhance wet-web stretch and dry strength. Compared with conventional deinked pulp, enzymatically deinked pulp will lead to better paper machine runnability without a loss in product quality. Strength and runnability may be enhanced even more after refining the enzymatically deinked pulp. However, additional studies are needed to understand fully the effects of refining on the runnability of enzymatically deinked pulp.

12.3 Effect on Sheet Properties

Use of secondary fibre to the furnish may affect the sheet properties in following ways:

- Residual contamination can affect sheet properties.
- Shorter fibres and higher fines content of re-pulped recovered paper can also affect sheet properties.

Screening and cleaning can remove only a few contaminants, and some will move forward to the headbox with the flow of the secondary fibre stock. The dirt count of the sheet will increase if this contamination is in the form of visible suspended particulate materials. The increased dirt detracts from the visual quality of the paper, particularly if it is to be printed. If the dirt contains a lot of stickies, adjacent wraps of a roll can adhere to each other. This will cause breaks in the pressroom, carton manufacturing and other converting operations. If the dirt contains a lot of metal, it can set off metal detectors in inspection lines for food packaging.

Several methods have been explored to reduce the number of contaminant particles and their size:

- Deactivating the surface of dispersed latex-type particles to limit the growth of large particles from collisions of smaller particles.
- Attaching small particles to the surfaces of fibres and fines via charge modification, usually in conjunction with deactivation, using chemical means to disperse large particles to a size suitable for emulsification.
- Agglomerating particulate materials into particles large enough to be removed by pressure screens and centrifugal cleaners (Smith & Bunker, 1993).

Sheet holes or light spots are caused when hard contaminants are carried with the sheet into the press section, where hydraulic forces, acting in a lateral direction, move fibres away from the solid. Sheet holes are intolerable regardless of their origin and have to be corrected. Wire contamination can be prevented by the following:

- Aqueous polymer treatments
- Wire-cleaning systems that use heated chemical solutions
- Using yarns produced from polymers designed to repel stickies.

Sheet holes originating in the press section create a more difficult problem. Solutions include the following:

- Continuous felt cleaning
- Continuous treatment of the face side of the felt with an aqueous polymer to prevent stickies from adhering to the yarn.

Surface picking results from a buildup of contaminants on lump breakers, centre rolls on binip presses, top press rolls on conventional single felt presses, dryer cans, breaker stack rolls or calender stack rolls. It detracts from the visual quality of the sheet, particularly in printing and writing grades. Surface picking is aggravated by the presence of fines or recycled fibres on the sheet surface, because they are too short to become mechanically entangled in the fibre web. Roll doctors, the aqueous polymer treatment of lump breakers and press rolls, and routine cleaning of dryer fabrics and cans will all mitigate the effects of secondary fibre on picking.

Contaminants, such as visible ink specks, reduce brightness and change the shade of newsprint and printing and writing grades. The maximum drop in brightness occurs with very small amounts of small ink particles. The losses are greater with higher initial brightness of the paper. Although a printing ink proportion of 0.1% in a newsprint causes a decrease of 16%, the brightness of a white paper falls by a total of 35%. A doubling of the amount of residual ink therefore results in only minor optical losses (Jordan & Popson, 1993). The reduction in brightness also depends on the size of the residual ink particles. Smaller particles have a greater specific surface area than larger particles, so they have a greater influence on brightness losses caused by light absorption. In addition to this optical phenomenon, the so-called light-spreading effect causes further light absorption with a drop in particle size. According to this theory, light penetrates below the printing ink particles because of the scattering within the paper structure and is absorbed there too.

Larger ink particles affect the optical homogeneity of the recycled fibre pulp and make it appear contaminated. The particles that are visible as specks are due to ink residues as particles or agglomerates detached from the fibres but not removed and due to printing ink still attached to the fibres. Difficult-to-deink printing inks, especially non-impact inks, contribute to a greater speck concentration and impair the optical quality of the finished product. Besides optical effects, residual ink particles can also have a detrimental effect on runnability (Selder, 1997). Conventional offset and rotogravure printing inks lead to deposits in the wet end of the paper machine. The effects of toner particles from non-impact printing processes are more evident in the dry end as coated felts or fabrics and rolls. The consequences can be measured not only by the decreased efficiency of process stages but also by faults in the paper that can lead to web breaks.

Secondary fibre contaminants also affect sizing. Because the re-pulping stages of many stock preparation systems are alkaline, the conventional sizes used in the original recovered paper are lost and must be replaced. Another sizing effect observed in sheets containing secondary fibre can be attributed directly to contamination with waxy substances. Residual contaminants in paper made from secondary fibre can also affect strength development, although the effect of fibre properties on sheet strength is far more important.

Sheet properties are also affected by fibres and fines. These have the greatest effect on mechanical properties. The contribution of fibre bonding to strength development and the role of fibre swelling in the formation of interfibre bonds has been reported. Swollen fibres collapse and come into intimate contact with adjacent fibres, which promotes the formation of hydrogen bonds. Excluding chemical dispersion, the most highly swollen fibres are found in never-dried pulp furnishes refined under slightly alkaline conditions. Furnishes derived from dried then re-pulped paper produce a sheet with poorer strength properties, because lamellae are brought together as the fibre cell wall loses water in the initial drying process. These lamellae are bonded in planes so tight that some of the regions will not admit water molecules when the fibre is later submerged in liquid water. The macroscopic effect is that the fibre never recovers its original swollen diameter, so it is not as flexible or conformable as before drying. This lack of ability to swell completely and to form intimate contacts with neighbouring fibres, namely to develop tightly bonded crossings, limits the strength of paper and board made from secondary fibre.

Sizing is affected by fines and fibres when they bring adsorbed anionics to the system. Chemical changes gradually occur in the fibres during the conversion of paper and board into printed matter, folding cartons, corrugated containers and other products; during the exposure of these products to heat, sunlight and water vapour; and through other conditions related to their end use. Residual alum and acids provide the catalysts for hydrolysis of newsprint. Losses can be expected in tensile strength, edgewise compression strength, interlaminar shear strength, plybond strength and burst strength. The stiffness and modulus of elasticity decrease. To offset losses in strength and stiffness, papermakers have devised several strategies:

- Chemical swelling in the fibre by using alkaline pulping conditions
- Fractionating the stock to recover the longer fibres for grades that require higher strength, increasing the refining energy per tonne
- Increasing the press loading
- Modifying the forming and pressing areas so they deliver a dried sheet to the first dryer. Although this also increases the sheet density, it is an economical way to restore lost strength properties and has the side benefit of reducing energy cost per tonne for drying.

Chemicals can be used selectively that will do the following:

- Improve drainage
- Increase retention
- Fibre bonding
- Reduce the amount of anionic dissolved solids in the system.

The organic and inorganic chemicals added in recovered paper treatment processes are transferred to the paper machine and its whitewater system. Not only do they trigger the formation of deposits, or secondary stickies, on the paper machine, but they also contribute to sheet properties related to surface chemistry. This is especially true with surfactants and soaps of the deinking process (Freeland & Gess, 1995; Guest, 1996). Besides the formation of stickies, an excessive carry-over of deinking soaps can interfere with fibre bonding and may relate to low surface friction that can increase layer-to-layer slippage when winding a reel. If low kinetic friction (sliding) of newsprint coincides with a high density of a wound reel, it might lead to crepe wrinkles (Lucas & Williams, 1997). Calcium soaps of the flotation process also adhere to fibres and are retained through a combination of mechanical entrapment and adsorption, resulting from the balance between van der Waals, electrostatic and hydrodynamic forces. These retention phenomena were studied by Paprican researchers (Fernandez & Garner, 1997) with pure fatty acid soaps and fibres. It has been reported (Haynes & Marcoux, 1997; Holmback, 1995; Krook, 1997) that soap carry-over with DIP is 0.6–13 kg/tonne. The average is about 2 kg/tonne. The design and operation of the deinking process, pH and deinking chemistry influence the carry-over. Sometimes a change from calcium soap chemistry to fatty acid emulsions or non-ionic surfactants in neutral deinking has been explained by an intention to avoid carry-over to the paper machine. Part of the great variation in carry-over reported in the literature probably comes from different methods of analysis.

A large variety of wet-end chemicals are used in modern papermaking. These chemical can remain in the finished products and finally return with the recovered paper to the papermaking system. In particular, cationic polymers are attached to furnish solids, although their exact location and mode of attachment are not known. An interesting question is whether the chemicals added initially stay on fibres in the deinking process and affect the functioning of fresh chemicals added in later papermaking stages. Probably the most common case is the use of cationic wet-end chemicals as fixatives to bind anionic trash to the sheet.

Cationic polymers such as starch or alkyl ketene dimer size can often also remain adsorbed on fibres during recycling (Sjostrom & Odberg, 1997). Some cationic long-chain dry-strength agents can lose their binding potential in a subsequent papermaking cycle even when they are well retained in the sheet (Grau, Schuhmacher, & Kleemann, 1996). The mode of attachment of those long polymers to fibres had probably changed significantly. Because of the increasing use of cationic additives in papermaking and the good preservation of their functional effectiveness during recycling, progressive accumulation can also be expected in recycled fibre. If a system gradually becomes over-cationised, these additives can also act as dispersing agents in suspensions because of cationic electrostatic repulsion.

Kruger, Gottsching, and Monch (1997) investigated the release of model substance (lignosulphonate) originally fixed by cationic additives during paper disintegration. Detachment of anionic substances was found to be very small at neutral pH, but the release was considerable above pH 9. This is a normal pH in alkaline deinking and bleaching. Washout of fixed anionic trash and old additives therefore seems to be possible in the deinking process if water management applies the countercurrent principle. The model substances became reattached to the fibres when the pH decreased (Kruger et al., 1997). This suggests that the interactions between fibres and still positively charged old cationic polymers can be restored when there is a decrease in the repulsive effects caused by high pH.

References

Ackermann, C., Gottsching, L., & Pakarinen, H. (2000). Papermaking potential of recycled fibers. In L. D. Gottsching & H. Pakarinen (Eds.), *Papermaking science and technology* (Vol. 7, pp. 357). Helsinki, Finland: Fapet Oy.

Ackermann, C., Putz, H. J., & Gottsching, L. (1996). Origin and content of adhesive impurities and stickies in graphic waste papers. *Wochenblatt für Papierfabrikation*, *124*(11/12), 508.

Bajpai, P. (2006). *Advances in recycling and deinking*. UK: Pira International. (180pp.)

Bajpai, P. (2012). *Biotechnology for pulp and paper processing*. New York, USA: Springer Science + Business Media. (414pp).

Bennett, C. (1985). Control of microbial problems and corrosion in closed systems. *Paper Technology and Industry*, *26*, 331–335.

Bhatt, G., Heitmann, J. A., & Joyce, T. W. (1991). Novel techniques for enhancing the strength of secondary fiber. *TAPPI Journal*, *74*(9), 151.

Bjorklund, M. (2000). Alternatives to conventional biocides in the pulp and paper industry. *IPPTA*, *12*(4), 1–4.

Bjorklund, M. (2002). Process water cleaning and slime control in tissue and food grade mills. *Seventh international conference on new available technologies*, Stockholm, pp. 216–218.

Blanco, A., Negro, C., Monte, C., & Tijero, J. (2002). Overview of two major deposit problems in recycling: Slime and stickies. Part 1: Slime problems in recycling. *Progress in Paper Recycling*, *11*(2), 14–25.

Blanco, M. A., & Gaspar, I. (1998). Microbial aspects in recycling paper industry. In M. A. Blanco, C. Negro, & J. Tijero (Eds.), *COST E1: Paper recycling: An introduction to problems and their solutions* (pp. 151). Luxembourg: Office of Official Publications of the European Communities.

Blanco, M. A., Negro, C., Gaspar, I., & Tijero, J. (1996). Slime problems in the paper and board industry. *Applied Microbiology and Biotechnology*, *46*, 203–208.

Blankenburg, I., & Schulte, J. (1997). An ecological method for slime and deposit control. *Pulp and Paper International*, *39*(6), 67–69.

Bouchard, J., & Douek, M. (1994). The effects of recycling on the chemical properties of pulps. *Journal of Pulp and Paper Science*, *20*(5), J131.

Brown, M. R. W., & Gilbert, P. (1993). Sensitivity of biofilms to antimicrobial agents. *Journal of Applied Bacteriology (Symposium Supplement)*, *74*, 87S.

Chaudhary, A., Gupta, L. K., Gupta, J. K., & Banerjee, U. C. (1997). Study of slime forming organisms of a paper mill-slime production, characterization and control. *The Journal of Industrial Microbiology and Biotechnology*, *18*, 348–352.

Danielsson, K. (1998). *Analysis of detrimental substances in two newsprint mills*. MSc thesis, Turku.

Delagoutte, T. (2005). Management and control of stickies. *Progress in Paper Recycling*, *15*(1), 31.

Fernandez, C., & Garner, G. (1997). Retention of fatty acid soaps during recycling. Part I: A study using packed beds of pulp fibres. *Journal of Pulp Paper Science*, *23*(4), J144.

Freeland, S. A., & Gess, J. M. (1995). What is the effect of residual deinking chemicals on the wet-end chemistry of the papermaking process and on paper properties? *Progress in Paper Recycling*, *4*(2), 107.

Fuentes, J. L., & Robert, M. (1986). French Patent 2,604,198.

Grau, U., Schuhmacher, R., & Kleemann, S. (1996). The influence of recycling on the performance of dry strength agents. *Wochenblatt für Papierfabrikation*, *124*(17), 729.

Guest, D. (1996). Chemical deposits should be tracked down and destroyed. *Pulp and Paper*, *1*(4), 18.

Haapala, A., Liimatainen, H., Korkko, M., Ekman, J., & Niinimaki, J. (2010). Runnability upgrade for a newsprint machine through defect identification and control – A mill case study. *Appita annual conference and exhibition*, Melbourne.

Haynes, D., & Marcoux, H. (1997). Evaluation of fatty acid carryover in North American newsprint deinking mills. *CPPA fourth research forum on recycling*, Quebec.

Holmback, A. (1995). *Analys Av Fettsyratvalar Och Vedharts I En Returpappersanlaggning.* MSc thesis, Turku.

Howard, R. C. (1990). The effects of recycling on paper quality. *Journal of Pulp and Paper Science*, *16*(5), J143.

Howard, R. C., & Bichard, W. (1992). The basic effects of recycling on pulp properties. *Journal of Pulp and Paper Science*, *18*(4), J151.

Hubbe, M. A., Rojas, O. J., & Venditti, R. A. (2006). Control of tacky deposits on paper machines – A review. *Nordic Pulp and Paper Research Journal*, *21*(2), 154–171.

Hubbe, M. A., & Rojas, O. J. (2008). Colloidal stability and aggregation of lignocellulosic materials in aqueous suspension: A review. *Bioresources*, *3*(4), 1419–1491.

Jackson, L. S., Heitmann, J. A., & Joyce, T. W. (1993). Enzymatic modifications of secondary fiber. *TAPPI Journal*, *76*(3), 147–154.

Jokinen, K. (1999). Paper machine microbiology. In J. Gullichsen, H. Paulapuro, & L. Neimo (Eds.), *Papermaking chemistry: volume 4 of papermaking science and technology series* (pp. 252–267). Helsinki, Finland: Fapet Oy.

Jordan, B. D., & Popson, S. J. (1993). Measuring the concentration of residual ink in recycled newsprint. *Paper presented at the second CPPA research forum on recycling*, Ste-Adele.

Kallio, T., & Kekkonen, J. (2005). Fouling in the paper machine wet end. *TAPPI Journal*, *4*(10), 20.

King, V. M. (1990). Microbial problems in neutral/alkaline paper machine systems. *Neutral-alkaline papermaking short course*, Orlando, FL, (pp. 211–216).

Klungness, J. H., & Caulfield, D. (1982). Mechanisms affecting fiber bonding during drying and aging of pulps. *TAPPI Journal*, *65*(12), 94.

Kokko, S., Niinimaki, J., Zabihian, M., & Sundberg, A. (2004). Effects of white water treatment on the paper properties of mechanical pulp – A laboratory study. *Nordic Pulp and Paper Research Journal*, *19*(5), 386–391.

Krook, K. (1997). *Stickies and wood resin in a deinking plant and on a paper machine*. MSc thesis, Turku.

Kruger, E., Gottsching, L., & Monch, D. (1997). The behaviour of fixed disturbing substances in the recycling process. *Wochenblatt für Papierfabrikation*, *125*(20), 986.

Kulkarni, A. G., Mathur, R. M., Jain, R. K., Gupta, A. (2003). Microbial slime in papermaking operations: Problems, monitoring and control practices. *IPPTA convention issue*. Mumbai, (pp. 121–125).

Linhart, F. (1988). Detrimental substances – New aspects of their action, new practical solutions of the problem. *Wochenblatt für Papierfabrikation*, *116*(9), 329.

Lucas R. G., & Williams, R. T. (1997). Winding problems associated with recycled fibers and additives in today's papers. *Tappi finishing and converting conference*, Atlanta, GA.

Lumiainen, J. (1994). Refining – A key to upgrading the papermaking potential of recycled fibre. *Paper Technology*, *35*(7), 41.

McKee, R. C. (1971). Effect of repulping on sheet properties and fiber characteristics. *Paper Trade Journal*, *155*(21), 34.

Nazhad, M. M., & Paszner, L. (1994). Fundamentals of strength loss in recycled paper. *TAPPI Journal*, *77*(9), 171.

Phipps, J. (1994). The effects of recycling on the strength properties of paper. *Paper Technology*, *35*(6), 34.

Pommier, J. C., Goma, G., Fuentes, J. L., Rousset, C., & Jokinen, O. (1990). Using enzymes to improve the process and the product quality in the recycled paper industry. Part 2: Industrial applications. *TAPPI Journal*, *73*(12), 197–202.

Roisum, D. R. (1990a). Runnability of paper. Part 1: Predicting of runnability. *TAPPI Journal*, *73*(1), 74.

Roisum, D. R. (1990b). Runnability of paper. Part 2: Troubleshooting web breaks. *TAPPI Journal*, *73*(1), 101–106.

Rutledge-Cropsey, K., AbuBakr, S. M., & Klungness, J. H. (1998). Performance of enzymatically deinked wastepaper on paper machine runnability. *TAPPI Journal*, *81*(2), 148–151.

Safade, M. L. (1988). Tackling the slime problem in a paper mill. *Paper Technology and Industry*, *29*, 280–285.

Salzburger, W. (1996). Paper and additives. In E. Heitz, H. C. Flemming, & W. Sand (Eds.), *Microbially influenced corrosion of materials* (pp. 415–427). Berlin, Germany: Springer-Verlag.

Sarkar, J. M., Cosper, D. R., & Hartig, E. J. (1995). Applying enzymes and polymers to enhance the freeness of recycled fiber. *TAPPI Journal*, *78*(2), 89.

Scott, W. E. (1993). A survey of the various contaminants present in recycled waste paper white water systems. *Paper presented at Tappi deinking short course*, Indianapolis, IN.

Selder, H. (1997). Improving the cleanliness of secondary fibre stocks. *Papier*, *51*(9), 455.

Sihvonen, A. L., Sundberg, K., Sundberg, A., & Holmblom, B. (1998). Stability and deposition tendency of colloidal wood resin. *Nordic Pulp and Paper Research Journal*, *8*(1), 233.

Sjostrom, L., & Odberg, L. (1997). Influence of wet-end chemicals on the recyclability of paper. *Papier*, *51*(6A), V69.

Smith, W. E., & Bunker, B. L. (1993). The impact of secondary fiber on paper machine. In R. J. Spangenberg (Ed.), *Secondary fiber recycling (pp. 29)*. Atlanta, Georgia, USA: TAPPI Press.

Smith, W. E., Dorn, W. P., & Orr, T. W. (1992). Optimizing wet end chemistry. *Paper presented at Tappi*, Atlanta, GA.

Song, Z., & Rys, L. (2004). Keep on running: How innovation in fixatives can help runnability. *Pulp and Paper International*, *46*(12), 25.

Stork, G., Pereira, H., Wood, T. M., Dusterhoft, E. M., Toft, A., & Puls, J. (1995). Upgrading recycled pulps using enzymatic treatment. *TAPPI Journal*, *78*(2), 79.

Stork, G., & Puls, J. (1995). *Proceedings of the sixth international conference biotechnology in the pulp and paper industry: Recent advances in applied and fundamental research.* Vienna, (p. 145).

Tay, S. (2001). Effect of dissolved and colloidal contaminants in newsprint machine white water surface tension and paper physical properties. *TAPPI Journal*, *84*(8), 43.

Tremont, S. R. (1995). Impact of residual deinking chemicals on newsprint machine runnability. *CPPA 81st annual meeting*, Montreal.

Wathen, R. (2007). Requirements for improved runnability at papermachines and in printing. *Ipw*, *5*, 58–64.

Wearing, J. T., Barbe, M. C., & Ouchi, M. D. (1985). The effect of white water contamination on newsprint properties. *Journal of Pulp and Paper Science*, *11*(4), J113–J121.

Yang, R., Chen, K., Chen, X., & Zhao, Y. (2002). Biocides: Solve the microbial contaminant problems in papermaking process. *China Pulp Paper*, *21*(2), 49.

13 Control of Stickies*

13.1 Introduction

Recycling of paper took off in the 1970s when the use of paper increased sharply and the need for cheaper raw material increased. Use of recovered paper as raw material for new paper is an important means of reducing waste and effectively using resources. It is therefore of considerable social importance. One of the major consequences of fibre recycling is dealing with stickies that are natural components of the recycled fibre used by the mills. According to Patrick (2006), stickies will become a growing process problem in the years to come with the increased use of recycled fibres and the closing of the mill water loops. Many significant operational and quality problems are caused by stickies in pulp and papermaking systems. Cleaning up fouled sections of the paper machine causes valuable machine downtime, which diminishes paper quality and reduces output, all costing millions of dollars per year.

Stickies are tacky, hydrophobic, pliable organic materials found in recycled paper systems (Delagoutte, 2005; Doshi, 1999; Doshi & Dyer, 2002). Stickies found in recycled fibre can be the following:

- Adhesives
- Styrene–butadiene rubber
- Vinyl acrylates
- Polyisoprene
- Polybutadiene
- Hot melts

Stickies have a broad range of melting points and different degrees of tackiness depending upon their composition. Table 13.1 lists classification of stickies.

The control of stickies involves mechanical and chemical methods. Mechanical methods use screens, cleaners, dissolved air flotation (DAF) and washing stages to remove the stickies from the stock. Depending on the nature of the stickie, its size, melting point, conformability, etc., it may be removed by one or more of the various types of equipment. This variability means that although some stickies will be removed in one step of the process, there is not one piece of equipment, nor even a combination, that will remove all the stickies, all of the time.

Mechanical methods can be made more efficient by manipulating process parameters such as temperature and pH. Process conditions can be set so that the stickies

*Some excerpts taken from Bajpai (2006) with kind permission from Pira International, UK and Bajpai (2012) with kind permission from Springer Science & Business media.

Recycling and Deinking of Recovered Paper. DOI: http://dx.doi.org/10.1016/B978-0-12-416998-2.00013-1

Table 13.1 Classification of Stickies

Macro-Stickies
These are stickies that cannot pass through a 100 μm slotted laboratory screen plate. This category of stickies is associated with the fibrous fraction of the pulp.
Micro-Stickies
The size of these stickies is below 100 μm and above 5 or 1 μm. That means that they pass through 100 μm slots in a screening plate. Micro-stickies are linked to the fine element fraction of the pulp.
Colloidal Stickies
These are the smallest stickies, less than 1 or 5 μm. In fact they belong to the dormant category of stickies. They could be potentially harmful if, owing to system upsets, imbalances are created leading to their precipitation. The perturbation of wet-end chemistry by a change in pH or ionic concentration could lead to a change in their state.

are in an optimum form to be removed by cleaners, screens, etc. However, once again the variability of the stickies means that not all of the stickies will be in the optimum form or that there even exists a single set of conditions that will allow the mechanical removal of all stickies.

Even with the best equipment and conditions, mechanical methods will not remove all of the stickies. This is where chemical methods come in. Chemicals are the second line of defence used by paper mills to control stickies. Chemical control methods have mainly involved dispersants, polymers and/or absorbents. The various chemicals used have two main modes of action:

- The first method is to tie up or passivate the stickie. These are stock applications.
- The second mode of action is 'point of problem' type applications. These passivate a surface or clean off the stickies after they have deposited on equipment.

The removal of a sticky material, particularly the specific point or piece of equipment where it is removed, depends on the nature of the sticky itself. For example, the size, melting point and conformability of a particular material determine where removal occurs during the process.

Factors such as process temperature and pH can be set or adjusted so that these unwanted materials are maintained in an optimum form for their removal by cleaners or screens.

Because difficulties in controlling stickies with existing systems continue, they will remain a cause of paper quality problems and lost production. As previously stated, there is no reason to believe that the problems will go away or get any better. As mills increasingly close up their water and fibre systems, there are fewer places where contaminants can be completely removed from the system. The trend towards increased mill closure is expected to continue, presenting additional challenges to the papermaker.

Table 13.2 Problems Caused by Stickies

Deposits within the paper machine can cause the following:
– Web breaks owing to weak spots and localised sticking to moving rolls
– Picking of fibres
– Lengthy non-productive cleaning times owing to build-up on wires, felts, rolls and doctor blades
Deposits in or on the paper can cause the following:
– Increased downgrades owing to contribution to dirt count
– During converting processes at high temperature and pressure, migration of stickies from inside the paper reaches outside the paper. This is particularly costly when goods are returned after printing, varnishing or laminating
– Converting problems owing to adhesion between sheets. Problems during unwinding of a reel causing breaks, surface blemishes or misfeeds in a sheet-fed process

13.2 Problems

The problems caused by stickies are many, varied and can occur at practically every stage in the papermaking and paper conversion chain including printing. Finding a suitable way to treat them can be difficult.

The problem is complicated by the fact that there are two types of stickies.

1. Primary stickies, which are insoluble components that are tacky under normal conditions and are present in the incoming recycled paper and coated broke, for example hot melts or pressure-sensitive adhesives, inks, waxes, plastics, synthetic adhesives, resin size, wet-strength resins and binders.
2. Secondary stickies, composed of the soluble components of adhesives and chemicals used in papermaking and converting. Secondary stickies are formed as a consequence of sudden changes in wet-end chemistry, temperature or pH, when they react to produce insoluble or gel-like tacky particles. They can be more difficult to handle than primary stickies. The problems caused by these tacky impurities are familiar. Deposits occur on stationary parts of a processing system such as pipes and chest walls and in the approach flow system. This includes the headbox itself because of the adhesive character of these materials. Deposits primarily influence moveable parts such as wires, felts or rolls. Stickies will end up in the final paper produced and cause quality problems aesthetically and in its subsequent use (Table 13.2).

A good discussion of the issues caused by stickies can be found in Doueck (1997). When troubleshooting a stickies issue it is important to use some standard of measurement. There are several measurement methods available for macro- and micro-stickies. A good reference for these methods can be found in Doshi, Dyer, Heise, and Cao (2000a,b, 2003). There are new methods being introduced, some of which are on-line. Effective on-line measurement of stickies will go a long way to assisting the pulp and papermaker deal with them. When addressing a stickies issue, one question always appears. This is whether the problem is caused by macro- or micro-stickies. In some cases the nature of the problem will answer the question. The best example would be if the problem is visual stickies on the surface of the paper. This almost always has to be macro-stickies. In many other cases it is not so clear cut. Is the problem being caused by macro-stickies or micro-stickies that have, over time, deposited onto a surface and then broken loose?

Measuring stickies throughout the system, especially before and after unit operations that remove stickies, is the first step in troubleshooting. If the required equipment is available then both macro- and micro-stickies should be measured. The data from the 'stickies audit' and a study of the location and nature of the problem will help indicate the most effective control solution.

13.3 Control

The control of stickies is achieved by three main mechanisms. These are recovered fibre selection, mechanical removal and chemical control products. Depending on the mechanism, the goal is to make larger stickies smaller, to make the stickies less sticky or tacky and/or to tie up the stickies so that they do not deposit (Delagoutte, 2005; Doshi, 1999; Doshi & Dyer, 2002; Hamann & Strauss, 2003; Lamot, 2004; McKinney, 1995; Putz, 2000). Methods for control of stickies are listed in Tables 13.3 and 13.4.

13.3.1 Control of Waste Paper Quality

By controlling waste paper quality, stickies problems can be prevented. The criteria for acceptable and unacceptable waste paper should be established and given to waste paper dealers for quality control of the incoming paper. Depending on the type of the furnish, final product and specific problems or customer needs, it may be desirable to measure the concentration of stickies, plastics or clay as well as the

Table 13.3 Methods for Control of Stickies

- Control of waste paper quality
- Pulping and deflaking
- Screening
- Cleaning
- Flotation
- Dispersion and kneading
- Use of chemical additives
- Use of esterase enzymes

Table 13.4 Chemical Additives Used for Stickies Control

- Talc
- Bentonite
- Diatomite
- Dispersants mixed with solvents
- Zirconium compounds
- Alum-sequestering agents
- Cationic polymers
- Surfactants

brightness, freeness, groundwood content or fibre length distribution. One or more of these quality indicators may be used, depending on a mill's circumstances (Hisey, 1986; Jackson, 1989; Sisler, 1980).

13.3.2 Pulping and Deflaking

The recovered paper is pulped after selection of the desired quality of the paper. The objective of pulping is to defibrise paper without significant disintegration of contaminants (Hoffmann & Ala-Jaaski, 2000). Most modern pulpers have supplementary equipment to remove contaminants before they are broken down into small pieces. This equipment includes a ragger to remove wire and string, a junker for large contaminants and a secondary pulper. A stream is bled off at the secondary pulper and is subjected to mild defibrising. High-density contaminants accumulate in a chamber with a double-valve arrangement while stock is sent back to the pulper or is screened in a rotary screen. Many older pulpers are being retrofitted with these accessories, and most newer pulpers come equipped with secondary pulpers.

The drum pulper is gaining in popularity, particularly for newsprint deinking (Borchardt, 2003; Patrick, 2001). The drum pulper does not have any cutting action, therefore many contaminants, such as plastics and book bindings, remain almost intact and are rejected by the associated rotary screen. Quimipel and Falcao Consultoria cooperated to develop a new re-pulping technology that allows stickies control at the beginning of the waste paper defibration process and is economically viable (Almeida Falcao, 2003).

Bouchette (1991) and Nagy (1991) proposed steam explosion pulping for defibring waste papers. Loose waste papers are subjected to steam at about 400 psi and 400°F. When the pressure is released, the material explodes and defibres. Contaminants, stickies and inks are also dispersed in this pulper.

13.3.3 Screening

Coarse screens with holes and fine screens with slots are used to remove contaminants, primarily based on their size (Goldenberg, 1989; Halonen & Ljokkoi, 1989; Koffinke, 1989). Heise, Kemper, Wiese, and Krauthauf (2000) has reported that 5–30% of primary stickies may pass through the screening slots. As slot size decreases, contaminant removal increases but so does fibre loss (Eck, Rawlings, & Heller, 1995; Lerch & Audibert, 1995).

Several factors influence the screening efficiency of sticky contaminants (McKinney, 1995). These include the following:

- Stock temperature
- Contaminant consistency
- Contaminant size
- Nature of the contaminant
- Slot or hole size of the screen basket
- Average stock velocity through the slot or hole
- Slot or hole profile
- Screen rotor type and speed

Most pressure screens operate with mass reject ratios of 15–30%. To reduce the fibre loss, there are second, third and sometimes fourth screening stages, and their arrangement is very important for contaminant removal efficiency. The elastic, deformable nature of stickies, especially pressure-sensitive adhesives (PSAs), allows them to extrude through fine slotted screens (Pikulin, 1997). Mill data indicate that stickies in the screen accepts were significantly larger than the screen slot size, sometimes an order of magnitude larger.

Mechanical screening is the most efficient tool for sticky separation in the industrial process. This means that stickies have to be as large as possible to be screenable. A particle size of macro-stickies above 2000 μm, as determined by INGEDE Method 4, is favourable to guarantee complete removal in a state-of-the-art recycling process.

13.3.4 Cleaning

Cleaners are classified as high-, medium- or low-density cleaners. They remove the contaminants based on their density differences. Their use depends on the density and size of the contaminants they are removing.

- High-density cleaners, or forward cleaners, are used to remove nuts, bolts, paper clips and staples. They are usually located immediately after the pulpers.
- Smaller-diameter hydrocyclones are used for medium-density contaminants. As the hydrocyclone diameter decreases, it becomes more efficient at removing small contaminants. For practical and economic reasons, the 3 in. (75 mm) diameter cyclone is the smallest cleaner used in the paper industry.
- Reverse hydrocyclones or through-flow cleaners are used to remove low-density contaminants. A disadvantage of reverse hydrocyclones is that 55% of the flow is in the reject stream, and secondary and tertiary stages are needed to recover the usable fibre (Bliss, 1985). This problem does not occur in through-flow cleaners. The accepts and rejects come out at the same end in through-flow cleaners. The reject stream is only 10% by volume and 2% by mass (Bliss, 1985; Chivrall & McCool, 1983; Flynn, 1984). However, the contaminant removal efficiency of hydrocyclones is usually higher than for through-flow cleaners. Another problem with through-flow cleaners is that they are somewhat prone to plugging because of the narrow gap at the exit and the lower pressure drop.
- Another lightweight cleaner, Gyroclean, functions as a rotating centrifuge and is effective at removing low-density contaminants (Marson, 1990). The reject ratio of these cleaners is so low that the second or third stage is not necessary. Lightweight cleaners generally become inefficient as the particle density exceeds 0.95 g/cm^3. A new design modifies the hydrocyclone's length, diameter, cone angle, etc. to increase the residence time (Maze, 1997). Researchers claim this new design is 95% efficient in lightweight contaminant removal, compared with 63% for conventional through-flow cleaners under the same operating conditions.

13.3.5 Flotation

The removal of stickies from recycled papers by flotation has been reported by several researchers (Chin, Hipolit, & Longhini, 1996; Chou, 1993; Delagoutte, Brun, & Galland, 2003; Doshi et al., 2000a,b, 2003; Hsu & Dauplaise, 1996; Johansson, Wikman, Lindström, & Österberg, 2003; Li, Hipolit, & Longhini, 1996; Ling, 1993a;

Nerez, Johnson, & Thompson, 1997). For efficient flotation there is a need to select a flotation aid that minimises any reduction in the surface hydrophobicity of the stickies but still generates a sufficiently stable froth for their flotation (Ling, 1994).

Johansson et al. (2003) have reported that flotation may remove over 70% of micro-stickies in a pulp. A study in a deinking mill showed 66% efficiency for micro-stickies removal (Delagoutte & Brun, 2005). Flotation efficiency depends on the shape, size and surface properties of the stickies and on the hydrodynamic parameters of the flotation. Its main advantage, especially compared with screening, is its ability to remove micro-stickies from the pulp suspension (Glover, Fitzhenry, & Hoekstra, 2001; Heise et al., 2000; Lee & Kim, 2007).

Doshi et al. (2000a,b) and Hsu and Dauplaise (1996) have reported that the nature of the stickies plays an important role. Wax and hot-melt adhesives are quite well removed by flotation, whereas waterborne PSAs are not. This is because these two types of adhesive present different surface properties. The waterborne PSA has a more hydrophilic character than the hot-melt adhesive. Chemical additives may significantly change the ability of flotation to remove stickies. The removal of waterborne PSAs may be improved by the addition of cationic polymers (Hsu & Dauplaise, 1996). These polymers may induce aggregation of the PSA particles into a size range more favourable for flotation removal. Polymer fixation may modify the surface properties, which could also favour interaction between the sticky particles and the air bubbles. Heise et al. (2000) reported that the concentration of surfactant also plays an important role. High surfactant concentration can decrease stickies removal because the initial hydrophobicity of the stickies is reduced owing to high amounts of surfactant, which decrease the attachment force of stickies to air bubbles.

Doshi et al. (2003) studied froth flotation to remove wax and stickies from repulped old corrugated container (OCC). Trials at a pilot plant used a conventional OCC stock preparation process with and without froth flotation. Additional washing and DAF were also evaluated. Including flotation in the OCC stock preparation system significantly improved stickies removal and promoted a decrease in the area of wax spots in handsheets. Flotation was more effective in removing wax and stickies than through-flow cleaners. Analysis revealed that three stages of flotation in an OCC system was sufficient and there was no significant loss of yield. Efficient water clarification was achievable using an effective polymer programme and DAF.

13.3.6 Dispersion and Kneading

This process step breaks up contaminants so they are not visible in the final product. Important dispersion parameters are the following:

- Consistency
- Temperature
- Pressure

Dispersion uses consistencies of 25–30%; temperatures range from 160°F to 180°F (71–82°C) at atmospheric pressure, and higher temperatures are sometimes used (Cochaux, Carre, Vernac, & Galland, 1997; Gilkey & Mark, 1987; Matzke & Selder, 1986; Rangamannar & Silveri, 1989). One application of dispersion is for breaking up waxes, hot-melt glues and bitumen in recovered papers to minimise problems of spots and sticking between sheets. Most European suppliers recommend non-pressurised units (95–100°C). In contrast, American suppliers generally advocate dispersion under pressure at high temperature (Fetterly, 1992) for more efficient dispersion of stickies.

Fergusson and McBride (1994) studied the Shinhama kneader. They found that no steam or pressure was needed for satisfactory stickies reduction. Reduction of the thermoplastic material from fibre was found to be more effective at low temperatures than at high temperatures, where the contaminants smeared onto the fibre. However, according to Mannes (1997), hot dispersion is needed to ensure a high level of product quality and to avoid runnability problems on the paper machine. He also reported that effective and reliable stickies treatment is possible with high-speed dispersers operating at high stock consistencies and high temperatures.

Kneading as a processing stage has been studied for its effect on stickies. It modifies the shape of the stickies, causing them to become more spherical (Julien Saint Amand, Perrin, Vinson, De Luca, & Cremon, 1994; Julien Saint Amand, Perrin, & De Luca; 1998). Also, sticky particles are stripped of adhering fibres, so they are no longer 'hairy'. Therefore, a kneading stage is recommended before screening, cleaning and/or flotation to enhance stickies removal (Seifert, 1997).

Julien Saint Amand et al. (1994, 1998) investigated shape changes of stickies during dispersion to improve their removal efficiency in subsequent processes. They found that kneading caused stickies to become more spherical. The shape modification was more significant, even on small particles, when kneading was done at high temperatures. This change in shape increased the removal efficiency of screening and improved the removal efficiency of cleaning and flotation, owing to the reduction in size of the stickies. Kneading at high temperature can have a harmful effect, as sticky particles can be degraded to sizes that are not effectively removed by the other mill processes (Galland, Carre, Cochaux, Vernac, & Julien Saint Amand, 1998; Julien Saint Amand et al., 1994, 1998; McKinney, 1999).

13.3.7 Chemical Additives

Many types of chemical additives have been used to control stickies (Delagoutte, 2005; Doshi, 1999; Doshi & Dyer, 2002; McKinney, 1995; Putz, 2000). These are listed in Table 13.3.

13.3.7.1 Talc

It has a hydrophobic surface and hydrophilic edge (Doshi, 1999; Doshi & Dyer, 2002; McKinney, 1995; Putz, 2000). The hydrophobic surface has an affinity for stickies whereas the hydrophilic edge allows easy dispersion of the talc in water. It is a hydrous magnesium silicate. To be effective, it should be as pure as possible. Impurities will reduce the affinity of the surface for organic material. Talc is not effective on stickies that are not tacky at the headbox temperature but become tacky at the dryer temperature. Most of the talc should be retained in the sheet to avoid excessive concentration in the whitewater. Talc is often introduced at the inlet of the kneader or disperser to favour its mixing with the pulp components. Few strategies have been proposed to control stickies deposition with talc.

Maat and Yordan (1998) examined the effect of talc in removing tackies from hot melt containing polyvinyl acetate. In their experiment, application of 0.5% talc (on an oven-dried pulp basis) gave a 77% reduction in the amount of tacky stickies detected using the Doshi microfoam method. Application of 2.5% talc reduced tacky stickies by 96%.

Sharma, Kortmeyer, Spivey, and Lasmarias (2002) have proposed modified talc in which the hydroxyl groups on the edges are modified by a cationic polymer for stickies control. The cationic charge carried by this modified talc is believed to favour its attachment to negatively charged stickies and its retention in the paper sheet. Luzenac Asia developed a talc solution that can be used for stickies removal from pulp containing recycled materials (Hayakawa and Williams, 2004).

Biza, Kaiser, and Gaksch (2002) has reported that when stickies become coated with talc, it increases their density so they can be removed more efficiently by high-density cleaners. Stickies removal efficiency increased from almost zero (in the absence of talc) to more than 50% when 2% talc was added before the cleaning step. Stickies on the paper surface were passivated by spraying talc slurry onto the paper sheet in the forming section of the paper machine (Biza, Gaksch, & Hang, 2001). Tests in three paper and pulp mills revealed that 1 tonne of recycled pulp treated with 9 kg of talc has 70% fewer deposits in the final pulp product compared with the untreated pulp and no negative effects on the resulting paperboard quality. This talc treatment has a better cost performance than cationic polymer treatment. No stickies control can be achieved with the cationic polymer using the same budget as for talc treatment.

13.3.7.2 Bentonite

Several researches have studied the use of bentonite to control stickies (Beaudoin et al., 1998; Tay, Shields, & Loucks, 1998). A retention chemistry based on a combination of bentonite, enhancer and polyethylene dioxide was introduced in 1996 (Tay et al., 1998) and had an incredibly positive effect on the paper machine. This retention programme was extended to the other two machines and had comparable results.

Beaudoin et al. (1998) used bentonite with polyethylene dioxide and found an increase in retention, drainage and sheet solids content. They also found lower turbidity of the whitewater and lower stickies deposits. Furthermore, the sheet quality and water removal characteristics could be more easily optimised and the paper machine fabrics were cleaner.

13.3.7.3 Diatomite

It has been increasingly successful at controlling hardwood pitch, white pitch, stickies from recycling and other organic contaminants. Diatomite has worked successfully on various grades of paper and with various organic contaminants. In addition, because it is added in small amounts, there have been no detectable effects on paper properties or adverse effects on machine runnability. It is a unique mineral made from the skeletal remains of single-celled aquatic plants called diatoms (Vogel, 2002;

Williams, 1987). Diatomite powders used in the paper industry for stickies passivation have exceptionally high silica content. Case studies demonstrate how diatomite has improved machine cleanliness at a mill producing text and cover paper from virgin pulp, improved productivity at a mill producing coated and uncoated recycled paper, and improved converting efficiency at a mill manufacturing 100% recycled brown towelling (Vogel, 2002).

13.3.7.4 Dispersants Mixed with Solvents

Mixing dispersants with appropriate solvents can help to defibre wet-strength papers (Dykstra & May, 1989; Elsby, 1986; Hoekstra & May, 1987; Miller, 1988; Moreland, 1986). Environmental concerns such as toxicity, odour and flammability need to be addressed when selecting solvents and dispersants.

Most dispersants are sensitive to pH, contaminants, temperature and the presence of other chemicals. They should be selected to be compatible with the existing system. Non-ionic dispersants are generally long-chain molecules with one hydrophobic end and one hydrophilic end. When a secondary fibre slurry is mixed with a non-ionic dispersant, the dispersant's hydrophobic end will attach to one of the stickies, leaving its hydrophilic end exposed to the water. The hydrophilic end has no affinity for stickies, which is how the stickies are prevented from agglomerating. Anionic dispersants keep small stickies suspended in a slurry by inducing a negative charge on them, which repels other negatively charged stickies.

13.3.7.5 Zirconium Compounds

These compounds consist of soluble polymer chains that can be anionic, cationic or neutral, depending on their charge. The active zirconium-containing groups react with functional groups on the stickies to reduce the negative effects of the stickies (Putz et al., 2000). Zirconium salts have been used in the paper industry for several years. Laboratory tests have shown that the tack and tacky area of PSAs and hot-melt adhesives in model fibre stock can be significantly reduced by treatment with solutions of zirconium compounds at 0.1–0.3% on recycled fibres (Lobbes & Forester, 1996). By adding up to 0.2% of these zirconium compounds, four paper mills considerably reduced their stickies problems.

13.3.7.6 Alum-Sequestering Agents

Alum can contribute to stickies. Effective control of stickies therefore requires alum to be kept to a minimum (Ormerod & Hipolit, 1990). It is the most commonly used chemical in paper production as it is needed in stoichiometric proportions for rosin sizing. It can also be used to improve the retention behaviour in the wet end of a paper machine. Excessive use of alum can cause colloidal particles to agglomerate into larger stickies, increasing the potential for stickies problems.

13.3.7.7 Cationic Polymers

Several researchers have studied cationic polymers to control stickies (Delagoutte, 2005; Doshi, 1999; Doshi & Dyer, 2002; Putz, 2000; Renaud, 2001; Taylor, 2001). These polymers can be added at the thick stock chest, fan pump or headbox. They help to avoid stickies building up in the whitewater system as the stickies leave the system in the paper sheet.

The function of these polymers is to act as fixing agents, helping stickies and the other colloidal particles in the pulp slurry to be adsorbed onto fibres (Woodward, 1996). Cationic polymer wire sprays are also used. A thin film of polymer attracts anionic materials, including stickies, which are then washed away when the paper machine wire is cleaned. The thin polymer film prevents stickies from coming into close contact with the wire. The polymer eventually ends up in the whitewater and serves as a retention aid. Proper use of showers to keep wires and felts clean will help maintain product quality and reduce downtime for the paper machine. Some products are also available for spraying on felts.

Angle (2002) has reported that Hercules Inc. has combined an amphoteric, surface-active structured protein with a highly charged cationic polymer. The structured protein increases the stability and reduces the tackiness of colloidal contaminants. This approach combines three different methods of control:

1. Stabilising colloidal material
2. Reducing the tackiness of the colloidal material
3. Removing contaminants from the system

It has been considerably more effective on commercial paper machines than conventional methods, according to laboratory measurements of zeta potential, contact angle after washing, colloidal stability and detackification. The method was successfully applied to problems of stickies and pitch deposition at a mill producing 100% recycled corrugating medium, problems of wet- and dry-end breaks and a blue-tinted deposit in the stock chests and whitewater systems of an integrated alkaline fine paper mill, and problems of stickies at a mill producing 100% recycled bleached and unbleached towel and napkin grades.

Rys and Song (2004) and Song and Rys (2004) used diallyldimethylammonium chloride (DADMAC) homopolymers and co-polymers. These products have high fixation power for controlling pitch and stickies for improved runnability. They performed very well compared with the best products for pitch and stickies in furnishes of thermomechanical pulp (TMP), deinked pulp (DIP), coated broke and blends thereof. They include the high-molecular-weight structure poly-DADMAC as well as co-polymers of DADMAC and a co-monomer that can boost the polymer's fixation power, and terpolymers of DADMAC co-polymerised with two co-monomers that improve the fixation power. Vacuum drainage filtrate turbidity tests were conducted on the new polymers and the existing commercial fixatives. Their performance was tested on coated broke at a US paper mill, on a DIP recycled furnish at a UK newsprint mill, and on TMP, DIP and a TMP/DIP mixture at a Canadian newsprint mill.

DADMAC gave significantly improved performance over existing commercial fixatives in removing turbidity for deposit control.

A joint study by Darmstadt Technical University and BASF showed that the effectiveness of fixative agents can be examined by the rotating wire mesh test and the pitch counter method (Hamann et al., 2004). Results from these two tests correlated well for coated broke. The tests showed that polyvinylamines with cationic charges of 3–5 milliequivalents per gram and molecular weights greater than 400 kg per mole were the most effective, whereas poly-DADMAC and polyamines were less effective. Tests on coated broke showed that the molecular weight and linearity of a polymer had a greater effect on the polymer's fixative properties than its density or cationic charge. In addition, polyvinylamines and polyethylene imines with high cationic charges were effective fixatives for DIP contaminants.

13.3.7.8 Surfactants

Surfactants were initially developed to control pitch deposits in paper mills using virgin fibre as raw material, but were later applied to reduce clogging by stickies in paper mills using recycled fibre. Passivators to reduce the tacky character of stickies have been based on particular non-ionic surfactants (Fogarty, 1992; Hall & Nguyen, 1998; Ling, 1993b; Moreland, 1995), which modify the surface properties of sticky particles but do not favour their removal, so they are still free to accumulate in the circuits. Other proposals include stickies removal (Back, 1995; Bossaer, 1999; Coffey, 1999; Crowe & Landstra, 2001; Doshi, 1999; Severtson, Coffey, & Nowak, 1999; Ward, Hensel, & Patterson, 1994). Non-ionic surfactants are ethoxylated alkyl phenol and ethylene oxide–propylene oxide block co-polymers.

Surfactants may also be used to treat paper machine clothing such as felts or wires. Combined treatments based on dispersants and cationic fixing agents have also been studied. Dispersants are usually added at the pulping stage for dispersing and passivating the stickies. Cationic fixing agents are added near the paper machine to fix the dispersed stickies onto fibres, so they can be eliminated via the paper output. Another combined treatment involves dispersing the wax at low temperature using a formulation that includes a wax melting-point depressor and an anionic dispersant. The dispersed paraffin is then separated from the fibres by a washing or thickening stage. Finally, the resulting water, containing dispersed wax, is treated by microflotation with a specific chemistry that removes wax particles.

13.3.8 Enzymes

Enzymes have been used in a variety of industries, including pulp and paper, for many years. Enzymes are catalysts that drive the chemical reactions in all living things. Enzymes have the following properties:

- Effective in very small amounts – a few enzyme molecules will catalyse thousands of reactions per second
- Not consumed in the reaction
- Reduce the activation energy of a reaction and therefore increase the speed of reaction

- Very specific to a reaction
- Specific pH and temperature range that they are active in.

The fact that enzymes are specific to a certain reaction allows enzymatic products to be tailored to specific needs. A study of stickies revealed that many have a large number of ester type chemical bonds that link the basic building blocks of the sticky. Therefore, an esterase is required to break down stickies. Several esterase-type enzyme mixtures have been studied to find one with the ability to break down a high percentage of stickies. Not only will the enzyme break the ester bonds in the sticky and reduce their size, but also, as the sticky compound is broken down into smaller components, there will be a greatly reduced chance that the stickies can reagglomerate further down the process (Anonymous, 2003a,b, 2005; Jones, 2005; Jones & Fitzhenry, 2003; Fitzhenry J., Hoekstra, P. & Glover, D., 2000; Fitzhenry, J., Hoekstra, P. M., & Glover, D., 2000; Toland, 2003; Van Haute, 2003).

Buckman Laboratories, Canada, has developed enzymes that control stickies. These enzymes have several advantages compared with conventional stickies control technologies (Covarrubias & Eng, 2006; Eng & Covarrubias, 2005; Kimura, 2006; Sokol & Huszar, 2005) (Table 13.5). More mills now have single-stream recycling, which means that old newspapers (ONPs) probably includes old magazines (OMGs) and OCC. The increased use of single-stream recycling processes has resulted in more stickies. Buckman manufactures Optimyze, which contains an esterase that breaks up the ester bonds in polyvinyl acetate stickies. This is particularly effective in the pH range 6.5–10 and temperature range 25–60°C (Covarrubias & Jones 2005). Buckman's Optimyze Plus range also includes an enzyme that acts on wood pitch. Buckman has developed an enzyme product combined with a dispersant that would disperse and stabilise the stickies. The enzyme is then more likely to penetrate the stickies rather than just acting on the surface (Jones, 2008). In all the early work, esterase was added to the stock, but Jones (2005) showed that applying the enzyme directly onto paper machine clothing also reduced stickies deposition. Positive results were obtained by applying enzyme to forming fabric and press felts. The enzyme was applied through a shower bar, and full coverage was important. The enzyme detackified the stickies, so the cleaning showers could remove them more effectively. In 2004, the US Environmental Protection Agency gave Optimyze a Presidential Green Chemistry Challenge Award. Optimyze is now used in many mills around the world.

Several mill-scale studies related to the production of tissues, newsprint, writing and packaging papers from recovered or recycled paper have been reported (Sokol & Huszar, 2005). The use of enzyme has been found to enhance the screening efficiency, reducing the amount of rejects. It resulted in the reduction of stickies by

Table 13.5 Benefits of Using Enzymes

Reduced paper machine downtime
Decreased cleaning chemical costs
Increased machine-clothing life
Reduced life of forming fabrics

70–90%, generated clean cellulose fibre and big contaminant particles that were easily removable from the system. Other mill studies with enzymes produced results such as 90% reduction in stickies and increased brightness across the deinking facility, production performance improvements, reduction in the use of cleaning solvents and a 95% reduction in downtime caused by stickies (Covarrubias & Eng, 2006).

In many other studies involving almost every grade sector, from multi-ply coated boxboard to packaging grades, stickies have been reduced dramatically with enzymatic control. This has resulted not only in higher-quality products and fewer customer complaints, but significant reductions in culled production and increased use of lower-quality and much less costly recovered fibre (Patrick, 2004).

The mills using Optimyze enzyme to control stickies observed significant gains over traditional chemical approaches they were using. On treatment of ONPs and OMGs with an esterase-type enzyme in a mill trial, a dramatic reduction in the size of the sticky particles was observed (Jones & Fitzhenry, 2003). The stickies content of all the sizes and the total stickies were much less on enzyme treatment compared with those without enzyme (pre-trial results). The larger size stickies were totally absent in the recycled fibre treated with enzyme When the recycled fibre from mixed office waste (MOW) was treated with esterase enzyme in a mill trial, the total stickies content was reduced appreciably (Jones & Fitzhenry, 2003). Without enzyme treatment of MOW, it was not possible to increase the recycled fibre content in the final furnish beyond 50%. Even then, the total stickies were more than 250 ppm. However, on treating the recycled fibre from MOW with enzyme, total stickies content could be reduced to around 100 ppm, and that was with a higher content (60%) of recycled fibre.

In another MOW mill, the recovered fibre was treated with enzyme to reduce the percentage of high brightness virgin pulp in the final furnish without compromising on brightness of the finished stock. With addition of 35% high-bright pulp, the normal brightness gain from coarse screen to finish stock was only 12–14 points when no enzyme treatment was given to the recovered fibre (Jones & Fitzhenry, 2003). However, after esterase enzyme treatment of MOW recovered fibre, it became possible to obtain a brightness gain of more than 15 points, even when the high-bright pulp content was only 15–20%. The results of mill trials with esterase in two MOW processing facilities and one tissue mill are presented in Figures 13.1–13.3.

In one US mill producing coated paperboard from 100% OCC for conversion to food boxes, problems with stickies were causing significant deposition problems on the mill's paper machine and related runnability and off-quality problems in the sheet. The high level of paper machine breaks was causing significant operating downtime. Kerosene and other harsh chemical solvents were being used to remove the deposits. After launching an Optimyze enzyme programme, paper machine deposition was reduced 75% and machine breaks/downtime were reduced accordingly. Off-quality product tonnage dropped and the mill's use of cleaning chemicals was practically eliminated. With holes and other stickies-related problems rectified, customer complaints fell to almost none. This coated paperboard mill experienced a US$1.33 million annual return by eliminating stickies (based on Patrick, 2004).

A mill in Brazil producing 270 tonnes/day of linerboard using 100% OCC was having significant downtime problems owing to stickies. Stickies in the sheet were

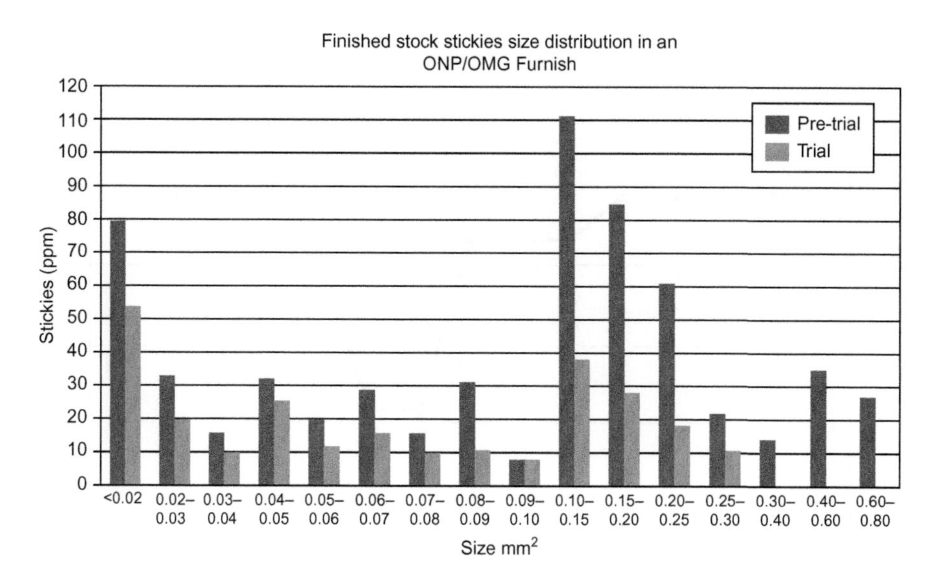

Figure 13.1 Stickies size distribution derived by using the enzyme system on a furnish of ONP/OMG waste paper.
Source: Jones and Fitzhenry (2003). Reproduced with permission from RISI.

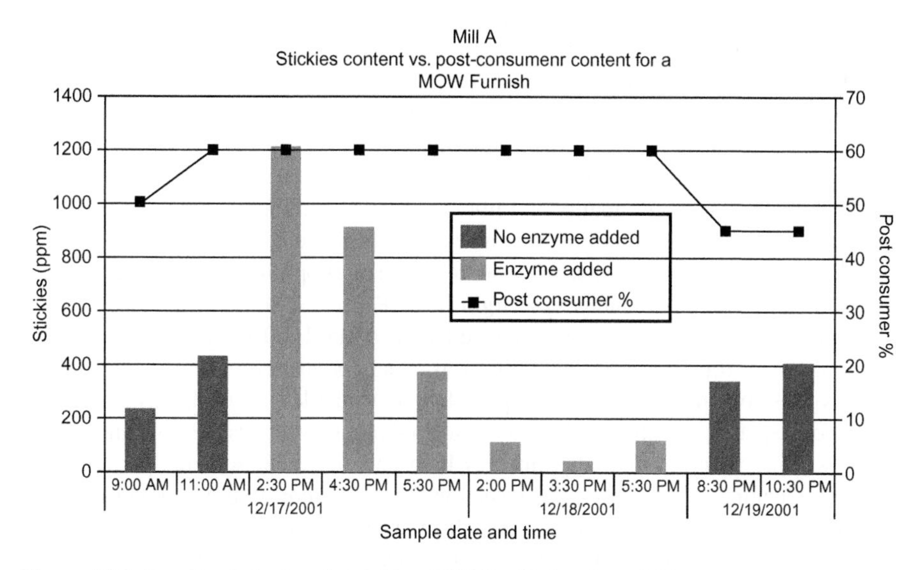

Figure 13.2 Results of plant-scale trial in a MOW mill showing stickies content vs. post-consumer content.
Source: Jones and Fitzhenry (2003). Reproduced with permission from RISI.

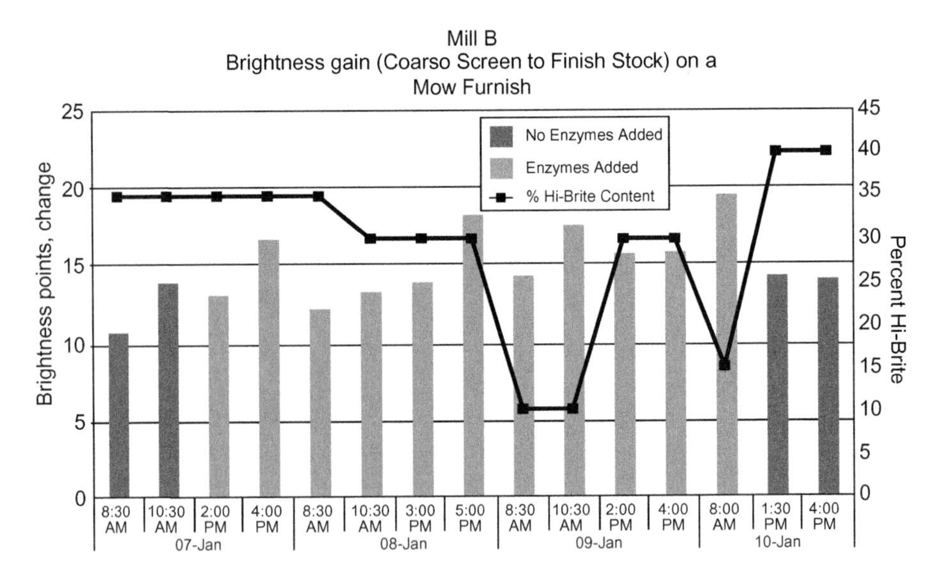

Figure 13.3 Results of plant-scale trial in a MOW mill showing brightness gain.
Source: Jones and Fitzhenry (2003). Reproduced with permission from RISI.

also causing breaks at the converting operations of the mill's customer. The mill was having to use a higher quality, more costly OCC in an attempt to minimise these problems. When the mill switched to Optimyze technology, product quality improved immediately. Production breaks were almost totally eliminated (reduced by 30 per month), giving a much more efficient operation. In fact, the mill achieved record production after the switch. In addition, it was able to use a poorer quality OCC, cutting production costs even further. The stickies were eliminated after the Optimyze programme was started. In other sample cases, use of Optimyze technology allowed one US mill to produce tissue and towelling products that easily met a fast-food chain's stringent quality and fibre content requirements (at least 65% post-consumer fibre). In fact, this mill was able to boost recycled fibre content to 100% and still meet the requirements, whereas other mills supplying the restaurant chain had to use much high percentages of virgin fibre (Patrick, 2004).

Another mill in the USA that produced more than 1000 tonnes/day of various grades of tissue from 100% recycled fibre was experiencing severe stickies deposits in its press felts. When the deposition approached a critical level, which was happening much too frequently, production had to be stopped and solvents used for cleaning. The mill was using several hundred gallons of kerosene and other solvents per day during these cleaning cycles. Also, press section showers were used continuously in an attempt to keep this machine's press felts clean. The water used in these showers was recycled within the paper mill system, which meant that this source too was building up high levels of stickies and re-depositing them in the cleaned felts and on other paper machine components. Clearly, another solution was needed. The mill added Optimyze to the water used in the felt cleaning showers. As a result, downtime for solvent cleaning was reduced from 1.81 h per week before the treatment to an average of 0.73 h per week. This represented a 60% reduction in downtime, and the amount of solvent used for cleaning the machine was reduced by a corresponding amount. Similar problems were occurring on another paper machine in this mill and the same strategy was used to solve them. With this machine, the average downtime for solvent cleaning dropped from an average of 1.6–0.46 h per week, representing a 70% reduction in downtime and solvent use. This 70% gain would have been closer to 80% if, during the 3 weeks of the test, a cleaning shower had not been inadvertently turned off (Patrick, 2004).

Buckman has developed its Next Generation Stickies Control enzyme. There are two new products that contain the same proven enzyme from the first generation with the addition of a second enzyme with similar but not identical action against stickies. NGESC-1 is for in-stock application and NGESC-2 is for paper machine clothing. In one application of NGESC-1, a 15% reduction in breaks was realised compared with the first-generation product. Application was in the stock. In a second case, NGESC-1 was applied in an OCC mill. The result was a marked reduction in sticky counts and reduced customer complaints. There is a further next generation product in development that contains an enzyme and a dispersant (Jones, Glover, & Covarrubias, 2010).

Zeng et al. (2009) used lipase- and esterase-type enzymes to degrade stickies in DIP and chemithermomechanical pulp. These enzymes removed stickies by over 30% in DIP and 75% in chemithermomechanical pulp. There was a remarkable change in the molecular weight of the stickies in DIP. Enzyme application had no negative effect on the properties of either pulp.

13.4 Future Prospects

The use of recycled fibre continues to grow. As demand increases, the quality of the recycled fibre will decrease, which will result in more stickies. Mechanical and traditional chemical control approaches have worked but they are not the complete answer. Enzymes are a new approach in the battle against stickies and have proved very effective. They will continue to play a major role in stickies control. The challenge is to develop new enzymes that work in different temperature and pH ranges and on all types of stickies. The Optimyze product line from Buckman is successfully helping to minimise or eliminate the problems caused by stickies in many mills around the world.

References

Almeida Falcao, J (2003). A new concept of stickies control. *Papel*, *64*(12), 71.

Angle, C. D. (2002). A superior new approach to paper machine contaminant control. *56th Appita annual conference*. Rotorua.

Anonymous, (2003a). Buckman: Preventing stickies. *Papermaking and Distribution*, *2*(4), 14.

Anonymous, (2003b). Buckman: Expansion geared towards better serving customers. *Pulp and Paper Canada*, *104*(1), 32.

Anonymous, (2005). Combatting and avoiding stickies in waste paper processing. *Internationale Papwirtschaft*, *5*, 35.
Back, E. L. (1995). Principles for wax removal from OCC after alkaline hot dispersion. Autoredispersible waxes for recyclable packaging paper. *TAPPI Journal*, *78*(7), 161.
Bajpai, P. (2012). *Biotechnology for pulp and paper processing*. New York, USA: Springer Science + Business Media. (414 pp.).
Bajpai, P. (2006). *Advances in recycling and deinking* (pp. 1–2). UK: Pira International.
Beaudoin, R., Dube, G., Lupien, B., Lauzon, E., Vines, M., & Blais, S. (1998). Increased retention and drainage and controlling stickies deposition with bentonite in combination with polyethylene oxide and a cofactor – A mill experience. *CPPA 84th annual meeting*. Montreal.
Biza, P., Gaksch, E., & Hangl, M. (2001). Treatment of sticky particles by surface application of talc. *Second CTP-PTS packaging paper and board recycling international symposium*. Grenoble.
Biza, P., Kaiser, P., & Gaksch, E. (2002). Use of talcum to enhance the removal of stickies. *10th PTS-CTP deinking symposium*. Munich.
Bliss, T. (1985). Throughflow cleaners offer good efficiency with low pressure drop. *Pulp and Paper*, *59*(3), 61.
Borchardt, J. K. (2003). Recent developments in paper deinking technology. *Pulp and Paper Canada*, *104*(5), 32.
Bossaer, P. (1999). The control of stickies by the means of dispersants and fixation aids. *First packaging paper recycling international symposium*. Grenoble.
Bouchette, M. P. (1991). Steam explosion technology and fibre recycling. *TAPPI focus 95. Landmark paper recycling symposium*. Atlanta, GA.
Chin, L., Hipolit, K., & Longhini, D.A. (1996). "Removal of stickies and electrostatic inks using flotation process," in *1996 TAPPI Recycling symposium*, New Orleans, March 1996.
Chivrall, G. B., & McCool, M. A. (1983). The uniflow centrifugal cleaning philosophy in light contaminant removal. *TAPPI pulping conference*. Houston.
Chou, C. S. (1993). Repulpable pressure sensitive adhesives designed for paper recycling. *TAPPI recycling symposium*. Atlanta, GA.
Cochaux, A., Carre, B., Vernac, Y., & Galland, G. (1997). What is the difference between dispersion and kneading? *Progress in Paper Recycling*, *6*(4), 89.
Coffey, M. J. (1999). Dispersion of wax coatings at neutral pH and typical process temperatures *Preparing for the next millennium*. Atlanta, GA, USA: TAPPI Press. (p. 221).
Covarrubias, R. M., & Eng, G. H. (2006). Optimyze: Enzymatic stickies control developments. *Paper Asia*, *22*(8), 31–34.
Covarrubias, R. M., & Jones, D. R. (2005). Optimyze: Enzymatic stickies control developments. *91st annual meeting Pulp and Paper Technical Association of Canada* (pp. A107–A116), Book A. Montreal, 8–10 February.
Crowe, P., & Landstra, M. (2001). Successful stickies and tackies program implemented at an Australian paper shoalhaven giving reduced machine production cost. *55th Appita annual conference*. Hobart.
Delagoutte, T. (2005). Management and control of stickies. *Progress in Paper Recycling*, *15*(1), 31.
Delagoutte, T., & Brun, J. (2005). Drying section deposits; origin, identification and influence of the recycling processes, deinking and packaging lines comparison. *Review ATIP*, *59*, 17.
Delagoutte, T., Brun, J., & Galland, G. (2003). Drying section deposits: Identification of their origin. *IPE international symposium new technological developments in paper recycling*. Valencia, June.

Doshi, M., Dyer, J. M., Heise, O. U., & Cao, B. (2000a). Removal of wax and stickies from OCC by flotation. Part 1: Green bay packaging trial. *Progress in Paper Recycling*, *9*(4), 71.

Doshi, M., Dyer, J. M., Heise, O. U., & Cao, B. (2000b). Removal of wax and stickies from OCC by flotation. Part 2: Menasha corporation trial. *Progress in Paper Recycling*, *10*(1), 55.

Doshi, M., Dyer, J. M., Heise, O. U., & Cao, B. (2003). Removal of wax and stickies from OCC by Froth Flotation. *Fall technical conference*. Chicago, IL.

Doshi, M. R. (1999). Properties and control of stickies. In M. R. Doshi & J. M. Dyer (Eds.), *Paper recycling challenge* (Vol. 4, pp. 67). Appleton, WI, USA: Doshi & Associates.

Doshi, M. R., & Dyer, J. M. (2002). Overview, recent advances in paper recycling – Stickies. In M. R. Doshi (Ed.), *Paper recycling challenge* (pp. 1). Appleton, WI, USA: Doshi & Associates.

Doueck, M. (1997).. In M. R. Doshi & J. M. Dyer (Eds.), *Paper recycling challenge* (pp. 15–21). Appleton: Doshi & Associates.

Dykstra, G. M., & May, O. W. (1989).Controlling stickies with water soluble polymers. *TAPPI pulping conference*. Seattle, WA.

Eck, T. H., Rawlings, M. J., & Heller, P. A. (1995). Slotted pressure screening at southeast paper manufacturing company: *TAPPI deinking short course*. Vancouver, Canada (p. 149).

Elsby, L. E. (1986). Experiences from tissue and board production using stickies additives. *TAPPI pulping conference*. Toronto.

Eng, G. H., & Covarrubias, R. M. (2005). Optimyze: Enzymatic stickies control developments. *Proceedings of the seventh international conference on pulp, paper and conversion industry* (pp. 135–142). New Delhi, 3–5 December.

Fergusson, L. D., & McBride, D. H. (1994). De-inking sorted office waste. *Pulp and Paper Canada*, *89*, 62.

Fetterly, N. (1992). The role of dispersion within a deinking system. *Progress in Paper Recycling*, *1*(3), 11.

Fitzhenry, J., Hoekstra, P., & Glover, D. (2000). New measurement techniques and new technologies for stickies control. *Pira conference on scientific & technical advances in wet end chemistry* (p. 16). Barcelona, 19–20 June.

Fitzhenry, J., Hoekstra, P. M., & Glover, D. (2000). Enzymatic stickies control in MOW, OCC, and ONP furnishes. *TAPPI pulping, process and product quality conference*. Boston, MA.

Flynn, P. (1984). Coreclean reverse cleaner. *TAPPI pulping conference*. San Francisco, CA.

Fogarty, T. J. (1992). Cost-effective, common sense approach to stickies control. *TAPPI pulping conference*. Boston, MA.

Galland, G., Carre, B., Cochaux, A., Vernac, Y., & Julien Saint Amand, F. (1998). Dispersion and kneading. In M. R. Doshi & J. M. Dyer (Eds.), *Paper recycling challenge* (Vol. 3, pp. 13). Appleton, WI, USA: Doshi & Associates.

Gilkey, M. W., & Mark, E. L. (1987). Dispersing stickies at medium consistency. *American Papermaker*, *50*(3), 26.

Glover, D., Fitzhenry, J., & Hoekstra, P. (2001). Stickies removal across the float cell. *TAPPI pulping conference*. Seattle, WA.

Goldenberg, P. H. (1989). Recent developments in screening. *TAPPI pulping conference*. Seattle, WA.

Hall, J. D., & Nguyen, D. T. (1998). Stickies control using non-ionic polymers in systems with lower operating temperatures. *TAPPI recycling symposium*. New Orleans, LA.

Halonen, L., & Ljokkoi, R. (1989). Improved screening concepts. *TAPPI pulping conference*. Seattle, WA.

Hamann, A., Gruber, E., Schadler, V., Champ, S., Kuhn, J., & Esser, A. (2004). Effect of fixative agents on stickies control. *Wochenblatt für Papierfabrikation*, *132* (3/4), 102.

Hamann, L., & Strauss, J. (2003). Stickies: Definitions, causes and control options. *Wochenblatt für Papierfabrikation, 131*(11/12), 65.

Hayakawa, S., & Williams, G. R. (2004). A cost-effective talc solution to stickies control in OCC pulps. *Japan Tappi Journal, 58*(9), 1214–1217.

Heise, O. U., Kemper, M., Wiese, H., & Krauthauf, E. (2000). A removal of residual stickies at haindl paper using new flotation technology. *TAPPI Journal, 83*(3), 73.

Hisey A. B. (1986).Quality control of recycled newspaper. *TAPPI pulping conference*. Seattle, WA.

Hoekstra, P. M., & May, O. W. (1987). Developments in the control of stickles. *TAPPI pulping conference*. Washington, DC.

Hoffmann, J., & Ala-Jaaski, T. (2000). Stickies removal in pulpers and screens. *Wochenblatt für Papierfabrikation, 128*(10), 666.

Hsu, N. N. C., & Dauplaise, D. L. (1996). Effectiveness of ionically charged chemicals as flotation aids in stickies removal during mixed office wastepaper recycling. *TAPPI recycling symposium*. Atlanta, GA.

Jackson, P. (1989). Wastepaper characteristics for quality control. *TAPPI Journal, 72*(5), 263.

Johansson, H., Wikman, B., Lindström, E., & Österberg, F. (2003). Detection and evaluation of micro-stickies. *Progress in Paper Recycling, 12*(2), 4.

Jones, D. R. (2005). Enzymes: Using Mother Nature's tools to control man-made stickies. *Pulp and Paper Canada, 106*(2), 23–25.

Jones, D. R. (2008). The next steps in enzymatic stickies control. *Pulp and Paper, 82*(6), 21.

Jones, D. R., & Fitzhenry, J. W. (2003). Esterase type enzymes offer recycled mills an alternative approach to stickies control. *Pulp and Paper Canada* (February), 28–31.

Jones, D. R., Glover, D. E., & Covarrubias, R. M. (2010). Enzymatic stickies control, next plus generation. *Proceedings of TAPPI peers conference and ninth research forum on recycling* (Vol. 2, pp. 1282–1315).

Julien Saint Amand, F., Perrin, B., & De Luca, P. (1998). Stickies removal strategy. *Progress in Paper Recycling, 7*(4), 39.

Julien Saint Amand, F., Perrin, B., Vinson, K., De Luca, P., & Cremon, P. (1994). Use of surface and rheological properties in stickies removal and control. Part I: Modification of stickies size and shape. *Final report, CR CTP 3174*. January.

Kimura, M. (2006). Stickies control agent for recycling paper by Optimyze. *Japanese TAPPI Journal, 60*(7), 35–41.

Koffinke, R. A. (1989). Screening and cleaning. *TAPPI pulping conference*. Seattle, WA.

Lamot, J. (2004). Effective chemistry for improved recycled fibre quality. *Eighth Pira International conference on paper recycling technology*. Prague.

Lee, H. K., & Kim, J. M. (2007). Quantification of macro and micro stickies and their control by flotation in OCC recycling process. *Appita Journal, 59*(1), 31.

Lerch, J. C., & Audibert, S. W. (1995). Slotted screen system evaluation at garden state paper: *TAPPI deinking short course*. Vancouver, Canada (p. 161).

Li, C., Hipolit, K., & Longhini, D. A. (1996). Removal of stickies and electrostatic inks using flotation process. *TAPPI recycling symposium*. Atlanta, GA.

Ling, T. F. (1993a). The effects of surface properties on stickies removal by flotation. *CPPA second research forum on recycling*. Montreal.

Ling, T. F. (1993b). Modifying surface properties of stickies materials through polymer/surfactant adsorption. *Pulp and Paper Canada, 94*(10), 273.

Ling, T. F. (1994). The effects of surface properties on stickies removal by flotation. *Pulp and Paper Canada, 95*(12), 109.

Lobbes, T. J., & Forester, W. K. (1996). The identification of stickies. *Progress in Paper Recycling, 5*(2), 78.

Maat, P., & Yordan, J. L. (1998). Use of the microfoam (doshi) method for assessing the effectiveness of talc in controlling stickies. *TAPPI recycling symposium*. New Orleans, LA.

Mannes, W. (1997). *Dispersion, an important process stage for reducing stickies problems*. Ravensburg: Voith Sulzer.

Marson, M. (1990). New lightweight cleaner units solve mill's plastic problems. *Pulp and Paper*, *64*(6), 93.

Matzke, W., & Selder, H. (1986). Additional benefits from pressureless high-consistency disperging of secondary fibre stock. *TAPPI pulping conference*. Toronto.

Maze, E. (1997). New cleaner development for lightweight contaminant removal. *Progress in Paper Recycling*, *6*(3), 40.

Mckinney, R. W. J. (1995). Waste paper preparation and contaminant removal. In R. W. J. Mckinney (Ed.), *Technology of paper recycling*. London, U.K: Blackie. (pp. 401).

McKinney, R. W. J. (1999). Ink removal following disc dispersion and kneading. *TAPPI recycling symposium*. Atlanta, GA.

Miller, P. C. (1988). Chemical treatment programs for stickies control. *TAPPI pulping conference*. New Orleans, LA.

Moreland, R. D. (1986). Stickies control by detackification. *TAPPI pulping conference*. Toronto.

Moreland, R. D. (1995). Stickies control by detackification: *TAPPI short course on deinking*. Washington, DC: Vancouver.

Nagy, J. (1991). Steam explosion technology. *Wastepaper II conference*. San Francisco, CA.

Nerez, R., Johnson, D. A., & Thompson, E. V. (1997). Laboratory repulping and flotation studies of three pressure-sensitive adhesives. In M. R. Doshi & J. M. Dyer (Eds.), *Paper recycling challenge* (Vol. 2, pp. 94). Appleton, WI, USA: Doshi & Associates.

Ormerod, D. L., & Hipolit, K. J. (1990). Aluminum control prevents stickies problems. In J. M. Coleman (Ed.), *Recycling paper from fibre to finished product* (Vol. 2, pp. 489). Atlanta, GA, USA: TAPPI Press.

Patrick, K. (2001). Advances in paper recycling technologies. *Paper Age*, *117*(7), 16.

Patrick, K. (2004). Enzyme technology improves efficiency, cost, safety of stickies removal program. *PaperAge*, 22–26.

Patrick, K. (2006). Stickies still a critical concern for today's recycling plants. *PaperAge*, *September/October*, 28–31.

Pikulin, M. A. (1997). Stickies and their impact on recycled fibre content of fine papers. In M. R. Doshi & J. M. Dyer (Eds.), *Paper recycling challenge* (Vol. 2, pp. 89). Appleton, WI, USA: Doshi & Associates.

Putz, H. J. (2000). Stickies in recycled fibre pulp. In L. D. Gottsching & H. Pakarinen (Eds.), *Papermaking science and technology* (Vol. 7). Helsinki, Finland: Fapet Oy.

Rangamannar, G., & Silveri, L. (1989). Dispersion – An effective secondary fibre treatment process for high-quality deinked pulp. *TAPPI pulping conference*. Seattle, WA.

Renaud, S. (2001). New measuring technique enhancing fixative control. *TAPPI papermakers conference*. Cincinnati, OH.

Rys, L., & Song, Z. (2004). New innovation in fixatives for runnability improvement. *Eighth Pira International conference on paper recycling technology*. Prague.

Seifert, P. (1997). Deinking technology basics. In M. R. Doshi & J. M. Dyer (Eds.), *Paper recycling challenge* (Vol. 2, pp. 139). Appleton, WI, USA: Doshi & Associates.

Severtson, S. J., Coffey, M. J., & Nowak, M. J. (1999). Wax dispersion during recycling of old corrugated containers. *TAPPI Journal*, *82*(12), 67.

Sharma, S., Kortmeyer, J. C., Spivey, A., & Lasmarias, V. B. (2002). A cost-effective talc solution to stickies control in OCC pulps. *TAPPI fall technical conference and trade fair*. San Diego, CA.

Sisler, G. (1980). Quality control for fine paper deinking. *TAPPI pulping conference*. Atlanta, GA.

Sokol, A., & Huszar, L. (2005). Optimyze modern enzymatic programme for stickies control in papermaking process. *Progress '05: 15th international papermaking conference* (11 pp.). Warsaw, 28–30 September.

Song, Z., & Rys, L. (2004). Keep on running: How innovation in fixatives can help runnability. *Pulp and Paper International*, *46*(12), 25.

Tay, S. C. H., Shields, G., & Loucks, K. (1998). New cost effective retention system based on polyethylene oxide for pitch control and improved machine operating efficiency with DIP in furnish. *Coating and papermakers conference*. New Orleans, LA.

Taylor, T. E. (2001). Novel chemistry for white pitch and stickies control. *TAPPI papermakers conference*. Cincinnati, OH.

Toland, J. (2003). Developments in deinking: Rounding up some of the latest trends in the recovered paper sector. *Pulp and Paper International*, *45*(4), 25.

Van Haute, E. (2003). Optimyze: Enzymatic stickies control products. *Invest. Tec. Pap.* *40*(149), 47–51.

Vogel, J. H. (2002). Diatomite for stickies pacification. *TAPPI Fall technical conference and trade fair*. San Diego, CA.

Ward, J., Hensel, D., & Patterson, J. (1994). A new approach to the control of stickies. *TAPPI recycling symposium*. Boston, MA.

Williams, G. R. (1987). Physical chemistry of the absorption of talc, clay and other additives on the surface of sticky contaminants. *TAPPI pulping conference*. Washington, DC.

Woodward, T (1996). Controlling contaminants in the production and use of deinked pulp. *Wastepaper VII conference*. Chicago, IL.

Zeng, X., Fu, S., Yu, J., Li, K., Zhan, H., & Li, X. (2009). Study on the degradation of stickies in the pulps by complex enzymes. *China Pulp Paper*, *28*(2), 1–4.

14 Water Reuse, Wastewater Treatment and Closed-Cycle Operation

14.1 Water Reuse

The pulp and paper industry is one of the heaviest users of water. With water used in nearly every step of the manufacturing processes, pulp and paper mills produce large volumes of wastewater and residual sludge waste, presenting several issues in relation to wastewater treatment, discharge and sludge disposal. However, increasingly advanced treatment technologies, including innovative strategies geared towards water reuse and resource recovery, present viable solutions for pulp and paper manufacturers in terms of wastewater and solid waste management. Contaminants inherent to pulp and paper waste streams include effluent solids, sediments, absorbable organic halides (AOX), chlorinated organic compounds, and chemical oxygen demand (COD) and biological oxygen demand (BOD) contaminants. According to industry experts, approximately 85% of the water used in the pulp and paper industry is used as process water, resulting in relatively large quantities of contaminated water and necessitating the use of on-site wastewater treatment solutions. Treatment options include primary treatment such as clarification to remove solids and particulate matter, and secondary biological treatment processes for removing biodegradable organic matter and decreasing the effluent toxicity. Additionally, tertiary treatment technologies such as membrane filtration, ultraviolet disinfection, ion exchange and granular activated carbon can also be used to treat effluent water further to higher qualities (Tempesta, 2007).

The water loops in processing recovered paper generally decrease fresh water consumption. Process water is reused many times (Figure 14.1) (European Commission, 2001). The basic assumption applied is the backward process water flow in the systems countercurrent to the fibre flow. Recovered paper mills use whitewater (untreated, fibre enriched) from the paper machine for stock dilution in the mixing pump before the paper machine and in the stock preparation line similar to virgin-fibre-based paper mills. Some whitewater is clarified in save-alls using filtration, flotation or sedimentation to reuse the clarified water for the replacement of fresh water at the showers used for cleaning machine clothing (Hamm, 2000). Excess clarified process water then goes to the wastewater treatment plant. In some mills, purified wastewater is partly reused as process water after an additional polishing step.

Advantages of reduced fresh water consumption and water loop closure are as follows: (i) Lower volumes to external effluent treatment, i.e. the treatment plant can

Recycling and Deinking of Recovered Paper. DOI: http://dx.doi.org/10.1016/B978-0-12-416998-2.00014-3

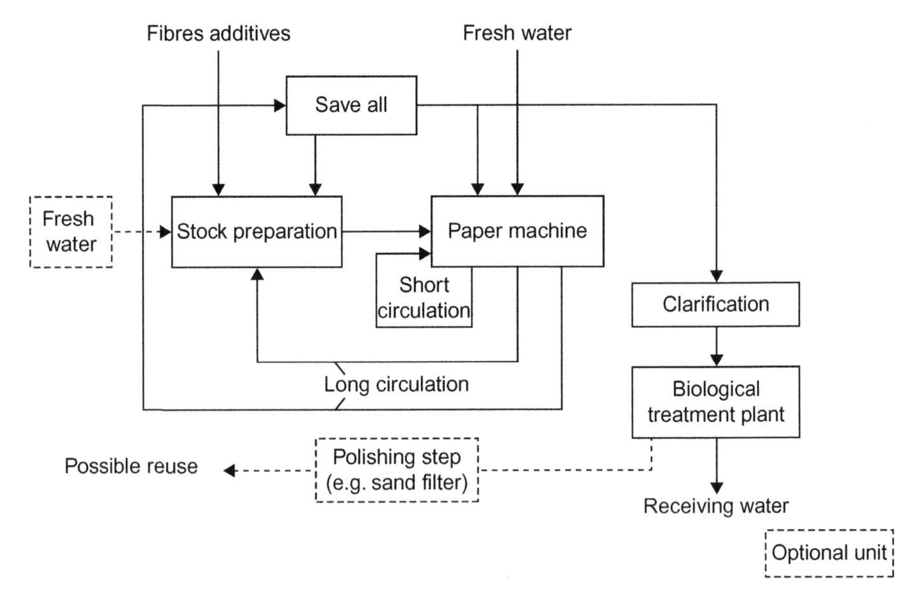

Figure 14.1 Water loop in recovered paper processing.
Source: Based on Hamm (2000)

be built with smaller hydraulic capacities and lower investment costs. Contaminants are more concentrated in the effluent, which often contributes to higher removal effectiveness. (ii) Lower costs of raw water. (iii) Lower losses of fibres and fillers. (iv) Lower energy consumption. (v) Higher temperatures in the process water systems result in faster dewatering on the paper machine wire.

The need for fresh water in a modern recovered paper preparation plant can be reduced to approximately 1 m^3/tonne. Consequently, only about 10% of the total fresh water demand in a modern integrated newsprint mill based on recycled fibres is consumed in the recovered paper preparation plant. However, in a tissue mill sometimes a lot of fresh water is required for efficient paper machine felt washing. In recycled-fibre-based mills without deinking, no fresh water is needed in the stock preparation.

The separation of water loops by insertion of thickeners leads to considerable change in the composition of the paper machine water loop with respect to the levels of organic and inorganic substances. This will lead to a changed additive regime in the wet-end that, in turn, will have its effect on the COD levels. In certain applications, the insertion of an extra water loop may drop the temperature in the paper machine loop, for example, when the paper machine loop is separated from the disperger and refiner that acts as a heating device for process water.

The main principle, whitewater flowing backwards in the system, countercurrent to the product, can be applied only in integrated pulp and paper/board mills. Recovered paper plants are, with very few exceptions, integrated with a paper mill. In stock preparation, the cleanness of the process water is less critical than in the whitewater of the paper machine. Therefore, the stock preparation uses, to a very large extent, excess clarified whitewater from the paper mill and internally clarified whitewater. However, potential drawbacks of water system closure need to be controlled. Otherwise, the build-up of suspended solids as well as dissolved organic and inorganic substances in the whitewater system may cause negative effects. The costs of this measure depend on the number and nature of water circuit rearrangements necessary and the type of additional installation needed. The driving force for recycling process water is to decrease the wastewater load from integrated recycled-fibre-based mills (CEPI, 1997; Pöyry, 1994; Senhorst, Zwart, Boulan, & Luttmer, 1997).

Recycled fibre processing has two main groups. The first is processes with exclusively mechanical upgrading, namely without deinking. This includes products such as test liner, corrugating medium, board and carton board. The second is processes with mechanical and chemical unit processes, namely with deinking and bleaching (Bajpai, 2006). This includes products such as newsprint, hygienic paper, printing and copy paper, magazine papers or market deinked pulp (Hamm, 2000). Wastewater loads from processes of the first group are slightly higher than those that must provide additional fibre processing such as deinking. The wastewater contains dissolved and undissolved substances of an organic and inorganic nature. Among the undissolved substances are fibres, fines, fillers and coating pigments. Removing these from the wastewater causes no problems using today's cleaning technology. The dissolved wastewater substances originate from different sources. Small amounts enter with the fresh water into the production process. Various substances are released from the virgin pulp. These are carbohydrates, lignin, inorganic salts and extractives. Of much greater importance are those coming from the recovered paper and the chemical additives used in paper production. The effect of chemical additives on the wastewater load is very complex. They form an extremely heterogeneous group and occur in different concentrations in a dissolved or dispersed state in the wastewater.

The composition of the wastewater is generally described by COD, BOD, total organic carbon, adsorbable organic halogen-containing compounds and suspended solids. The concentrations of these parameters depend on the specific fresh water consumption, which is in turn determined by the programme of grades produced in a paper mill. More indicative are the specific loads, which are the loads in relation to the unit of weight of the produced paper given in grams per tonne or kilograms per tonne. The specific emissions of recovered paper processing mills vary over a wide range. This has proved useful in distinguishing between mills with and without deinking. Packaging paper and board manufacturers working with recovered paper do not usually deink the recovered paper. Although no deinking chemicals are used, the untreated wastewater sometimes reveals a higher load of dissolved organic compounds than the wastewater of deinking mills. This is due to the greater share of strength-enhancing additives, especially starch, in the recovered paper and in the production process and to steam-volatile organic acids. This is primarily acetic acid formed in tightly closed water circuits by microbial transformations. This can be an important part of the organic content of wastewater.

The treatment processes in general will address removal of contamination arising from increased use of recovered fibres: adhesives, micro-organisms, food residues, etc.; microflotation and filtration for recovery of raw materials and reducing fresh water use; removal of dissolved and colloidal materials. The process water from recovered paper pulp recycling operations may also contain low percentages of hydrophobic contaminants, such as stickies: pressure-sensitive/hot-melt adhesives (Klungness, Tan, Gleisner, & Abubakr, 1998), food residues and micro-organisms that create problems to papermakers.

Contaminants could be pressure-sensitive adhesives and hot-melt adhesives in the process waters of paper mills using old corrugated container in manufacturing. Micro-stickies and anionic trash will also be present in the process waters (Klungness et al., 1998). Mills have to take action against the contaminants before they cause production and quality problems. Paper production has the problem of chemicals escaping into circulation water, which disrupts production and affects paper quality. The trend now is to reduce water consumption, which is leading to a further increase of these substances in process water (Bajpai, 2008).

Recovered fibre may also be contaminated with food residues and micro-organisms, which will eventually find its place in the process water circuit. Various well-proven and effective techniques are available for reducing the contaminant concentrations in mill water circuits, including optimising the water circuit design, passivation and/or fixation of detrimental substances and removal from the circuit with the paper produced, dissolved air flotation (DAF), filtration, evaporation and biological treatment methods. All these methods have a non-selective character in common.

Food residues and other dissolved organic compounds can be removed by biological treatment (Jimenez & Arrieta, 2004; Wirth, Kosse, & Welt, 2005): anaerobic (Bajpai, 2000), aerobic or sequential anaerobic–aerobic (Schabel & Hamm, 2007), whereas microbial contaminants can be removed by either sand filtration (Quadri, 2007; Wirth et al., 2005) or DAF (Huhtamaki, 2002; Rooks & Bottiglieri, 2003) processes.

Production of catalase enzyme during the metabolism of aerobic organisms and catalysing the breakdown of hydrogen peroxide has been reported to be one of the problems associated with hydrogen peroxide bleaching in the pulp and paper industry, particularly with deinking processes (Lenon, Fourest, & Vernac, 2000). Significant improvements have been introduced to control catalase production in deinking plants, notably the increased use of peroxide, the thermal treatment of the pulp or process waters and the addition of specific chemicals to deactivate or inhibit catalase. Specific treatments that have proved successful in controlling catalase generation include the following: catalase deactivation with sodium azide, 3-amino-1,2,3-triazole, treatment with sodium hypochlorite, glutaraldehyde, increased rate of peroxide introduction, treatment with chlorine and hypochlorite, treatment with chlorine dioxide, treatment with biocides, treatment with peracetic acid and thermal treatments. Centre Technique du Papier, France, has conducted a study over 2 years devoted to catalase-related problems in deinking plants (Lenon et al., 2000).

Microflotation is a well-known method of particle separation. In microflotation, also called DAF, microbubbles are created by dissolving air into water under pressure. When the air-saturated pressurised water is released, microbubbles are formed. Suspended solids and colloids attach to one another, or to air bubbles, owing to chemical, physical and electrical forces. These particle flocks then float to the water surface and are scraped off. Coagulation and/or flocculation chemicals are used to improve and enhance separation. DAF has long been used in the pulp and paper industry for different applications such as separating ash, fines, resin and other detrimental substances from the process, recycling water and/or fibre, and in all stages of process and effluent water treatment (Thurley, Niemczyk, & Turner, 1996). Microflotation is used in the pulp and paper industry for deinking, water purification and few other applications. The purpose of microflotation is to purify the internal circulation waters used in stock preparation. Microflotation is also used to improve the brightness of the pulp, remove ash and decrease the effluent flow to the wastewater purification plant. Filtrate from the pulp wash is used as feeding water for DAF. The purified water is recycled to stock preparation, where it is used for pulp dilution. The sludge from the microflotation process is led to the mill's wastewater purification plant for further treatment. Microflotation DAF systems are available in circular as well as rectangular configurations. Circular systems are preferred for recovery of larger particles like fibre, whereas rectangular ones are generally used for removing smaller and colloidal particles. There are several vendors who supply the microflotation system. Examples include FlooDaf™ microflotation of EIMCO Water Technologies; OptiDaf™ microflotation of Metso Paper (Strengell, 2004); Deltapurge NG™ microflotation of Voith Paper's partner Co., Meri Water Equipment, Germany (Zippel, 2003); and Flootek™ microflotation of Valmet. Screen baskets can be used to recover the fibre from the whitewater. Metso Paper has developed an innovative screen basket, called NimClear, designed to enhance fibre removal and water cleaning. The Kadant Petax™ fine filtration system is the first commercially available unit that removes particles from 1 to 20 μm from whitewater streams with up to 2000 parts per million and can produce water of a quality (less than 20 parts per million) that is acceptable for use in critical services (Koepenick, 2006). Petax™ removes this high quantity of particles without chemical flocculants, without the use of a precoat or sweetener stock and without a vacuum drop-leg. It is a highly engineered system that uses a series of submerged rotary discs and is available in multiples of five dick configurations. It can process flows of up to 250 m^3/h in a single unit with inlet solids concentrations up to 2000 parts per million. Visy Smithfield Paper, New South Wales, Australia, installed a Petax water recycling system supplied by Kadant Lamort to cut down effluent discharge costs and improve mill water quality (Koepenick, 2006). The Petax fine filtration system uses eight discs covered with a fine filter medium through which passes the filtrate into hollow collection voids in each disc and thence to the central hollow rotor core. The cakes of filter are doctored off, the filters are scrubbed and the clean filtrate is recirculated through the filter medium. Filters need replacing every 6 months, the process taking about 2 h. Each disc can process 7–12 m^3/h and installations of up to 25 discs are operating. Visy Smithfield uses the filtrate for a wide range of applications. All of the high-pressure showers serving the forming section are fed through the filter. It also handles the retention aid dilution and supplies the shower for the suction couch roll. Water consumption at Visy Paper has been reduced by 800,000 L a day, reducing the burden on the municipal drinking water supply.

A 240-tonne per day paper mill in the northern India (belonging to Century Pulp and Paper) uses recycled fibre. The deinking plant with a double loop and a paper machine was commissioned in February 2007. For water conservation, it uses poly-disc save-all for filtration of machine backwater and DAF for clarification of its wastewaters (Joshi, Thareja, Sharma, Aggarwal, & Aggarwal, 2008). All excess machine backwater is filtered through a poly-disc filter which gives three types of filtrates: cloudy, clear and super clear. Cloudy filtrate (80–100 parts per million total suspended solids) is recirculated again to the inlet of the poly-disc. Clear water (20–30 parts per million total suspended solids) is stored in a buffer tank, which is used for wet- and dry-end pulpers, centricleaners, consistency dilution, flushing pulp lines and the deinking plant. The super clear water as such (15–20 parts per million total suspended solids) or after further filtration (8–12 parts per million total suspended solids) is used in de-aeration tanks, top and bottom wire low-pressure showers, vacuum boxes and deckle lubrication systems, etc. Water from the vacuum system is filtered through a 60 mesh inclined screen and taken to backwater tank. The first loop of the deinking plant includes a pulper, protector system, screening, pre-flotation and washing, and dispersing. The second loop consists of oxidative bleaching, post-flotation, and disc and disperser. Backwaters of loops I and II are clarified and recycled through DAF-I and DAF-II with the help of coagulant and flocculants. There is significant reduction in total suspended solids and the turbidity of the water. Centricleaner rejects, ink sludge coming out from pre- and post-flotation, and sludges from DAF-I, -II and -III are collected in a sludge tank and from there taken to the sludge dewatering unit. Dewatered sludge is further squeezed through a screw press to reduce the moisture level to more than 50% before disposal. Water removed at the dewatering unit is clarified further through DAF-III before disposal to ETP. This way, the specific water consumption is 23 m^3/tonne of paper. Scaling, slime, stickies, wire and felt plugging are the common problems with water recycling. These are overcome by following various precautionary and preventive/remedial measures such as the use of anti-sticky chemicals at the disperser inlet and high-density towers/polytetrafluoroethylene (PTFE) coating on the first dryer and an updated cleaning system for screens (which remove stickies from the screen surface), deinking with enzyme, use of soft water as fresh water, an effective biocide programme for slime control and on-line wire and felt cleaning.

14.2 Wastewater Treatment

Aerobic and anaerobic biological processes are used at recovered paper processing mills (Hamm, 2000). Anaerobic technologies have grown in importance during the past two decades (Bajpai, 2000). They find use mostly with aerobic process technologies because an exclusively anaerobic clarification alone does not reach the specified emission limits. Anaerobic technologies presuppose a COD over 1500 mg/L in the wastewater to be treated. For this reason, anaerobic technologies have been applied almost exclusively to treat wastewater from the production of corrugating medium and test liner from recovered paper. More recent developments have been oriented towards the treatment of wastewater from tissue production and for partial streams from deinking systems.

14.2.1 Aerobic Biological Treatment

This technique has been used to treat effluents from recycled paper mills for over three decades to remove oxygen-consuming organic substances and other specific organic compounds. In aerobic biological processes, organic matter is oxidised to CO_2 and H_2O by micro-organisms in the presence of oxygen (Bajpai, 2001). A fraction of the organic matter removed is synthesised to form new bacterial cells. The micro-organisms are a mixture of bacteria, algae, fungi, protozoa, etc. Bacteria are the predominant species and are primary consumers of the organics. Factors affecting aerobic oxidation efficiency include the concentration of dissolved oxygen, pH, temperature and nutrients. To ensure aerobic conditions, it is generally accepted that the dissolved oxygen must be maintained above 2 mg/L. The optimum pH lies between 6.8 and 8.0, and often the pH of the mill effluent requires some adjustment. Systems currently used in pulp mill effluent treatment operate in the mesophilic temperature range (35°C). Because the temperature of most process streams exceeds this value, most mill effluents require cooling (from 60°C to 35°C) before entering biological treatment systems. Thermophilic operation (50–60°C) of a treatment system is, therefore, attractive and is the subject of research (Barr, Taylor, & Duff, 1996). Nutrients include nitrogen and phosphorus as well as micronutrients (trace metals). To support aerobic bacterial growth, nutrients with a BOD:N:P ratio of 100:5:1 by mass are considered suitable. Because mill effluents are normally deficient in nitrogen and phosphorus, an additional amount of these two nutrients needs to be added, particularly nitrogen. The amounts of trace metals or micronutrients in the effluent are normally considered sufficient. Preferred processes for the aerobic treatment of mill wastewater are activated sludge, trickling filters, fixed beds and aerated lagoons.

The activated sludge process is a high-rate biological process adapted largely from sanitary waste treatment. In it, there is a sludge settler after the aeration basin. The function of the settler is to separate the sludge from the treated effluent so that it can be recycled to the aeration basin and the bacterial concentration in the aeration basin can be maintained at a high level (2000–5000 mg/L). The high biomass concentration increases the rate of treatment, so the required hydraulic retention time for treating the same effluent is much shorter than in an aerated stabilisation basin, the required size of which is also greatly reduced. Two major activated sludge processes used in paper mills are air and pure oxygen activated sludge treatment systems. The activated sludge process has been used by the pulp and paper industry when the available land space is small and/or a low suspended solids concentration of the treated effluent is required. Activated sludge processes have been adopted in many paper mills around the world. Several full-scale processes are used to treat various pulp mill effluents, including those from kraft, paperboard, deinking, thermomechanical pulp and chemithermomechanical pulp, sulphite and newsprint mill operations (Bajpai, 2001; Buckley, 1992). Activated sludge processes are generally reported to remove much higher quantities of AOX than aerated lagoons. Removal efficiencies ranging from 14% to 65% have been reported.

In activated sludge treatment, the biomass, namely the activated sludge, and the wastewater are mixed to give a homogeneous suspension in the tank. The sludge–water suspension is removed from the aeration tank and directed to a secondary clarifier where the sludge settles. Some sludge is pumped back into the aeration tank. The excess is removed and recycled or disposed. Manufacturers of board, corrugating medium and test liner frequently reintroduce this excess sludge with the sludge from primary clarification into production.

A precondition for good degradation is sufficient residence time of the wastewater in the aeration tank and a sufficient amount of activated sludge. For this reason, the size of the aeration tank and the separating efficiency of the secondary clarifier have special importance. Important parameters for activated sludge treatment processes include the following: volume load, which usually refers to the BOD or COD in the inlet; sludge load, also referring to the BOD or COD in the inlet; and sludge volume index. This last parameter reflects the settling properties of activated sludge in the aeration tank. The target value should be less than 300 mL/g. If the sludge volume index exceeds 300 mL/g, bulking sludge with a very poor settlement behaviour can occur. Another parameter is the oxygen content (mg O_2/L) in the aeration tank, which should be above a minimum level to ensure efficient functioning of the process.

In addition to one-stage activated sludge treatment processes, two-stage processes have frequent use in the paper industry. The first stage involves a low oxygen content in the aeration tank with a high sludge loading. The second stage is then a conventional activated sludge process designed according to the type of wastewater being treated. Two-stage systems are less sensitive to load fluctuations and bacterial-toxic wastewater components. The reduced sludge load of the second stage allows easier removal of poorly degradable substances. The COD reduction efficiency of two-stage plants is superior to that of one-stage treatment. Table 14.1 shows conditions for one- and two-stage activated sludge treatment plants at paper mills (Hamm, 2000; Mobius & Cordes-Tolle, 1993).

The oxygen supply necessary for the biomass is usually provided by mechanical air input or pure oxygen in special cases. In most plants, the input air is used simultaneously for the mechanical mixing of the aeration tank. Besides controllability,

Table 14.1 Conditions for One- and Two-Stage Activated Sludge Treatment Plants at Paper Mills

Stage	Sludge Load (kg BOD_5/kg d)	Oxygen Content (mg/L)	Sludge Volume Index (mL/g)
One-stage aeration plant	0.25–0.30	2.0–3.5	<300
Two-stage aeration plant *Stage 1*	1.4–2.00	0.4–0.8	<100
Two-stage aeration plant *Stage 2*	0.12–0.22	2.5–3.5	<200

Source: Based on Mobius and Cordes-Tolle (1993) and Hamm (2000)

the decisive factor for the choice of aeration systems for recovered paper processing mills is the operating reliability. This primarily concerns the blocking of pressure-type ventilators and injectors caused by precipitation of calcium carbonate. Heavy-duty, pressure-type aerators are especially effective. The specific energy consumption related to 1 kg BOD eliminated is 0.6–3 kW h/kg. Especially efficient designs of the aeration systems are necessary to reduce this to less than 1 kW h/kg.

The nutrient content in the wastewater of recycled paper mills is generally low. So, to ensure efficient operation of the biological system, controlled addition of phosphorus and nitrogen is necessary. To control eutrophication problems, overdosing of nutrients and unnecessary emissions of nutrients should be prevented. This is applicable to new and existing mills. Removal efficiency is in the range 95–99% for BOD_5 and 75–90% for COD. Tables 14.2–14.5 show effluent properties with and without deinking before and after biological treatment (Hamm, 2000).

The aerated lagoon is a low-rate aerobic biological process. It is the oldest and simplest type of aerobic biological treatment system to construct and operate. The system consists of an aeration system for supplying dissolved oxygen. The wastewater to be treated is continuously fed into the aeration lagoon where bio-oxidation of the organic matter occurs, then directly flows out to the receiving environment. In some cases, a settling pond is installed after the aeration basin to remove biological and other solids from the treated wastewater. A large portion of the sludge

Table 14.2 Characterisation of Wastewater of RCF Mills Without Deinking Before Biological Treatment

Paper Grade	Specific Wastewater Volume (m^3/tonne)	BOD Load (kg/tonne)	COD Load (kg/tonne)
Board	3.6	1.8	3.5
Board	10	5.0	10.0
Board	–	4.7	8.4
Test liner fluting	3.4	9.7	19.4
Test liner fluting	–	6.1	12.3

Source: Based on Hamm (2000)

Table 14.3 Characterisation of Wastewater of RCF Mills Without Deinking After Biological Treatment

Paper Grade	BOD Load (kg/tonne)	COD Load (kg/tonne)
Board	0.06	0.77
Board	–	1.2
Test liner fluting	–	0.7
Test liner fluting	0.1	1.0

Source: Based on Hamm (2000)

Table 14.4 Characterisation of Wastewater of RCF Mills with Deinking Before Biological Treatment

Paper Grade	Specific Wastewater Volume (m^3/tonne)	BOD Load (kg/tonne)	COD Load (kg/tonne)
Newsprint, tissue	9	10	20
Newsprint	8	–	18
Newsprint	7	–	–
Newsprint	7.7	–	–
Recycled graphic paper	10	–	–
Newsprint, tissue	15	27	–
Tissue	8	–	–

Source: Based on Hamm (2000)

Table 14.5 Characterisation of Wastewater of RCF Mills with Deinking After Biological Treatment

Paper Grade	BOD Load (kg/tonne)	COD Load (kg/tonne)
Newsprint	–	3.1
Tissue	–	–
Newsprint, tissue	0.1	3.1

Source: Based on Hamm (2000)

produced settles in the lagoon or settling pond, where it subsequently undergoes auto-oxidation or endogenous respiration, which not only reduces the sludge production but releases and reuses the nutrients from the sludge. Because aerated lagoons do not recycle the biomass, they normally require a long hydraulic retention time (volume/volumetric flow) of 5–10 days, and the concentration of micro-organisms in the lagoon is too low (<200 mg/L). In both Canada and the USA (Turk, 1988; Wilson & Holloran, 1992), most of the early secondary treatment systems in pulp and paper mills, where available land space was not limited, were aerated lagoons. In China, India and other developing countries, lagoons are the major process for treating pulp mill effluent.

Aerated lagoons are less suitable for treating the wastewater of recovered paper processing mills. This system is primarily used when the BOD_5 concentration in the wastewater to be clarified is less than 100 mg/L.

Many successful applications of trickling filters in recovered paper processing mills have been reported. The development of special plastic support media considerably reduced the risk of blocking and allowed operation at high-volume load. Degradation occurs through an aerobic biofilm that forms on the support media sprinkled with the wastewater. Compared with activated sludge processes, trickling filters have a lower energy consumption and usually a higher operating reliability.

The temperature of the treated wastewater is usually lower. This is a desirable side feature. Trickling filters are not efficient when used alone to purify highly concentrated wastewater. For this reason, they operate primarily as the first stage of a two-stage biological process in which the trickling filter is followed by activated sludge treatment, usually without intermediate clarification. Odour development can be a disadvantage of trickling filters. Odours can be due to the anaerobic decomposition of wastewater or the evaporation of odorous substances contained in the wastewater. In both cases, total enclosure of the trickling filter may be necessary. In one German newsprint mill using 100% recycled fibres, two highly loaded trickling filters are integrated. The flotation plant that follows the biological clarification (trickling filter, activated sludge treatment and secondary clarification) acts as a 'police filter' to ensure observance of the strict legislative stipulations in the biologically purified wastewater effluent.

Biofilter processes are fixed-bed processes. These separate suspended solids and provide biochemical degradation of dissolved organic substances. The plastic or porous ceramic filter bed remains aerobic by atmospheric oxygen input. The bed height is 2.5–4 m. The decisive factors for controlled operation of biofilters are the rinsing and rinsing sequence. Care is necessary to ensure that the entire filter bed is completely aerated. The main advantages of biofiltration over activated sludge treatment are that the process is immune from bulking problems and that its energy demand is only one-third that of an activated sludge plant of the same capacity.

Special forms of aerobic fixed-bed reactors are biorotors. These reactors are normally used for small plants. They consist of a fixed medium in cylindrical or disc form rotating on a horizontal shaft in a tank. The biofilm on the medium is alternately in contact with wastewater and then with air. Usually, three to five rotors are connected in series.

Removal efficiencies of combined anaerobic/aerobic biological treatment are usually slightly higher (Bajpai, 2001). The overall treatment efficiency is over 99% for BOD_5 removal and over 95% for COD removal. However, up to now, the application of anaerobic treatment has mainly been limited to recovered paper mills without deinking. In the Netherlands, the wastewater of three recycled fibre processing paper mills is treated simultaneously by an anaerobic/aerobic treatment system. One of these mills is a deinking plant. However, there are some promising trials at a laboratory scale to apply combined anaerobic/aerobic biological treatment to deinking plants, too. Recycling of a part of the water after biological treatment seems possible. At least one newsprint mill using 100% recovered paper as raw material is using about 10% of the treated effluent (activated sludge plus sand filter) without any problem.

Usually the main effluent parameters are measured daily or at least a few times a week. Additional measurements to control the activated sludge system are necessary, for example oxygen content, sludge volume index, water flow and analyses of the biomass. During aerobic wastewater treatment, excess sludge is produced which has to be thickened, dewatered and further treated. A typical value for activated sludge plants is in the range 0.6 kg excess sludge per kilogram BOD_5 eliminated (as dry solids) generated during treatment. Thus, depending on the quality of recovered paper and on the process design, about 10 kg excess sludge from biological treatment per tonne of paper (on a dry basis) may be expected. Electrical energy is needed to aerate the active biomass (activated sludge) and for pumps. The specific consumption of energy for degradation/elimination of 1 kg BOD_5 amounts to 0.6–3 kW h/kg BOD_5 eliminated. If the system is well designed, a value of <1 kW h/kg BOD_5 eliminated can be achieved (Möbius, 1997). This value can also be used to compare expected operating costs of different wastewater systems.

Especially during the summer period, the wastewater treatment plant of recycled fibre paper mills may emit annoying odours. If the wastewater treatment is well designed and controlled, these can be avoided. Aerobic biological treatment for effluents from recovered paper mills has been successfully used for over 20 years (Bajpai, 2000). The phenomenon of bulking sludge also has to be controlled.

Numerous aerobic wastewater treatment plants are operated in European paper mills producing paper based on deinked recovered fibres (CEPI, 1997; Dutch Notes on BAT, 1996; Möbius, 1997).

14.2.2 Anaerobic Biological Treatment

Anaerobic biological treatment has been used for a long time to reduce the amount of organic pollutants in pulp and paper mill effluents, and extensive experience of this method is available (Bajpai, 2000). The reasons why anaerobic treatment is now becoming a viable alternative are its general advantages over the aerobic process. The most important of these are as follows: much lower electrical power demand especially for highly concentrated effluents, lower nutritional demand, production of energy rich biogas, lower sludge production and the biological sludge may be stored during relatively long shut-down periods without serious deterioration. Unlike aerobic biodegradation, anaerobic digestion of organic matter is a multistage, complex biological process that occurs in the absence of oxygen by different groups of bacteria. It is often divided into a four-step process: hydrolysis, acid formation, acetogeneses and methanogenesis. In hydrolysis, non-soluble organic compounds are hydrolysed by enzymes excreted from acidifying bacteria. Because the rate of this process is rather slow, it is often regarded as the rate controlling step for the entire anaerobic treatment. In acid formation, the hydrolysed compounds are converted into organic acids such as lactic acid, butyric acid, propionic acid and acetic acid by acid-forming bacteria, as well as into alcohol, hydrogen and carbon dioxide. In acetogenesis, organics from the previous step are converted into acetic acid, hydrogen and carbon dioxide. In methanogenesis, methane-forming bacteria convert the products from the previous step into methane as follows:

a. $CH_3COOH \rightarrow CH_4 + CO_2$
b. $4H_2 + CO_2 \rightarrow CH_4 + 2H_2O$

A very small part (<10%) of the degraded organic matter is converted into new bacterial cells.

The principal factors affecting the rate of anaerobic digestion of a wastewater are anaerobic conditions, good mixing for intimate bacteria–substrate contact, temperature, pH and alkalinity, presence of toxic substances, nutrients, trace metals, solids retention time, volatile solids, loading rate and hydraulic retention time. All the anaerobic treatment systems currently used in the treatment of pulp mill effluent operated in the mesophilic temperature range (35–38°C). The optimal pH for the growth of methanogenic bacteria is between 6.8 and 7.5, although methane production is possible in the range 6.0–8.5. Alkalinity in an anaerobic system must be sufficient to neutralise the volatile acids produced during the process to maintain an optimum pH (6.8–7.5). Commonly, a minimum of COD:N:P ratio of 100:1:0.5 is provided to ensure a slight excess of nitrogen and phosphorus.

The major anaerobic processes currently used for the treatment of pulp mill effluents include anaerobic lagoons, anaerobic contractor, upflow anaerobic sludge blanket, anaerobic fluidised bed and anaerobic filter (Bajpai, 2001).

As the optimum temperature for the growth of mesophilic anaerobic microorganisms is 35–38°C, for the anaerobic purification of the wastewater of recovered paper processing mills with tightly closed process water circuits, heat exchangers must cool the wastewater to this temperature range. The optimum pH of 6.8–7.2 is usually present in process waters of recovered paper processing mills. Controlled addition of acid and alkaline solutions is necessary to prevent pH fluctuations that would disturb the activity of the bacteria in the wastewater treatment plant.

It has been agreed at an international level to base the process technology rating of anaerobic reactors on the COD volume load. This rating criterion depends largely on the different processes. In contact sludge reactors, the COD volume load is 2–5 kg COD/m^3/day. Fixed-bed reactors operate with COD volume loads of 5–20 kg COD/m^3/day. Considerably higher volume loads are possible for the upflow anaerobic sludge blanket reactor. Usually a value of 5–15 kg COD/m^3/day is chosen. Newly developed pellet reactors can be loaded with 30 kg COD/m^3/day. With all reactor types, COD reductions of 70–80% are possible. The reductions of CODs are 80–95%. As part of this process, about 0.4 m^3 biogas is produced per kilogram of COD eliminated.

The produced biogas comprises 70–80% methane, 20–30% carbon dioxide, less than 2% hydrogen sulphide (H_2S) and traces of other gases. The heating value of the biogas is 22–30 MJ/m^3. In the anaerobic decomposition process, hydrogen sulphide forms from sulphur-containing organic compounds and sulphates. Caused by sulphate-reducing bacteria, the reaction is unavoidable because it is thermodynamically favoured over methane formation. Any dissolved H_2S that is present has a toxic effect on the entire bacteria population and can seriously disturb the methane formation process. Because sulphate is usually present in the wastewater of recovered paper processing mills, problems in the anaerobic treatment are often due to high sulphate concentrations.

Biogas must usually be cleaned before using for energy purposes. The most important consideration is elimination of H_2S because H_2S and SO_2 produced by incineration cause corrosion in a combustion plant. Caustic soda washers are primarily used for desulphurisation before burning the biogas. An interesting development is a bioreactor in which sulphide contained in the wash water is oxidised to elemental sulphur by micro-organisms. The sulphur is separated and can be reused elsewhere.

Problems can occur in anaerobic plants if the calcium concentration in the wastewater to be treated is more than 400 mg/L (Hamm, 2000). By reaction with carbon dioxide at a neutral pH, higher calcium concentrations lead to calcium carbonate precipitation. With processes using biomass carriers, precipitation can cause blocking of the growth surfaces. An additional risk exists for blocking the reactor, especially with fixed-bed reactors. To minimise calcium carbonate precipitation, running the anaerobic reactor in a lower acid pH range and accepting lower COD reduction levels is best. This is especially suitable where an aerobic treatment stage follows the anaerobic reactor.

14.3 Closed Cycle in Paper Mills Utilising Recycled Fibres

For zero discharge operation, paperboard and roofing felt mills using recycled fibre as furnish are good candidates (Bajpai, 2008). The reason is that their product quality standards are less demanding than for many other paper products, and typically they do not have extensive treatment facilities already in place. Several recycled paperboard mills have already implemented the closed whitewater systems. Stuart and Lagace (1996) have reviewed the typical process modifications required to implement zero discharge operation at paperboard mills. The items used to achieve this are not sophisticated and generally include primary treatment, adequate water storage, stainless steel metallurgy and often a segregated cooling water system. Chemical programmes must achieve consistently high first-pass retention of fines, clays, size and colloidal contaminants such as pitch. Recycled water must replace fresh water for most applications, including gland water and paper machine shower water. In a typical effluent treatment installation for a closed-cycle paperboard mill, the loop comprises removal of suspended solids by either a primary clarifier, DAF unit or screening equipment. Often, the sludge recovered in this treatment stage can be returned to the pulper to be incorporated into the board product. Clarified effluent is then sent to a large storage tank, which is sized to hold the mill. Finally, recovered water is passed over sidehill screens to protect downstream equipment from clarifier upsets. Screened water is then used throughout the mill to replace fresh water.

Most effluent treatment facilities at existing zero discharge mills can be divided into three main categories: (i) mills with facilities able to remove suspended solids effectively and achieve lean whitewater qualities under 200 mg/L suspended solids; (ii) mills with small or non-existent primary treatment facilities, whose total suspended solids in the recycled water are often above 1500 mg/L and can be as high as 5000 mg/L (mills in this category typically have problems such as plugged showers); (iii) mills that have biological treatment facilities.

St.-Laurent Paperboard Inc., a corrugated medium mill, situated at Matane, Quebec, Canada, successfully attained zero process effluent discharge (Rousseau & Doiron, 1996). The process effluent flow was steadily decreased from 6000 m^3/day to zero in October 1995. The mill was able to meet environmental regulations. Chemical dosages and injection points were optimised. Although the chemical dosages increased, the incremental costs were only a fraction of what it could cost to operate a secondary treatment plant. Furthermore, zero effluent resulted in an old corrugated containers yield increase from 85% to 92%, related to the higher overall retention of fines/clay into the paper. The reuse of process water also resulted in energy savings of about 5%.

With a closed-cycle operation, water leaves the process by dryer evaporation, with wet solid wastes such as from the rejects handling system, and as moisture in the paper product. To achieve zero discharge, less water must be added to the process on a continuous basis than is required to make up losses from the paper machine dryers. To maintain the overall balance, zero discharge mills segregate rain water, cooling water and non-contact pump seal water. These streams are not part of the water balance and the allowable fresh water make-up can be used for more critical applications. On average, slightly more than 1 m^3/tonne fresh water make-up to process is required to compensate for losses, mainly because of evaporation in the dryers. This implies that larger production mills have a greater fresh water make-up requirement with zero discharge operation. Some of the critical applications that must remain for fresh water, such as trim squirts and felt cleaning showers, are not directly related to production level, so it is easy for larger production mills to achieve zero discharge.

Green Bay Packaging Inc. implemented a closed-cycle system in the 1970s using reverse osmosis as the method for purging inorganic matter (Young, 1994). However, the mill is still operating in a closed-cycle mode without reverse osmosis or any other expensive technology. The mill achieved systems closure primarily by extensively replacing fresh water with process water and proper selection of chemical additives. The Haltown Paperboard Company mill used a similar approach for mill systems closure (Anonymous, 1993; Bajpai & Bajpai, 1999). In cases where high degrees of system closure were achieved, corrosion and bacterial growth in the paper machine area were observed. Several mills are attempting to achieve complete closure by treating the effluent stream biologically and reusing the treated effluent as the fresh water source.

Habets, Hooimeijer, and Knelissen (1997) described a biological treatment concept that has been in operation in a German zero discharge paper mills since September 1995. This mill is producing 400 tonnes/day of corrugated medium and test liner from water furnish and is being expanded to 1000 t/day. The biological treatment plant consists of anaerobic and aerobic stages. After the anaerobic process, where most of the COD is removed, the water is polished in an aerobic plant to remove residual BOD, sulphides and precipitated $CaCO_3$. The methane gas produced is scrubbed to remove the H_2S and is used for steam generation in a boiler. An indirect cooling facility brings the water temperature down from 55°C to 35°C. A buffer/pre-acid tank is used for pre-conditioning the water with nutrients before feeding it into the upflow anaerobic sludge blanket reactors of 720 and 1250 m^3 volume, respectively. The aerobic polishing is done in two parallel carousel-type aeration tanks, each with a volume of 900 m^3. In this way, 28 tonne/day of incoming COD is reduced to 1.0 tonne/day of soluble COD and 1.0 t/day of solids. The sludge solids produced during the aerobic polishing are added to the stock preparation to be recovered in the product. The total investment for the in-line biological plant was about 5%

of the total investment for the new machine. Part of the in-line biological treatment plant was started up in September 1995 for treating the process water of the existing machine. The new machine was started in October 1996. Within a few weeks of the start, the process water quality had changed drastically. The levels of COD, sulphate and calcium were reduced by 80%, 60% and 80%, respectively. The pH of the process water increased from 6.5 to about 7.5, which allowed the use of more effective papermaking chemicals at higher pH levels. Volatile fatty acid levels in the product were reduced to a much lower level than those usually found in similar mills with an open circuit.

Based on two surveys that characterised manufacturing conditions and water use practices at zero discharge paperboard mills (Barton, Stuart, Lagace, & Miner, 1995), the following conclusions were drawn. (i) Considerable whitewater surge capacity is needed for zero discharge operation, especially where white lined or coloured grades are manufactured. (ii) All but one of the mills visited by National Council for Air and Stream Improvement (NCASI) maintained zero discharge by providing either no external treatment or external treatment in the form of gravity clarification only. Only one of the mills used biological treatment. (iii) Operational difficulties, possibly encountered owing to closure of the mill water cycle, included (a) fogging in the machine area, (b) wet-end chemistry changes, (c) reduced felt efficiency, (d) reduced vacuum pump efficiency, (e) development of slime holes or other imperfections in the sheet and (f) odour, either in the sheet or emanating from the external treatment process. Furthermore, accelerated corrosion rates associated with zero discharge dictated the need for stainless steel process equipment and piping. Zero discharge implementation was shown to have a negative impact on the efficiency of a wet-end retention aid programme (Lagace, Pruszynski, Stuart, Anderson, & Warrick, 1997).

Methods were developed to reduce fresh water consumption in a mill for deinking old newspaper. Excess whitewater from a neighbouring paper mill was reused. The impact on pulp quality and loss of fine fibre after closing circuits was determined through simulation (Savu, Sarrailh, Marechal, & Paris, 2001). The deinking mill comprised an alkaline and an acid loop in sequence with a daily capacity of 150 t of deinked pulp. Closure strategies were developed from either the original procedure or a completely open procedure. Distribution of fines, fibres, total dissolved solids, ashes and some ions for each method in permanent use was determined by simulation. Varying degrees of closure, up to complete closure, were studied and strategies evaluated for feasibility. Different options were suggested and consideration given to possible increased consumption of chemical products and feasibility of use. A reduction of 86% in fresh water consumption was achievable. Effluents were more concentrated.

The catalase enzyme produced during the metabolism of aerobic organisms which catalyses the breakdown of hydrogen peroxide was implicated in the early problems associated with hydrogen peroxide bleaching in the pulp and paper industry, particularly with deinking processes (Lenon et al., 2000). Significant improvements have been introduced to control catalase production in deinking plants, notably the increase of peroxide introduction, the thermal treatment of the pulp or process waters and the addition of specific chemicals to deactivate or inhibit catalase. Specific treatments that have proved successful in controlling catalase generation include the following: catalase deactivation with sodium azide, 3-amino-1,2,3-triazole, treatment with sodium hypochlorite, glutaraldehyde, increased rate of peroxide introduction, treatment with chlorine and hypochlorite, treatment with chlorine dioxide, treatment with biocides, treatment with peracetic acid and thermal treatments.

Durango-McKinley Paper, a 220,000 tonne/year producer of linerboard based in McKinley County, New Mexico, USA, and serving the Los Angeles region, received a national environmental performance award from the US Environmental Protection Agency in 2005 in recognition of its environmentally friendly completely closed water cycle (Aukia, 2005). This mill lacks both the water resources necessary to sustain a standard mill process and the waterways to discharge water into; all the water used in the plant has to be treated and reused. In the closed system, process water is treated to avoid accumulation of impurities and chemicals. Water from various processes is fed to the reclamation process in separate streams according to the level of contamination. A pretreatment stage removes dirt, fibre and other solids, which are recovered and sold for soil improvement. Further treatment stages comprise DAF, microfiltration and reverse osmosis. Water lost by evaporation is replaced by water purchased from the nearby Tri-State power plant, from which the mill also obtains steam. To manufacture of its linerboard, Durango-McKinley uses 100% recycled board and old corrugated containers supplied from various sources in New Mexico. The processes in the mill must be operated more carefully than those in an ordinary system, as the paper machine and water treatment plant are interdependent on each other. As a result, the company strives to use as few additives as possible, employing a biocide programme from Kemira. A pulsed biocide dose is used to deal with variations in microbial contamination in the raw material.

References

Anonymous (1993). *Development document for proposed effluent limitations, guidelines and standards for the pulp, paper and paperboard point source standard.* US EPA-821-R-93-014 (p. 610). October.

Aukia, J. -P. (2005). Managing a fully closed water cycle calls for expertise, proper biocide: Durango-McKinley Paper is one of a few mills to operate a completely closed water cycle and was recognized for its success this spring by the EPA. *Pulp and Paper*, *79*(11), 31–34.

Bajpai, P. (2000). *Treatment of pulp and paper mill effluents with anaerobic technology*. UK: Pira International.

Bajpai, P. (2001). Microbial degradation of pollutants in pulp mill effluents. In S. Neidleman & A. Laskin (Eds.), *Advances in applied microbiology* (Vol. 48, pp. 79–134). New York, NY: Academic Press.

Bajpai, P. (2006). *Advances in recycling and deinking* (180 pp.). UK: Smithers Pira.

Bajpai, P., & Bajpai, P. K. (1999). *Recycling of process water in a closed mill systems – An introduction* (104 pp.). UK: Pira International.1 85802 277 0.

Bajpai, P. K. (2008). *Water recovery in pulp and paper making* (170 pp.). UK: Pira International.

Barr, T. A., Taylor, T., & Duff, S. (1996). Effect of HRT, SRT and temperature on the performance of activated sludge reactors treating bleached mill effluent. *Water Research*, *30*(4), 799–802.

Barton, D. A., Stuart, P. R., Lagace, P., & Miner, R. (1995). Experience with water system closure at recycled paperboard mills. *TAPPI Journal*, *79*(3), 191–197.

Buckley, D. B. (1992). A review of pulp and paper industry experience with biological treatment process bacterial augmentation: *TAPPI environmental conference* (pp. 750–810). Atlanta, GA: TAPPI Press.

CEPI (1997). *BAT in the manufacturing of pulp*, Report carried out by J. Poyry Soil and Water Ltd.

Dutch Notes on BAT (1996). Ministry of Transport, Public Works and Water Management, Directorate-General for Public Works and Water Management and RIZA (Institute for Inland Water Management and Wastewater Treatment), Dutch Notes on BAT for Paper and Board from Recycled Fibres, 1996, Contact: W.J. Luttmer, RIZA, P.O. Box 17, 8200 AA Lelystad, The Netherlands.

European Commission (2001). Integrated Pollution Prevention and Control (IPPC). Reference document on best available techniques in the pulp and paper industry. Seville: Institute for Prospective Technological Studies.

Habets, L. H. A., Hooimeijer, A., & Knelissen, H. J. (1997). In-line biological process water treatment for zero discharge operation at recycled fibre board mills. *Pulp and Paper Canada*, *98*(12), 184–187.

Hamm, U. (2000). Final fate of waste from recovered paper processing and non-recycled paper products. In L. D. Gottsching & H. Pakarinen (Eds.), *Papermaking science and technology* (7, pp. 506). Helsinki, Finland: Fapet Oy.

Huhtamaki, M. (2002). Profitable water management solutions to meet tightening water requirements. *Fiber Paper*, *4*(4), 10–13.

Jimenez, I., & Arrieta, J. (2004). Biological treatment in closed system for mills producing recycled paper. *Papel (Spain)*, *110*, 74–76.

Joshi, H. C., Thareja, L. K., Sharma, P. K., Aggarwal, S. K., & Aggarwal, D. C. (2008). Recycling & conservation of water in recycled paper plant: A case study of century pulp and paper. *IPPTA Journal*, *20*(3), 139–142.

Klungness, J. H., Tan, F., Gleisner, R., & Abubakr, S. (1998). Process water recovery: Dissolved air flotation compared to high shear rate separation. *Proceeding of the 1998 TAPPI recycling symposium* (pp. 123–132). Atlanta, GA: TAPPI Press.

Koepenick, M. (2006). Saving water, saving money – A new recycling system at Visy Smithfield produces fibre-free water without concern for colloidal content. *Pulp and Paper International*, *48*(10), 31–32.

Lagace, P., Pruszynski, P. E., Stuart, P. R., Anderson, N., & Warrick, S. B. (1997). Impact of zerodischarge implementation on wet-end retention at recycled fiber board mills. *1997 TAPPI environmental conference and exhibit* (pp. 679–686). Minneapolis, MN, 5–7 May, Book 2, Atlanta, GA: TAPPI Press.

Lenon, G., Fourest, E., & Vernac, Y. (2000). Microbial problems related to water reuse: Catalase control in deinking mills (bibliographic survey). *Investigaction Y Tecnica del Papel*, *37*(146), 559–571.

Möbius, C. H. (1997). Pulp and paper industry wastewater. Biological treatment of wastewater from pulp and paper manufacturing (in German only: Abwasser der Papier- und Zellstoffindustrie. Biologische Reinigung von Abwässern aus der Erzeugung von Papier und Zellstoff), 2nd ed., Munich, Germany.

Mobius, C. H., & Cordes-Tolle, M. (1993). Produktspezifische Abwasserzusammensetzung und Konsequenzen fur die Abwasserreinigung. *DasPapier*, *47*(10A), V53–62.

Pöyry, J. (1994). Techno-economic study on the reduction measures, based on BAT of emissions (water, waste, air) from the Paper and Board Manufacturing Industry, European Commission–DG XI, 1994.

Quadri, R. (2007). Process water recycling with the help of sand. *Industria della Carta*, *45*(5), 40–44. (Italian).

Rooks, A., & Bottiglieri, J. (2003). Water treatment: A solution roundtable. *Solutions*, 19–21.

Rousseau, S., & Doiron, B. (1996). Zero process effluent discharge attained at St.-Laurent Paperboard Inc., Matane, Quebec. *Pulp and Paper Canada*, *97*(9), T302–T305.

Savu, E., Sarrailh, S., Marechal, F., & Paris, J. (2001). Impact of system closure in a deinking mill. *Sixth research forum on recycling* (pp. 155–158). Magog, Quebec, 1–4 October (Montreal, Canada: Pulp and Paper Technical Association of Canada, 2001).

Schabel, S., & Hamm, U. (2007). Effluent-free papermaking: Industrial experience in the German and European paper industry. *Proceedings of the eighth international technical conference on pulp, paper, conversion & allied industry (paperex 2007)* (pp. 43–54). New Delhi, 7–9 December.

Senhorst, H., Zwart, L., Boulan, R. P., & Luttmer, W. (1997). A broad approach to water loop closure at Papierfabriek Doetinchem (and Appendix Report), RIZA-document 97.188X, RIZA, 1997.

Strengell, M. (2004). Metso supplies clean water: The consumption of fresh water in paper making has decreased significantly in the past 30 years. *Papermak Distrib*, *3*(2), 26–27.

Stuart, P., & Lagace, P. (1996). Review of process modifications for implementation of zero discharge at recycled fiber board mills: *Proceedings of the TAPPI minimum effluent mill symposium*. Atlanta, GA: TAPPI Press. (pp. 327–333).

Tempesta, M. (2007). Ultrafiltration unit. *International Papwirtsch*, *9*, 17–18.

Thurley, D., Niemczyk, B., & Turner, G. (1996). The use of dissolved air flotation to clean process water. *Proceedings of the 50th Appita annual general conference* (Vol. 2, pp. 765–769). Auckland, NZ, 6–10 May (Technical Association of the Australian and New Zealand Paper Industry, Carlton, Australia 1996, 2 vols, 890 pp).

Turk, S. (1988). Option for treatment of CTMP effluents. Report WTC Bio-7-1988. Environment Canada, Ottawa.

Wilson, D. G., & Holloran, M. F. (1992). Decrease of AOX with various external effluent treatments. *Pulp and Paper Canada*, *93*(12), T372–378.

Wirth, B., Kosse, J., & Welt, T. (2005). The importance of water circuits in the production of corrugating base papers. *Wochenblatt für Papierfabrikation*, *133*(16), 974–978.

Young, J. (1994). Green Bay packaging begins third year with closed water system. *Pulp and Paper*, *68*(11), 105–111.

Zippel, F. (2003). Water management in paper mills in Central Europe. *56th ATIP annual meeting*. Grenoble, France, 7–9 October (5 pp.) (Paris, France: Association Technique de l'Industrie Papetiere, 2003).

15 Environmental Aspects of Recycling

15.1 Introduction

Paper production has massive effects on the environment. Among others, the use and processing of raw materials have several negative effects on it. However, there are technologies that can moderate these negative impacts and have a positive economic effect. One of these is recycling. One of the main benefits of recycling is a double decrease in environment loading, known as environmental impact reduction (Bajpai, 2006). From the first viewpoint, natural resources are conserved on the side of manufacturing process inputs; from the second, the amount of harmful compounds leaking to the environment decreases on side of the manufacturing process outputs. Paper production from recycled fibres consumes less energy, conserves natural resources (namely wood) and decreases environmental pollution. The conflict between economic optimisation and environmental protection has received wide attention in recent research programmes for waste management planning. This has also resulted in a set of new waste management goals in reverse logistics system planning. A mixed-integer goal-programming model to capture the interrelationships within the paper recycling network system has been proposed by Pati, Vrat, and Kumar (2008). Use of this model can indirectly benefit the environment and improve the quality of wastepaper reaching the recycling unit. In 2005, the total production of paper in Europe was 99.3 million tonnes. This generated 11 million tonnes of waste, representing about 11% of the total paper production. During the same period, the production of recycled paper was 47.3 million tonnes, generating 7.7 million tonnes of solid waste (about 70% of the total generated waste in papermaking). This represents 16% of the total production from this raw material (Confederation of European Paper Industries (CEPI), 2006). Consumption of recovered paper has continuously grown in recent decades. According to CEPI, the use of recovered paper was almost equal to that of virgin fibre in 2005. This development has been boosted by technological progress and the good price competitiveness of recycled fibre, but also by environmental awareness – both at the producer and at the consumer ends – and regulation that has affected the demand for recovered paper. The European paper industry suffered a very difficult year in 2009, during which the industry encountered more downtime and capacity closures as a result of the weakened global economy. Recovered paper utilisation in Europe decreased in 2009, but exports of recovered paper to countries outside CEPI continued to rise, especially to Asian markets (96.3%). However, the recycling rate expressed as 'volume of paper recycling/volume of paper consumption' resulted in a record high 72.2% recycling rate after reaching 66.7% the year before (CEPI, 2006; CEPI Annual Statistic, 2010;

Recycling and Deinking of Recovered Paper. DOI: http://dx.doi.org/10.1016/B978-0-12-416998-2.00015-5

European Declaration on Paper Recycling, 2010; Huhtala & Samakovlis, 2002; Hujala, Puumalainen, Tuppura, & Toppinen, 2010; CEPI, 2008).

Nowadays, production and use of recycled paper is well established and widely accepted. The necessary technologies are available and it is possible to produce all types and qualities of paper using wastepaper as raw material. Paper recycling induces numerous environmental benefits.

15.2 Environmental Benefits of Recycled Paper

Recycled paper is better for the environment than virgin paper. It helps preserve forests, because it reduces demand for wood. It conserves resources and generates less pollution during manufacturing, because the fibres have already been processed once. And it reduces solid waste, because it diverts usable paper from the waste stream. Rigorous scientific research supports the benefits of recycled paper, and government agencies, environmental groups and many other large purchasers have adopted policies mandating its use.

Recycling reduces the total number of trees that are cut down to make paper and can reduce overall demand for wood. However, more importantly, paper recycling saves forests. By substituting used paper for trees, recycling reduces the overall intensity of forest management needed to meet a given demand for paper, and the pressure to convert natural forests and ecologically sensitive areas like wetlands into tree plantations. With recycling, not only are fewer trees harvested to make paper, but those trees that are harvested can be produced using methods that have less impact on the environment. Thus, recycling helps preserve the full range of values that forest ecosystems provide, including clean water, wildlife habitat and biodiversity. Every tonne of recycled fibre that displaces a tonne of virgin fibre results in the following reductions in usage:

- Wood, 100%
- Total energy consumption, 27%
- Wastewater, 33%
- Air particulate emissions, 28%
- Solid waste, 54%

Producing recycled paper uses much less total energy than producing virgin paper (Table 15.1) (Vest, 2000). Depending on the grade, producing recycled paper may use more or less purchased energy in the form of fossil fuels and purchased electricity. Virgin freesheet grades require slightly less purchased energy to produce than recycled ones, because some of their energy needs are met by burning wood-derived process waste. Virgin groundwood papers, by contrast, require more purchased energy to produce than recycled groundwood papers.

Incorporating recycled fibre reduces the amount of virgin fibre (namely trees) needed to produce a given amount of paper. This helps to reduce pressure on forests and the environmental impacts of commercial forestry.

Table 15.1 Energy Required for Producing Different Types of Pulp and Paper

	Pulp Production (GJ/tonne)	**Paper Production (GJ/tonne)**
Wastepaper pulp	5	13–17
Wood pulp	15–25	30–37
Kraft pulp	26–45	35–54

Source: Based on Vest (2000)

Table 15.2 Pollutants Generated in Different Paper Production Technologies

Pollutants (kg/tonne)	**Wastepaper Pulp**	**Wood Pulp**	**Kraft Pulp**			
			Sulphate Pulp		**Sulphite Pulp**	
			Conventional Bleaching		**Chlorine-Free Bleaching**	
BOD_5	0.1–1.5	2–50	26–81	2–25	20–30	1–8
COD	0.8–5	3–90	70–290	20–35	72–120	22–65
AOX	0.012–0.2	<0.02	3.7–7.0	0	3.7–10	0.22–1.2

Source: Based on Vest (2000)

The production of pulp from wastepaper shows very low water pollution factors compared with the other processes (Table 15.2) (Vest, 2000).

Life-cycle analysis shows that even after the energy used to collect, transport and process used paper is accounted for, the recycled paper system uses much less total energy than the virgin paper system. This is because the energy needed to recover used paper and get it back to the mill is quite small in comparison to the energy saved by using recovered paper rather than trees to manufacture new paper. Making virgin paper also requires energy to cut, collect and transport trees to the mill, all of which is derived from fossil fuels. And although the distances are shorter, the magnitude is greater: between 2.2 and 4.4 tonnes of wood are cut and transported for every tonne of virgin pulp, compared with 1.4 tonnes of wastepaper for a tonne of recycled pulp. Apart from reducing total energy use and greenhouse gas emissions, switching to recycled paper cuts emissions of other air pollutants such as nitrogen oxides (which contribute to smog) and particulates (which contribute to respiratory problems). It also reduces the volume and improves the quality of wastewater from the paper mill. Virgin paper production requires substantially more water and yields wastewater that has significantly higher levels of major water pollutants than does recycled paper production.

Recycled mills do generate more solid waste, mostly in the form of sludge, than virgin paper mills. However, that increase is more than offset by the reduction in solid waste that comes from diverting paper from the waste stream. And the same inks, coatings and fillers present in recycling mill sludge would go into the ground anyway if the paper were landfilled instead of recycled. Finally, recyclers are increasingly finding ways to reclaim and reutilise some components of recycled paper sludge, which cannot happen if that paper goes to a landfill or incinerator.

Paper recycling conserves wood and other forest resources, and reduces environmental impacts – energy use, air and water pollution, and solid waste – during manufacturing. Finally, by reducing paper's contribution to landfills, recycling avoids releases of methane and other pollutants, and reduces the need to site additional landfills where such releases would occur. Methane, a gas with 21 times the heat-trapping power of carbon dioxide, is a potent greenhouse gas and contributor to global climate change. The US Environmental Protection Agency cites municipal landfills as the single largest source of methane emissions to the atmosphere and has identified the decomposition of paper as among the most significant sources of landfill methane.

The environmental impacts of papermaking are undeniably complex, but the basic conclusion is simple: recycled paper is better for the environment than virgin paper. Switch to recycled paper and use the highest level of post-consumer recycled content you can find. It is the right thing to do for the Earth. To achieve a higher use of recovered paper as a raw material in the paper and board industry, it is necessary to increase its availability, by examining existing or new sources.

Recycled paper can be used for the same purposes as paper from primary sources. There are no indications that recycled paper creates any harm to human beings resulting from contamination by pathogens, chemicals (e.g. formaldehyde dioxins and furans) or heavy metals (e.g. lead). Deinking sludge is waste of normal category that can be landfilled or incinerated together with municipal solid waste.

At a European level, about 50% of recovered paper is collected from industry and trade, 40% from households and 10% from offices (CEPI, 2006). These percentages can differ greatly among countries, as well as among the collection systems used. These are exploited very differently. First, sources of high quality and high generation by volume are exploited, followed by others, of lower quality and low volume, e.g. households. The future potential for recovered collection lies in households, as other sources – mainly industrial and trade – have, to a great extent, already been tapped. However, it is well known that an extended recovered paper collection is always detrimental to quality (Blanco, Negro, Monte, Fuente, & Tijero, 2004). The quality of recovered paper has been considered as the major threat to extended use of recovered paper in the paper industry (Miranda & Blanco, 2010; Miranda, Bobu, Grossmann, Stawicki, & Blanco, 2010). Without efforts on source segregation and separate collection from households, which are major prerequisites for sustainable recycling, it will be very difficult to extend the use of recovered paper as a raw material in the papermaking industry (Grossmann & Bilitewski, 2005). Consequently, the participation of households is key to achieving higher recoveries of paper and to providing raw material of high quality with little, if any, need for sorting owing to separation at source. The participation of citizens is thus considered the touchstone for the success of any recycling scheme (Bolaane, 2006). However, to join collection systems requires investment of time, storage space, money and effort from the participants (Nixon & Saphores, 2009; Sidique, Joshi, & Lupi, 2010).

15.3 The Negative Effects of Paper Recycling

Recycling paper saves energy, reduces pollution, preserves trees and conserves landfill space. However, it is a messy process that uses caustic chemicals and produces harmful by-products and emissions. The industry is making strides in the development of more Earth-friendly techniques. But the best way to reduce paper-related pollution and energy use is to reduce paper consumption, which will decrease the demand for new or recycled paper.

Although recycling paper saves 28–70% – depending on the facility – of the energy used for making virgin paper, this saving is controversial because of the type of energy used in these two processes. Paper recycling uses fossil fuels, whereas production of virgin paper fibre uses waste products from timber to supply a high percentage of its energy requirements. Moreover, recycled paper is less energy-friendly compared to plastic. The paper bag recycling process uses 98% more energy than that for recycled plastic bags.

The paper recycling process requires the removal of inks from the used paper. Recycling facilities use different processes and the chemicals they use range from detergents to caustic chemicals, such as chlorine. Print from copy machines and laser printers is particularly problematical because it is not really ink but rather a plastic polymer that the printer or copier burns onto the paper. Removal requires chemicals that are much more caustic than standard deinking chemicals. Similarly, printing inks contain heavy metals and other compounds that require strong solvents.

When recycling facilities remove inks from paper, the waste makes its way into the water stream. Metals from printing inks, including copper, lead, zinc, chromium and cadmium, enter the water stream. Wastewater from paper recycling often contains dioxins as well, although experts are unable to determine their precise origin.

Wastepaper reprocessing produces a sludge that contains solids including small fibres, ink from the deinking process and fillers. This waste, including the heavy metals from the inks, is often sent to landfills. Incineration is an alternative, but this process releases dangerous emissions, including dioxins and hydrocarbons, as well as the heavy metals from the inks. The ash that remains after incineration is also consigned to landfills.

Recycling programmes use effective advertising to convince consumers they can help the planet by recycling their waste. This advertising, perhaps inadvertently, sends the message that consumption does not matter so long as what is not used is recycled. Because recycling creates pollution, reducing consumption is the most effective way to help the environment.

Environmental issues, in particular the appropriate handling of waste, are particularly important concerns. These issues merit attention and cooperation from all members of society. A country's waste management and public cleansing laws establish a basis for developing measures for managing waste. Although people have long known that recovered paper can be recycled, they are often less aware that inappropriate handling of such paper can turn it into waste. Accordingly, it is necessary to educate and gain cooperation from all paper users: residences, offices, industrial users and so on. Effective recycling is also important in terms of protecting forest resources and reducing energy use. For this reason, governments and local authorities are tasked with developing systems, application technologies and distribution methods that support effective recycling.

The relationship between paper recycling and the carbon dioxide it produces is a little complicated. Wood fibre is largely divided into mechanical pulp, which is made from wood chips by mechanical force, and chemical pulp, which is made using chemicals. Using recovered fibre, compared with using mechanical pulp, can reduce the total amount of carbon dioxide produced, which is the total of carbon dioxide from biomass fuel and carbon dioxide from fossil fuel. However, a comparison with chemical pulp shows that although using recovered fibre produces less total carbon dioxide, chemical pulp produces less carbon dioxide from fossil fuel because black liquor can be used as energy from biomass. The main ingredient of black liquor is lignin, which is separated and extracted from wood chips with chemicals. Therefore, chemical pulp produces less carbon dioxide from fossil fuel than recovered fibre. To avoid increasing carbon dioxide in the atmosphere, it is also important to try not to lose accumulated carbon storage in forests.

Toxic heavy metals and solvents in inks pose the most significant environmental problems. Lead, cadmium and barium are a few of the highly toxic heavy metals found in some inks. These can easily seep into the environment, contaminating soil and groundwater. Concentrations of heavy metals in the human body have been linked to serious neurological disorders, particularly in developing brains. Petroleum products and solvents, typically consisting of alcohol or different hydrocarbons, are also common ingredients used in inks. Most solvents are toxic and nearly all release volatile organic compounds (VOCs) into the atmosphere as they dry. VOCs are a well-known cause in the problem of ozone depletion.

Non-toxic, biodegradable vegetable-based inks were in standard use before the 1960s, when petroleum products started becoming a major ingredient in new 'higher-performing' inks. It is now clear that many of the so-called advancements attributable to petroleum-based inks have adverse effects on the environment. Recent technological advances have not only created better performing agri-based inks, but also developed ways of capturing significant amounts of VOCs, thus preventing them from entering the environment.

Publishers have access to more eco-options than ever before. Inks are no exception, with a variety of choices beyond standard inks, ranging from recycled, non-toxic and soy to other smart agri-alternatives. Although most inks are still petroleum-based, some ink manufacturers now substitute renewable and biodegradable resources (such as soy, linseed and corn oils) for most or all of the petroleum. Vegetable oil replacement ink meets the same specifications as petroleum-based ink, and it does not release a significant volume of VOCs into the air when it dries.

Agri-based inks are produced without toxins and biodegrade effectively, eliminating environmental, health and safety hazards. When alternative materials such as agri-inks are not adequate or available, facilities can reduce emissions through 'control and capture technology'. Using ventilation systems, emissions are guided to devices that capture the VOCs. Captured emissions are either destroyed or reused on similar applications. Ask your printer about agri-based inks made with renewable oils, recycled inks and inks that are heavy metal-free. Using more eco-friendly inks – vegetable or soy-based inks – is simple.

15.4 Health Dangers Caused by the Use of Recycled Paper

The German Federal Environmental Office started a project in 1981 to evaluate the applicability of recycled paper in modern office use. No significant differences were found between paper from primary sources and recycled paper in its use as writing, copying or printing paper. Like paper from primary sources, quality differences among recycled paper depend more on the quality of wastepaper, the production process, the additives used and the finishing process. For very high-quality printing, few problems were observed with recycled paper. Small remaining particles of former glue, lacquer or synthetics were found to affect the printing quality.

Wastepaper can sometimes contain pathogens that are expected to cause diseases. So, there was apprehension that recycled paper might be hygienically intolerable. However, during production of recycled paper, the paper passes through certain stages where it is heated to high temperatures and so it is practically sterilised. Investigations in Germany showed that recycled paper is hygienically acceptable even for food packaging. A few years ago there were some reports that recycled paper contained a higher amount of formaldehyde, which degases during use. Formaldehyde can come from some special paper and board qualities, where it is used during the production process. However, because these papers are very rare in Germany (the situation might be different in other countries) and mixed up with other formaldehyde-free papers during the recycling process, formaldehyde is hardly traceable in recycled paper. Its content is well below any limit set by the environmental legislation. However, contamination of paper by dioxins and furans seems to be more serious (Vest, 2000). During bleaching with chlorine, several organic chlorine compounds are formed which include dioxins and furans. Dioxins in paper can also originate from wood preservation chemicals or certain printing colours apart from bleaching. It was found that some chlorine-bleached papers contain 30–50 ng of toxicity equivalent (TE) per kilogram (ng TE/kg) of dioxins and furans. Chlorine-free bleached paper, on the other hand, often contains less than 1 ng TE/kg. During the recycling of paper, chlorine-bleached paper and chlorine-free bleached paper are mixed. Because no chlorine bleaching is applied during the production of recycled paper, the intake of dioxins from chlorine-bleached paper will be diluted by the other more or less dioxin-free papers. Nowadays, this results in a dioxin content of some 3–4 ng TE/kg for standard recycling paper. This figure is far below any limit given by environmental legislation. The dioxin and furan content in recycled paper has decreased gradually during the past few years since the paper industry has moved over more and more to chlorine-free bleaching processes. At the beginning of paper recycling, sometimes 50–60 ng TE/kg were measured in recycled paper. When the reasons for the contamination were researched, certain sources of dioxins were identified. For example, carbon paper was identified as a major source of chloroparaffins, and some cardboard boxes for exotic fruits contained reasonable amounts of pentachlorophenol (PCP). Both substances may form dioxins and furans during the papermaking process.

In developing countries, the situation may be different from industrialised countries. The paper industry might still use chlorine-bleaching processes. Additionally, environmental control about the use of certain chlorine-containing chemicals – herbicides, fungicides, wood protection chemicals, paper and cardboard, etc. – might not be as severe. So, there are more possibilities for the intake of dioxins and furans into the recycling paper production. However, dioxin and furan contents in the vicinity of 60 ng TE/kg are not dangerous and meet accepted environmental standards. As regards the contamination of recycled paper, heavy metals, in particular lead, has been discussed. During former printing processes that used printing types from lead, traces of lead were deposited together with the ink. There was apprehension that during recycling, in particular, if paper was recycled several times, lead might accumulate in the recycled paper. Studies have shown that the lead content of such recycled paper never exceeded any critical limit. At the same time, deinking technology was developed which removed most of the lead together with the ink. Although enriched in deinking sludges, lead was never a problem. The concentration of lead was far below the limits of, for example, lead allowed in sewage sludge that is suitable as fertiliser on farmland. The heavy metal problem disappeared with the disappearance of old-fashioned printing technologies (Vest, 2000). Nowadays, deinking sludges can be landfilled on normal landfill sites or can be incinerated in normal municipal solid waste incinerators without any special precaution.

15.5 Organochlorine Compounds

Organochlorine compounds are present in most paper products. They are generated during chemical pulp bleaching with elemental chlorine and chlorine compounds such as chlorine dioxide and hypochlorite and chlorination of fresh water. They are present in chemical additives used for paper production and paper conversion, especially epichlorohydrin-based wet strength agents and in printing ink pigments. Because of the potential human and ecological toxicological hazards of these compounds and a frequently insufficient biological degradability, environmental protection organisations and others have called for their minimal release into the environment. This has had consequences for the pulp and paper industry, especially in the mid-1980s. Polychlorinated dibenzodioxins and dibenzofurans that have a high toxicological potential were found to occur in waste and effluent from pulp mills and in the pulp and paper produced. In some parts of the world, the entry of organochlorine compounds into the paper recycling loop resulting from pulp bleaching with chlorine and chlorine compounds has decreased significantly in the 1990s. It will decrease even more in the future as use of elemental chlorine-free and totally chlorine-free pulps in paper production occurs. This is because of the substitution of chlorine and chlorine compounds in pulp bleaching, integration of the deinking process into recovered paper processing and modifying of paper additives containing organochlorine compounds. From an analytical point of view, the presence of organochlorine compounds in paper and paper products can be quantified simply by the absorbable organic halides (AOX) parameter. Bromo- and iodo-organic compounds have little importance in the pulp and paper industry. The measured values of AOX are therefore due almost exclusively to organochlorine compounds. AOX encompasses a wide range of groups of organochlorine compounds such as dioxins, furans, polychlorinated biphenyls, chlorophenols, chlorinated homologues of the aliphatic carboxylic acids, alkanes, alkenes, catechols, cymenes, guaiacols, naphthalenes, syringols, vanillins, and azo dyes and azo pigments of printing inks. AOX is not a sufficient indicator of the presence of toxicologically relevant substances in paper. It covers an excessively wide range of organochlorine compounds varying from harmless, easily degradable to toxic, persistent substances. Dioxin is the frequent name for polychlorinated dibenzodioxins (PCDD). A total of 75 PCDD congeners can contain between one and eight chlorine atoms at the positions in the benzene rings. Closely related chemically with the category of polychlorinated dioxins are polychlorinated dibenzofurans (PCDF), of which 135 isomers exist. They are simply known as furans.

Dioxins and furans are extremely toxic and reveal clear carcinogenic effects in tests using rats, mice and hamsters. Especially important from the toxicological point of view are dibenzodioxins and dibenzofurans that are chlorine-substituted in positions 2, 3, 7 and 8. With the aid of toxicity equivalency factors, higher-chlorinated dibenzodioxins and dibenzofurans are converted to the toxicity of 2,3,7,8-chlorinated dibenzodioxins and dibenzofurans. Total toxicity equivalency factors are calculated by multiplying individual PCDD and PCDF concentrations with the toxicity equivalency factor and adding the resulting values.

The pattern of tetrachlorinated dibenzodioxins and dibenzofurans considered typical for chlorine bleaching plays only a minor role in all papers containing recycled fibres (Hamm, 2000). In extensive testing conducted by the Fraunhofer Institute in 1995, dioxin was detected in recovered paper used to produce paper for corrugated board. Within the framework of this project, the dioxin content had fallen on average by a factor of three compared with 1992. Most investigated paper additives, including starch-based adhesives, lignin-based additives, cationic starch, dyes and anti-foaming agents, have almost no traces of dioxin. The only relevant dioxin comes from the use of rosin sizes. Its value averages 0.16 ng international toxicity equivalent (I-TE)/kg of paper produced.

The concentrations of PCP in paper containing recycled fibres show broad fluctuations (Hamm, 2000). Higher contents occur in packaging papers containing recycled fibres. This is because the process water systems are very effectively closed in this product segment. This results in a wider use of anti-slime agents. The PCP contents of papers produced from deinked pulp are considerably lower. With these papers, including recycled graphic paper, newsprint, and hygienic paper, the flotation or wash deinking or both processes act as a PCP sink. Despite this, the PCP concentrations of the papers concerned are occasionally above the analytical detection threshold.

The possibility of a PCP contamination of paper by PCP impurities in chemical additives used in paper production is only slight. The manufacturers of packaging paper and board from recycled fibres usually work with the same chemical additives as those used to produce the corresponding papers from virgin fibres. In the latter papers, the PCP contents are usually all below the analytical detection threshold. A contamination of paper by the use of PCP in forestry is therefore also unlikely. If this source of contamination was relevant, higher PCP contents should also be detectable in virgin fibre papers.

The PCP problem in paper containing recycled fibres could concern the paper industry for some time to come. Up to now, the toxicologists on the Committee of Experts on Materials coming into Contact with Food of the European Council (Strasbourg, France) have taken the position that no PCP should be detectable in paper intended to come into contact with foodstuffs, in the interest of preventive consumer protection. Consequently, this makes the methods of detection and their detection threshold especially important. So far, no method of analysis has been agreed upon.

Investigations so far on the dioxin contamination of paper containing recycled fibres have clearly shown that a correlation exists between PCP and dioxin contents. A further reduction in dioxin contamination is only possible if PCP can be kept from the paper recycling loop. The example of PCP shows that national prohibitions are insufficient to influence the contamination of secondary base materials when these materials are traded internationally.

15.6 Carbon Footprints

Life-cycle analysis by the Waste and Resources Action Programme (WRAP) in the UK studied the carbon balances between paper recycling and the disposal options of landfill and incineration. This report – Environmental Benefits of Recycling (WRAP, 2006) – reviewed over 108 global life-cycle assessments for paper. It rejected 99 for several reasons including incomplete life-cycle review and studied nine in detail. The results, perhaps reflecting rather the variable quality of the life-cycle analyses, showed enormous variation for paper recycling. A few situations existed where both incineration and landfill were found to be preferable to recycling, but these typically involved extreme situations that are not typically found in practice (Stawicki & Barry, 2009). In most studies, recycling was the preferred option and for reduced greenhouse gas emissions, 73% of situations favoured recycling. Almost matching results were obtained by the European Environment Agency in a similar study. Based on the study, WRAP considers that it is reasonable to say that recycling 1 tonne of paper and cardboard will avoid 1.4 tonnes of carbon dioxide equivalent compared with landfill, and 0.62 tonnes of carbon dioxide equivalent compared with incineration. This alone would have meant a saving of 38–85 million tonnes of carbon dioxide equivalent a year with current volumes of recycling in Europe compared with incinerating everything and landfilling everything, respectively. Whereas recycling usually has other climate-change benefits in addition to avoided emissions from landfilling or incineration, such as the savings available from the substitution of primary raw materials with secondary materials plus the energy efficiency gains achieved through the reprocessing stage, calculating specific carbon dioxide savings from recycling is not straightforward. The difficulty in such calculations lies in the fact that paper manufacturing is extremely varied, much more than other industries, in terms of production processes, raw material inputs, energy sources, output products and locations. Some estimates do exist, such as that by International Energy Agency (IEA), which calculated that each tonne of recycled pulp offers a net energy savings potential of 10.9 GJ/tonne (IEA, 2007). IEA Energy Technology Perspectives (2008) notes that between 10 and 20 GJ/tonne can be saved per tonne of paper recycled, depending on type of pulp and the efficiency of the pulp production it replaces.

Another study by WRAP (2008) determined the long-distance haul of recovered paper and plastics to Asia. It concluded that the benefits of recycling might outweigh emissions from shipping to the Asian market. The study found that emissions from shipping recovered paper from the UK to China are less than one-third of savings from recycling, and less than one-tenth if the ship carrying the waste would otherwise have returned empty. CEPI has set out a framework for paper manufacturers to measure their carbon footprints. It will be applicable to the paper industry and individual companies, to determine this number based on their own unique circumstances. This work is in line with later developments in standardisation such as PAS 2050. However, the methodology for quantifying all benefits from paper recycling is still not ready, as one should be able to allocate carbon dioxide costs and savings between the phases of primary production and recycling. For example, the method for calculating carbon storage in products is not fully developed yet.

References

Bajpai, P. (2006). *Advances in recycling and deinking* (180 pp.). UK: Smithers Pira.

Blanco, A., Negro, C., Monte, C., Fuente, E., & Tijero, J. (2004). The challenges of sustainable papermaking. *Environmental Science and Technology*, *38*, 414–420.

Bolaane, B. (2006). Constraints to promoting people centred approaches in cycling. *Habitat International*, *30*, 731–740.

CEPI (Confederation of European Paper Industries). (2006). Special Recycling 2005 Statistics – European Paper Industry Hits New Record in Recycling. 27.02.2011. Available from <http://www.erpa.info/images/Special_Recycling_2005_statistics.pdf>.

CEPI (Confederation of European Paper Industries). (2008). 2007 CEPI Key Statistics, Brussels, Belgium. <www.vnp-online.com/index.cfm?firm=vnp&fuseaction…p>.

CEPI (Confederation of European Paper Industries). (2010). Annual Statistic 2009. 27.02.2011. Available from <http://www.erpa.info/download/CEPI_annual_statistics%20 2009.pdf>.

EDPR. (2010). <www.intergraf.eu/Content/.../2008-09_ERPC_AnnualReport_2007.p>.

Grossmann, H., & Bilitewski, B.(2005). Closing the material loops: Paper recycling in Germany and Europe. *COST E48 Meeting*. Brussels, Belgium, 28–29 November.

Hamm, U. (2000). In L. D. Gottsching & H. Pakarinen (Eds.), *Environmental aspects, Papermaking science and technology* (Vol. 7). Helsinki, Finland: Fapet Oy. (p. 592).

Huhtala, A., & Samakovlis, E. (2002). Does international harmonization of environmental policy instruments make economic sense? *Environmental and Resource Economics*, *21*(3), 261–286. ISSN: 0924-6460.

Hujala, M., Puumalainen, K., Tuppura, A., & Toppinen, A. (2010). Trends in the use of recovered fiber – Role of institutional and market factors. *Progress in Paper Recycling*, *19*(2), 3–11. ISSN: 1061-1452.

IEA. (2007). <http://www.iea.org>.

IEA Energy Technology Perspectives. (2008). <www.iea.org/techno/etp/etp_2008>.

Miranda, R., & Blanco, A. (2010). Environmental awareness and paper recycling. *Cellulose Chemical Technology*, *44*(10), 431–449.

Miranda, R., Bobu, E., Grossmann, H., Stawicki, B., & Blanco, A. (2010). Factors influencing a higher use of recovered paper in the European paper industry. *Cellulose Chemical Technology*, *44*(10), 419–430.

Nixon, H., & Saphores, J. D. M. (2009). Information and the decision to recycle: Results from a survey of US households. *Journal of Environmental Planning and Management*, *52*(2), 257–277.

Pati, R. K., Vrat, P., & Kumar, P. (2008). A goal programming model for paper recycling system. *Omega*, *36*, 405–417. ISSN: 0305-0483.

Sidique, S., Joshi, S., & Lupi, F. (2010). Factors influencing the rate of recycling: An analysis of Minnesota counties. *Resources, Conservation and Recycling*, *54*(4), 242–249.

Stawicki, B., & Barry, B. (2009). The future of paper recycling in Europe: Opportunities and limitations. <www.cost-e48.net/10_9_publications/thebook.pdf>.

Vest, H. (2000). Environmental aspects of waste paper recycling. Gate Technical Information E1e <http://www.gtz.de/gate/gateid.afp>.

WRAP. (2006). <http://www.wrap.org.uk/wrap_corporate/about_wrap/what_does_wrap_do/ environmental.html>.

WRAP. (2008). <www.wrap.org.uk/press-releases?page=14>.

16 Uses of Recovered Paper Other than Papermaking*

16.1 Introduction

According to American Forest and Paper Association statistics, between 4.8% and 7.4% of all recovered paper is utilised outside the paper industry for construction purposes, insulation, mailing bags and other end uses. For Europe, no accurate data is available to estimate the recovered paper utilisation for applications other than the paper industry. According to the Confederation of European Paper Industries, in addition to the paper industry use, over 8% of collected paper is used in other applications: construction materials, animal beddings, composting and energy. Because this figure includes energy use that according to current European definitions is not recycling, recovered paper recycling for other end uses must be significantly lower than 8%. If it is assumed that 4–5% of recovered paper in Europe is utilised for applications other than the paper industry, the actual recycling rate would be about 4% above the level that the official statistics show. The reason for this is that, so far, the recycling figures have included only the paper industry end uses.

Recycling is considered better than composting in the hierarchy of materials management (Stawicki & Barry, 2009). This is because most energy in the paper life cycle is spent producing paper from virgin pulp as opposed to growing trees, transporting the paper to the retailer, collection and processing. Thus, recycling for papermaking purposes saves most of the manufacturing energy. Most of the energy will be lost when the recovered paper is composted. Moreover, the avoided costs of other raw materials – water, bleaching chemicals, minerals – will be lost. Paper is a non-renewable resource and should be recycled, not composted. However, the recent changes in the energy market proved that this very much depends on the market's needs. Neither option is better than the other. Recycling or composting occur where it is economically feasible and the benefits or disadvantages of each must be considered on a site-specific basis.

From the point of view of papermaking, promotion of recycling should take greater importance than composting or other uses that remove material from the paper recycling loop. Old corrugated cardboard, which is a secondary raw material of high quality, is a good material to be recycled and utilised within the paper industry and should not be collected among the food scraps for composting. Finally, almost all types of paper-industry-related waste like deinking sludge, biological sludges and fibre-containing sludges from wastewater treatment plants are suitable for composting. Solid wastes from recovered paper processing are available for

*Some excerpts taken from Bajpai (2012) with kind permission from Springer Science & Business media.

Recycling and Deinking of Recovered Paper. DOI: http://dx.doi.org/10.1016/B978-0-12-416998-2.00016-7

composting to a limited degree. In this type of solid waste, some of the materials like non-paper components, plastics, glass, etc. have a harmful effect on composting. So efforts should be made to separate these materials.

16.2 Biorefinery

A biorefinery is a facility that integrates biomass conversion processes and equipment to produce fuels, power, heat and value-added chemicals from biomass (Axegard, 2005, 2007a,b; Axegard, Backlund, & Tomani, 2007; Bajpai, 2012; Farmer, 2005; Kamm, Gruber, & Kamm, 2006; Thorp, 2005a,b; Thorp & Raymond, 2005; Thorp, Thorp, & Murdock-Thorp, 2008; van Heiningen, 2006; Van Ree & Annevelink, 2007; Wising & Stuart, 2006). A biorefinery can in concept be considered equivalent to that of a petroleum/oil refinery, except that in this case it makes use of plant materials (renewable resources derived from photosynthesising plants), whereas an oil refinery uses non-renewable fossil-derived petroleum. A petroleum refinery is an industrial process plant where crude oil is processed and refined into more useful products such as petroleum naphtha, gasoline, diesel fuel, asphalt base, heating oil, kerosene and liquefied petroleum gas. In many ways, oil refineries use much of the technology, and can be thought of, as types of chemical plant. The International Energy Agency Bioenergy Task 42 on Biorefineries has defined biorefining as the sustainable processing of biomass into a spectrum of bio-based products (food, feed, chemicals, materials) and bioenergy (biofuels, power and/or heat). By producing multiple products, a biorefinery takes advantage of the various components in biomass and their intermediates therefore maximising the value derived from the biomass feedstock. A biorefinery could, for example, produce one or several low-volume, but high-value, chemicals, and a low-value, but high-volume liquid transportation fuel such as bioethanol and biodiesel (Kamm et al., 2006). At the same time, it can generate electricity and process heat, through combined heat and power technology, for its own use and perhaps enough for sale of electricity to the local utility. The high-value products increase profitability, the high-volume fuel helps meet energy needs and the power production helps to lower energy costs and reduce greenhouse emissions from traditional power-plant facilities. Although some facilities exist that can be called biorefineries, the biorefinery has yet to be fully realized. Future biorefineries may play a major role in producing chemicals and materials that are traditionally produced from petroleum (Axegard, 2005, 2007a,b; Axegard et al., 2007).

In a world hungry for food and fuel, it is a questionable practice to displace a feedstock resource (e.g. starch, sugar, seed oils) derived from agricultural crops such as cereals (wheat, rice), maize (corn), sugarcane, soybean, oil palm, vegetables (potato, cassava) and fruits (dates) for the production of energy and chemicals. Instead, a non-food material such as cellulose (the most abundant organic chemical on Earth) offers an alternative feedstock from which chemicals (bio-based products) can be derived, which would not deprive a feedstock for the production of bioenergy. The biorefining concept is now widely accepted throughout the world, and technology based on this concept can provide a wide range of bio-based products that include biofuels (bioethanol and biomethane), bio-materials (fibres, pulp for paper manufacture) and a host of biochemicals through downstream fermentation and refining processes. In the past, these products were traditionally derived from petroleum feedstocks. However, developed technology now allows them to be produced from renewable biological resources such as those arising from agricultural processing wastes (cereal straws, sugarcane bagasse) and forestry (wood, wastes from processing lumbar such as bark, foliage, sawdust). Thus, cellulose can serve as a substitute feedstock for conversion into bioproducts that adds value to plant biomass residues generated through processing agricultural crops for foods as well as forests.

Biorefining is still largely unexplored territory and presents many research and business opportunities for the production of bio-based products from agricultural and forest residues. There are distinct advantages and benefits of a biorefinery using renewable feedstocks compared with chemical refining of petrochemical feedstocks. Foremost, plant biomass (agricultural and forestry residues) constitutes a renewable resource, whereas crude fossil oil has a limited and finite supply, which could run dry during this century depending upon its increasing consumption as an energy source. Plant biomass will therefore reduce the dependency on crude oil (non-renewable), and will create new businesses that will lead to new jobs, and generate wealth. A forest-based bioproducts industry will therefore promote new markets that use the unexploited products from the forest. It should also reduce waste streams and minimise pollution, and, globally, assist against climate change by reducing carbon dioxide build up in the atmosphere.

The research initiatives of biorefining are not limited to value-adding of low-value forest biomass, but will also consider the economic benefits derived from forest microbiota including bacteria, actinomycetes, yeasts and fungi. These micro-organisms colonise and inhabit the forest soil, the decomposing forest litter, and the phytophyllosphere comprising epiphytes, endophytes and phytopathogens. The forest microbiota therefore comprises a rich, diverse ecosystem that can specifically produce high-value products, such as enzymes, antibiotics and pharmacologically active compounds in combating human and animal diseases including cancer, and species that can serve as agents to resolve environmental problems (such as bioremediation of polluted sites).

One concern in the biorefinery approach for producing chemicals and fuels to replace existing petrochemical products is the availability of biomass, namely the competition with food. Biomass sources can be considered under the following categories: agricultural residues and energy crops; forest biomass; biowaste streams including municipal solid waste, construction, demolition, packaging and household wood waste, market and garden waste, food processing waste and sewage sludge; marine biomass, keeping in mind that of the global primary biomass production half is terrestrial and half aquatic. In these categories, recycled fibres are generally considered a component of municipal solid waste, and not in the forest biomass where they primarily belong. In fact, pulp and paper mills are ideal sites for biorefineries (Connor, 2007). The reasons are as follows: paper and forest product companies are efficient growers, harvesters, transporters and processors of biomass; pulp and paper mills are located near numerous sources of biomass, such as forest and agricultural residuals, and energy crops, and have existing infrastructure to ship finished product; paper mills utilise over several million dry tonnes of wood per year as a raw material. Additionally, these mills have ready access to a roughly equal amount of forest residuals and an even greater amount of agricultural waste and energy crops. Pulp and paper mills are familiar with producing power from biomass: currently they produce 60% of their power from wood residuals and spent liquors. Pulp and paper mills also have a highly trained workforce capable of operating energy and biorefinery systems. Pulp and paper companies are considering following general biorefinery configurations. Many investigators are assessing biorefinery technologies and pathways (Chambost & Stuart, 2007; Wising & Stuart, 2006). However, these analyses are typically limited to techno-economics and lack analyses of product design and new customer linkages. Forest biorefinery design must consider a multidimensional and multidisciplinary approach, including customer/product and process concepts to make the biorefinery not only a capital project, but an initiative that results in a successful business transformation. A systematic methodology for helping industry executives decide which technologies to implement and which biorefinery products to manufacture is needed.

By integrating forest biorefinery activities at an existing plant, pulp and paper mills have the opportunity to produce significant amounts of bioenergy and bioproducts and to increase their revenues drastically while continuing to produce wood, pulp and paper products. Manufacturing new value-added by-products, such as biofuels, bulk and specialist chemicals, pharmaceuticals, etc. from biomass, represents for some forestry companies an unprecedented opportunity for revenue diversification. The biorefinery builds on the same principles as the petrochemical refinery. In a petrochemical refinery the raw material is normally crude oil, whereas in the forest biorefinery the raw material is wood/biomass. The raw material stream is fractionated into several product streams. The products can be a final product or a raw material for another process. New technology is being developed that could be integrated into an existing pulp and paper mill, transforming it into a forest biorefinery. There are still significant challenges associated with these new technologies, but several of them look promising. Research is focused on biorefinery technology development in North America and around the world (Closset, 2004; Mabee, Gregg, & Saddler, 2005). However, these process technology development activities alone do not address most of the significant risks associated with implementing the forest biorefinery.

Chemical pulp mills are among the largest producers of bioenergy as about half of the raw material is reused on site, which enables energy-neutral production of chemical pulps. Some specialist chemicals are also produced in some mills. Paper recycling and deinking mills also produce bioenergy from solid rejects and deinking sludge (incineration, combustion and heat plants). 'Biofuels' in different forms (like fuel pellets from coarse rejects using the Rofire process, methane from the wastewater treatment plant) and other by-products (like cement products produced from deinking sludge incineration processes) are also produced in some mills. The implementation of real biorefinery concepts in the pulp and paper industry sector means that, besides the main products (pulp and paper), the mill sites would produce significant biofuels and chemicals measured in terms of quantities, added value and carbon dioxide impact in the replacement of products derived from fossil oil.

Various biorefinery processes will have to be implemented in pulp and paper mills to produce such biofuels and chemicals from basically wood or recovered paper. The current pulp and paper mill uses logs and fibre, chemicals and energy to produce commodity pulp and paper products (Connor, 2007; Thorp et al., 2008). Future mills, integrated forest biorefineries, will import regional biomass instead of purchased energy. They will expand the industry's mission from simply manufacturing low-margin paper products. Typical biorefinery processes include chemical, thermochemical and biochemical processes. The main driver in the biorefinery approach is that forest biomass should first be used for products. The biorefinery aims to create more value from the bio-based raw material provided by the forest-based sector. Compared with direct use of wood or recovered paper for energy production, the production of pulp and paper creates several times more added value and new revenue streams by producing 'green' power and new, high-value products such as biofuels and biochemicals, all while improving the efficiency and profitability of core papermaking operations. Figure 16.1 shows possible products from a pulp mill biorefinery (Axegard, 2005; Samiei, 2007). Concepts of recycled fibre biorefineries have not been studied much. Clearly, the development of such a new approach where recycled fibres are reused for paper first and for energy last would retain jobs and add value, compared with the possibilities where biorefineries are developed on municipal waste, including recovered paper. In the recycled fibre biorefinery concept, the biomass raw material is essentially fibres from recovered paper, with

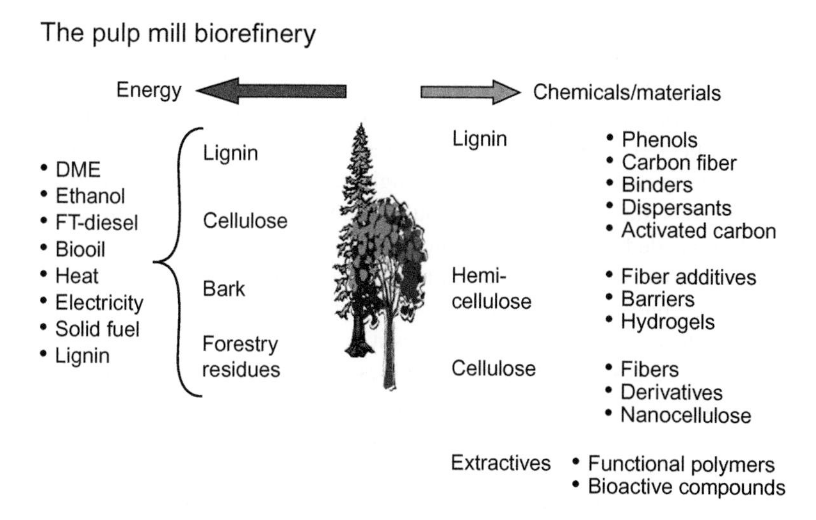

Figure 16.1 Possible products from a pulp mill biorefinery.
Source: Based on Axegard (2005) and Samiei (2007).

some other biomass like starch mainly from corrugated board as well as some non-renewable organic materials like plastics entering the paper recycling chain with the collection of packaging paper and board. Most of the large paper recycling and deinking mills already produce some bioenergy and other bioproducts, but not yet liquid biofuels or biochemicals, from their rejects. Coarse rejects can be incinerated to produce energy on site or secondary solid fuel, for example with the Rofire and FoxFire processes (Verhaevert, 2007), which are being used in Dutch mills. FoxFire is a Dutch company. It is turning paper mill waste into an energy product for the cement industry. Coarse rejects can also be used to produce various composite materials (Christmas, 2002). Deinking sludge incineration produces, despite the high ash content, some excess heat used on site. It is commonly used in the production of bricks and cement (Hanecker, 2007). A special incineration process (Verhaevert, 2007) treating sludge from different mills to produce cement has been reported in the Netherlands. Sludge can also be used to produce panel board for the building industry (Valtonen, Soukka, Martilla, & Hammo, 2000). Biogas is generated at the anaerobic effluent treatment plant, from the RP soluble biomass fraction. The biomethane produced in packaging paper recycling mills can lead to significant savings on gas consumption (Clumpkens, 2008). In addition to recycling and deinking rejects, RP sorting rejects should be included in the recycled fibre biorefinery.

16.3 Use of Recovered Paper in Construction and Manufacturing

Recovered paper can be also used for insulation in construction. The insulation is made from a composite of recycled paper consisting of newspaper, cardboard and telephone directories. Fibrous insulation can be used for building, renovation and re-insulation projects. These are more environmentally friendly than conventional fibreglass insulation and are denser and good for both sound reduction and increased airflow.

Old newspapers can play a major role in reducing the carbon footprint of buildings. Once collected, the newspaper can be processed using a combination of 'fibreising' milling and granulation to re-engineer the fibres of the newspaper into its original fibrous state. Then it is formulated so that it has the required structure and properties needed for the end application. Fibres can be created in forms that vary from an almost powder-like appearance (for use in paints) to longer fibres with excellent weave characteristics, which are essential in road aggregates or in sound insulation products. The insulation provides a complete seal to minimise heat loss, eliminating gaps, cracks or other cold bridges. It has very low embodied energy, requiring far less energy to produce than any other mainstream insulation material. Its fire resistance is improved through the addition of simple inorganic salts, enabling it to meet comfortably the fire protection standards required for timber-frame construction and conventional lofts. After additional treatment it is also resistant to biological and fungal attack, treated against insects and is unattractive to vermin. It is also safe to other common building components such as copper pipes, electric cabling and metal nail-plate fasteners. The traditional process involves grinding the paper in a hammer mill to produce a fibre fluff, which is treated with fire-retardant chemicals (boric acid or ammonium sulphate), then bagged and sold. The finished product usually consists of about 80% post-consumer waste, with the remaining 20% consisting of fire retardants and in some cases acrylic binding. When, eventually, insulation is removed from a building, it can be recycled again or incinerated safely without creating toxic waste or biodegradability problems.

Typical performance enhancements may include extreme resistance to fire (essential in house insulation), greater fluidity and dispersal, or the ability to absorb water more effectively. By weight, about 35% of municipal solid waste (before recycling) is paper and paper-related products. Recycling 1 tonne of newspaper eliminates $3\,m^3$ of landfill. Incineration of waste paper is also an option and is usually preferable because useful energy can be generated as a by-product. Organic materials, including paper, decompose in landfills, albeit sometimes slowly, releasing methane, a potent greenhouse gas. Many larger landfills now collect this methane for use as a biogas fuel. In highly urbanized areas, such as the northeastern USA and most of Europe, land suitable for landfills is scarce and must be used carefully. Fortunately, it is in such areas that collection of waste paper is also most efficient, as it creates more jobs for people and saves space in landfills.

16.3.1 Recycled Newspaper in Construction and Manufacturing

- Asphalt road surfaces – a development of traditional 'tarmac' road coverings – use old newspaper fibres to act as a thickener, which holds the liquid bitumen in place around the aggregate of the asphalt. Without it, the bitumen would drain away, allowing the aggregate to be more easily dislodged by traffic.
- Car brake linings similarly use recycled newspaper fibres to hold the lining material together.
- In concrete repair, recycled newspaper fibres are used as a carrier for an electrolyte solution, which enables concrete–steel reinforced structures to be re-alkalised or desalinated, both responsible for the corrosion of the steel reinforcement.
- Increasingly homes are being built using recycled fibres as wall and loft/roof insulation, providing levels of insulation far and above those of traditional houses. Passive house techniques and sustainable design principles make it possible for houses today to require no heating at all due to the high standards of insulation (provided there is adequate ventilation). Not only does this save money for the homeowner or tenant, but it vastly reduces the production of carbon dioxide gas normally generated by the heating boiler.

A further benefit of well-produced recycled newspaper fibres is that they remain inert throughout their lifetime, so that once their primary function has been completed they can still remain harmlessly within a manufactured component or product without posing any threat to the product's long-term performance or life expectancy.

16.4 Moulded Papers

Moulded paper is made from recycled paperboard and newsprint. It is considered one of the most sustainable packaging materials available today. It has enjoyed a renaissance due to the excellent environmental credentials of paper. Moulded pulp is produced from 100% recovered paper, with the process being tolerant of some contamination, typically more than a conventional sheet or board product. The method involves dispersing 100% recovered paper in water to form a homogeneous fibre suspension. This mixture is pumped into a vat containing a large revolving drum. Packaging moulds are mounted onto the surface of the drum. The liquid pulp is sucked into the moulds by a vacuum and water is simultaneously extracted. The wet packaging is then pressed and dried. The final product is subsequently post-pressed and printed or labelled. Machines used for this process very often work with a closed water system, in which the water is cleaned and used again within the process. Moulded pulp packaging can be shaped to suit a variety of applications. With the fibres being relatively soft and able to cushion items by absorbing any impacts, it has the ability to protect products and to promote their integrity, freshness and appearance. Small-scale, batch production of moulded pulp products for use in packaging has become technically and economically feasible with the advent of new moulded pulp technology. Over the years, the product range has considerably expanded. Today moulded pulp packaging is increasingly being used in electronic and consumer applications, from egg, meat, fruit and bottle packaging to disposable tableware, custom-made protective packaging and hospital disposables. Moulded pulp technology can also be used to manufacture high-end technical products (such as filter media) from material blends using virgin pulp fibres and synthetic fibres. Moulded fibre products are recyclable, biodegradable/compostable and easy to dispose of. They can be manufactured in different colours or special fibre mixes. Quality of recovered paper is very often not seen as an issue, although contamination with large solids such as compact discs in newspapers/magazines causes problems.

An increase in the number of companies manufacturing moulded pulp could be advantageous (and the market could be grown to accommodate their output), particularly as the process can utilise the lower grades that currently have little market value. There may in particular be an opportunity in regions where a proportion of paper collected for recycling is currently landfilled. A niche market may also exist in regions where there is an existing concentration of other (non-paper) manufacturing activities that could consume the moulded pulp packaging, such as electronics.

In recent years, moulded paper pulp packaging has undergone a transformation in terms of both manufacturing innovation and consumer acceptability. Owing to tough commercial packaging waste regulations and consumer-driven demand for greener packaging, moulded paper pulp packaging is now fast becoming a packaging medium that is not only acceptable and desirable but essential in terms of commercial environmental obligations. Waste paper is the main component of moulded paper pulp packaging, which is ideal as there is always a large amount of waste paper material available from the commercial paper recycling industry. Essentially the finished quality of a pulp moulding is determined by the quality of the waste paper material used in the manufacturing process. Printed waste paper destined for re-pulping is chemically deinked; this not only removes inks and other coatings but will also remove any other non-desirable substances that may be held within the waste paper. Corrugated cartons are a moulded paper pulp product. Corrugated fluting paper is manufactured from recycled paper pulp, which is moulded around giant rollers to create the corrugated effect. The average corrugated carton therefore contains between 70% and 80% recycled paper and always has done! Paper is the only raw material utilised in manufacturing moulded paper pulp packaging. For most pulp mouldings, waste paper is an acceptable material. Where waste paper is not acceptable for a pulp moulding, the moulded product must have a traceable raw material lineage. An example of this would be for a food container that is or has the potential to come into direct contact with the food itself. Environmental factors for this type of clean paper pulp moulding may demand that virgin grade raw material be sourced. Moulded paper pulp is in itself a fully recyclable product. When recycled through domestic or commercial waste packaging recycling programmes, moulded paper pulp is broken down and made back into waste-paper-based pulp slurry. When moulded paper pulp packaging is placed into landfill, it will biodegrade and compost within a matter of only a few months.

Moulded paper pulp is quickly becoming the raw material of choice for many types of protective packaging because of stricter waste regulations for plastics in packaging. Moulded paper pulp packaging is not only recycled but also fully recyclable. It is difficult to imagine a more efficient, timely and universally acceptable product. It is made predominantly from waste paper material that can be recovered, recycled and reused repeatedly. Its profile has slowly risen from protecting one of nature's most fragile creations, in the form of the humble egg box, to an internationally accepted medium for protective product packaging.

16.5 Recovered Paper as a Fuel

With the increase in paper consumption, more waste paper of poor quality is entering in the solid wastes circuit. For waste paper a better solution other than fibre recovery is its incineration (Gavrilescu, 2008). Although these residues are still largely dumped on landfill sites, there is a trend towards thermal processing. Mixed recycled paper represents a valuable source of energy for following reasons:

- Easy to separate from the waste stream
- Relatively homogeneous and mostly free from metals, putrescibles and other non-combustible materials
- Requires minimum processing to be converted into densified form of energy suitable for direct combustion
- Heating value is very high
- Low sulphur content
- Low nitrogen oxide (NO_x) emissions

One tonne of waste paper combusted as fuel will produce 9.8 GJ of thermal energy (Porteous, 2007). Table 16.1 shows the calorific values of some paper grades.

The moisture content of mixed paper waste lies between 6% and 28% by mass. If the moisture content of paper does not exceed 20%, heating values of mixed paper waste samples do not show big differences. Moisture contents exceeding 30% lead

Table 16.1 Calorific Values of Some Paper Grades

Paper Grade Mean Gross Calorific Value	kJ/kg (Dry Basis)
Newspaper	17,600
Kraft and cardboard	16,100
Boxboard	16,000
Tissue	15,200
Office paper	14,700

Source: Based on Gavrilescu (2008)

Table 16.2 Emissions Expected from the Combustion of Mixed Paper Waste

Particulates
Carbon dioxide
Carbon monoxide
Water
Oxygen
Nitrogen
Oxides of sulphur
Oxides of nitrogen
Ammonia
Hydrocarbons as hexane or methane
Aldehydes as formaldehyde
Organic acids as acetic acid

Source: Based on Gavrilescu (2008)

to low heating values for mixed paper waste and must be detrimental for shredding and for pelletizing of mixed paper waste. During paper combustion, all carbon in the organics is converted to carbon dioxide, with a trace amount of carbon monoxide. Hydrogen is converted into water vapour, and all chlorine is converted into hydrogen chloride and chlorine gas. Sulphur is converted to sulphur dioxide and sulphur trioxide. Nitrogen is converted into nitrogen oxides such as nitrogen monoxide or nitrogen dioxide, and nitrogen gas. Organics are decomposed into lower molecular mass species at high temperature in the flame zone. When some of these smaller species meet with oxygen and hydroxyl radicals, some combustion intermediates such as aldehydes, ketones, alcohols and acids are formed. In incomplete combustion, further oxidation of these intermediates is hindered owing to flame quenching by cold air, water or by a cold surface, or lack of additional oxygen. Therefore, these intermediates are emitted from the flame zone as products of incomplete combustion (Lyons & Kerstetter, 1990). Typical emissions that can be expected from the combustion of mixed paper waste are shown in Table 16.2.

The primary air pollution concern is the emission of particulate matter rather than gases and odours. A significant portion of particulate emission arises from the carry-over of mineral matter introduced with paper waste. The quantity of particulate

Table 16.3 Air Emission Limit Values (Solid Waste Incineration)

Pollutants	Concentration, mg/Nm^3 at 11% Oxygen
Total	10
VOC as C	10
HCl	10
HF	1
SO_x	50
NO_x	200
Cd	0.05
Hg	0.05
Heavy metals	0.5

Source: Based on EC Directive (2000)

emissions from an incinerator depends on the waste composition and the design and operation of the incinerator. Mixed paper waste, like all solid wastes and fuels, has substantial ash content. Some of the fly ash may settle in the bottom of the furnace or in other sections. Settlement reduces the emission rate depending on the settling rate, which is related to the size and mass of fly ash particles, flue gas properties and residence time in the furnace. The ash content of paper waste is a major factor influencing the particulate emission rates, and the amount of ash carried over generally ranges between 10% and 20% of the total (Lyons & Kerstetter, 1990). Directive 2000/76/EC of the European Parliament and of the Council of 4 December 2000 on the incineration of waste includes the limit values for incineration plant emissions (EC Directive, 2000). They concern heavy metals, dioxins and furans, carbon monoxide, dust, total organic carbon, hydrogen chloride, hydrogen fluoride, sulphur dioxide, nitrogen monoxide and nitrogen dioxide. The limits are listed in Table 16.3.

16.6 Recovered Paper for Use in Art and Handicraft

Paper has been used to an increasing degree as the basic ingredient for other forms of creative expression, in which the material aspect plays an important role. Recently, many artists have discovered the creative potential of paper as well as raw materials used for paper production, among those also recovered paper. Although this kind of activity will never provide a significant commercial outlet for recovered paper, it nevertheless has an important role to play in improving the consumer's perception of recovered paper as a valuable raw material and not as a waste product of society. Instead of throwing junk mail, catalogues and magazines, these can be used to make beads, paper designs and decoupage. It has been suggested to collect all kinds of papers, magazines and maps, and to recycle them to create incredible wrapping for ordinary objects that would otherwise go unnoticed.

References

Axegard, P. (2005). The future pulp mill – A biorefinery. *First international biorefinery workshop*, Washington, DC.

Axegard, P. (2007a). Swedish country report on biorefinery. *Second IEA bioenergy task 42, meeting*, Vienna, 4–5 October.

Axegard, P. (2007b). The kraft pulp mill as a biorefinery. *Third ICEP international colloquium on eucalyptus pulp* (pp. 6), Belo Horizonte, 4–7 March.

Axegard, P., Backlund B., Tomani P. (2007). The pulp mill based biorefinery. *Pulp paper 2007 conference. Biomass conversions* (pp. 19–26), Helsinki, 5–7 June.

Bajpai, P. (2012). *Biotechnology for pulp and paper processing*. Springer Science+Business.

Chambost, V., & Stuart, P. R. (2007). Selecting the most appropriate products for the forest biorefinery. *Industrial Biotechnology*, *3*(2), 112–119. Media, pp. 414.

Christmas, P. (2002). *Investigative Technical Paper*, *39*(148), 96–103.

Closset G. (2004). Advancing the forest biorefinery. *Presentation at forest products technobusiness forum*, Atlanta, GA, 26–27 October.

Clumpkens, M. (2008). The energy neutral mill. *COST strategic workshop improving energy efficiency in papermaking*, Amsterdam, 9–11 June.

Connor, E. (2007). The integrated forest biorefinery: the pathway to our bio-future. *International chemical recovery conference: efficiency and energy management* (pp. 323–327), Quebec City, QC, 29 May–1 June.

EC Directive. (2000). Directive 2000/76/EC of the European Parliament and of the Council of 4 December 2000, on the incineration of waste, Annex V, Official Journal of the European Communities, L 332/91, 28.12.2000, Brussels.

Farmer, M.C. (2005). The adaptable integrated biorefinery for existing pulp mills. *Presentation at TAPPI engineering, pulping, and environmental conference*, Philadelphia, PA, 28–31 August.

Gavrilescu, D. (2008). Energy from biomass in pulp and paper mills. *Environmental Engineering and Management Journal*, 537–546.

Hanecker, E. (2007). *Progress in Paper Recycling*, *16*(4), 34–37.

Kamm, B., Gruber, P. R., & Kamm, M. (2006). *Biorefineries – Industrial Processes and Products*. Weinheim, Germany: Wiley-VCH. ISBN: 3-527-31027-4.

Lyons, J.K., Kerstetter J.D. (1990). Mixed waste paper as a fuel: Environmental, economic, and market assessments, appendix A – Combustion of plastics, Report, A1. <http://www.p2pays.org/ref/24/23776.pdf>.

Mabee, W. E., Gregg, D. J., & Saddler, J. N. (2005). Assessing the emerging biorefinery sector in Canada. *Applied Biochemistry and Biotechnology*, *121–124*, 765–777.

Porteous, A. (2007). *Dictionary of environmental science and technology* (fourth ed.). New York, NY: John Wiley and Sons, Ltd.

Samiei, K. (2007). *Entropy analysis as a tool for optimal sustainable use of biorefineries*, Masters Thesis, University College of Borås, School of Engineering, BORÅS.

Stawicki, B., Barry B. (2009). The future of paper recycling in Europe: Opportunities and limitations. <www.cost-e48.net/10_9_publications/thebook.pdf>.

Thorp, B. (2005a). Forest biorefinery could open door to bright future for P&P industry. *PaperAge*, *120*(7), 16–18.

Thorp, B. (2005b). Transition of mills to biorefinery model creates new profit streams. *Pulp Paper*, *79*(11), 35–39.

Thorp, B., & Raymond, D. (2005). Forest biorefinery could open door to bright future for P&P industry. *PaperAge*, *120*(7), 16–18.

Thorp, B.A., Thorp B.A., Murdock-Thorp L.D. (2008). A compelling case for integrated biorefineries. <http://www.epoverviews.com/oca/Compellingcaseforbiorefineries.pdf>.

Valtonen, O., Soukka, R., Martilla, E., Hammo, S. (2000). Research Report EN B-135, Lappeenranta University of Technology, August.

van Heiningen, A. (2006). Converting a kraft pulp mill into an integrated biorefinery. *Pulp and Paper Canada, 107*(6), T141–T146.

Van Ree, R., Annevelink B. (2007). Status Report Biorefinery 2007, Agrotechnology and Food Science Group, U-Wageningen, November.

Verhaevert, J. (2007). Utilizing the energy potential of paper production by-streams. *KCPK international conference fibre raw material for paper & board*, 21 March, paper 12.

Wising, U., & Stuart, P. R. (2006). Identifying the Canadian forest biorefinery. *Pulp and Paper Canada, 107*(6), 25–30.

17 Future of Paper Recycling

17.1 Introduction

Recycled fibre has been an integral part of papermaking for many years (Bajpai, 2006). Many papermaking operations having relied on it as the sole or major part of the papermaking furnish. It has assumed even greater importance in recent times as the world has become environmentally conscious and a greater emphasis has been put on recycling (Grossmann, 2007). The material is fast becoming a globally traded commodity with countries that are rich in recycled fibre developing important export businesses around its trade. The main drivers for the use of recycled fibre continue to be availability at an economic price, legislative and voluntary agreements based on environmental pressure, and continuously improving deinking and other technologies for recovered fibre processing (Moore, 2002). Despite this increasing role as a raw material for the paper industry, issues such as price volatility of recovered paper, quality factors, food contact proposals and ever-increasing environmental pressures and considerations may cause a rethink in some sectors. This may result in some companies opting for virgin fibre, which greatly reduces issues related to quality and food contact. Quality and price issues can, to a certain extent, be influenced by the industry and its supply chain. Environmental considerations and any new proposals relating to food contact will require effective lobbying to ensure the overall industry is not disadvantaged.

It is estimated that in the very near future about half of the global amount of fibres used in papermaking will be recycled. There is no doubt that recovered paper as a raw material for the paper industry is here to stay. There should also be no reasonable doubt that utilising used paper products of an acceptable quality as raw material for new paper and cardboard is by far the most economical and ecological strategy compared with alternatives like thermal utilisation or even burning. An efficient and robust long-term strategy for recovery, reuse and recycling of materials is a cornerstone for improving the sustainability of the industry and society (Grossmann, 2007).

One of the decisive criteria for the future utilisation of recovered paper is its quality. This is a rather complicated function of the amount of undesirable components in the recovered paper, the state of the art of recovered paper treatment technologies, the accepted level of impurities in the finished pulp and the accepted level of residues produced during recycling (Grossmann, 2007). In the past the paper industry has always managed the problems associated with paper recycling very effectively: the menace of water-based inks, the problems of sludge disposal, the increasing problems caused by stickies, etc.

Recycling and Deinking of Recovered Paper. DOI: http://dx.doi.org/10.1016/B978-0-12-416998-2.00017-9

The main lessons on recycling may be presented briefly as follows:

- The quality of the individual fibre gets worse over time. There is no number of cycles by which the lifespan of a fibre can be characterised in a significant way. This is because the quality of a recycled fibre depends mainly on the wood species, the pulping process and its former individual history. On the other hand, an excessively recycled fibre might be of absolutely no use in a high-quality kraft paper whereas it still can be used as a natural filler in low-grade graphic paper.
- Recovering and re-using paper results in the establishment of a partly closed loop. The result is the build-up of concentrations of harmful matter, which is part of the loop depending on the degree of closure. To avoid this, purification processes have to be installed to keep the concentrations to a manageable level. This function is assumed by recovered paper treatment techniques. These techniques are still far from perfect.
- Increasing the utilisation rates beyond certain limits also requires the exploitation of resources that have not been used hitherto. These are, in many cases, low-quality or highly contaminated papers and cardboards. Even with highly sophisticated cleaning techniques, the quality of the recycled pulp produced from this raw material will be limited, yield will be low and the amount of residues produced during recycling, which have to been disposed of, will be high.

17.2 Issues and Trends Influencing the Availability, Quality and Quantity of Recycled Fibre

The trade in recovered paper has increased owing to globalisation of the world economy. China and other Asian countries will continue to import recovered fibre as domestic paper and board production continues to develop and expand and there is insufficient local, domestic collection to meet their fibrous raw material needs. Europe and North America will continue to develop to service this market demand. Demand for recycled fibre grades for use in the manufacture of packaging grades will continue to grow, with mixed paper and board grades being sourced along with old corrugated containers. The packaging sector is the biggest consumer of recovered paper and board in Europe. Recovered paper use rates are also high in newsprint, household and sanitary papers. Many European newsprint operations are already running on 100% recovered fibre. Although the printing and writing papers sector represents almost 40% of all paper and board production, it uses only limited volumes of recovered paper. Table 17.1 clearly indicates the two large groups of waste paper applications, which are packaging paper and information papers (printed material) (CEPI, 2009; Villanueva & Eder, 2011).

The potential for increasing paper collection and use exists both in developed and in developing countries. However, intensifying collection activities alone is not sufficient. Such activities have to be cost effective, efficient, acceptable and easy for end users (Moore, 2012).

The success of any efforts aimed at increasing collection rates is very much dependent on the public's outlook and environmental awareness. The green

Table 17.1 Amounts of Paper Consumed in CEPI Countries in 2008 (in Million Tonnes) and Breakdown by Use of Paper

Use of Paper	Paper Consumption in CEPI Countries (Million Tonnes), 2008	Share of Total (%)
Information: graphic papers	41.7	47
Packaging	36.0	41.0
Hygienic: household and sanitary	6.9	8
Others: special paper, etc.	3.4	4
Total	88.0	100

Source: Based on CEPI (2009)

awareness and focus of sustainability has supported the overall recycling culture. However, such awareness is not apparent in every country, and in Europe there are widely differing attitudes and support for recycling among the member states of the European Union. In Europe, recovery of waste paper and board continues to attract social and political attention. The European Commission is introducing several measures to implement such policies and is setting waste diversion and recycling targets that have to be met. This will mean that recovery will continue to be ahead of utilisation in most of the countries, which will lead to the continued development of export markets.

The recent economic recession has caused a change in paper use in many end-use applications, particularly in the printing and writing sector; paper demand has fallen and will not recover to pre-recessionary levels. The large decline in paper use experienced in printing and writing applications has been through the combination of economic issues but also, more importantly, the rise of electronic media. Paper is now part of a changing multimedia world of new technology. This ultimately means that overall paper demand will continue to decline and the availability of specific grades for recycling will be reduced. In contrast, paper for packaging applications has grown with improvements in overall economic performance, but also through the rise of environmental concerns which are favouring paper over plastic-based packaging.

Europe currently enjoys a good volume balance in paper production and consumption. It can safely be stated that for most grades the quality demands in the West European market are equivalent or superior to the products offered in any other market and are met by the European paper industry. The threat, in the graphic paper area, coming from electronic media seems to be far less than anticipated a few years ago. The main industrial threats for the European paper industry are probably large production units in tropical countries based both on fast growing hardwood (mainly eucalyptus and acacia) and recovered paper which is typically not available in those countries but has to be purchased from outside. Their products will be highly competitive partly also because of far lower costs for labour and energy. This results in a threat to the European paper industry, which does not only apply to the potential development of international prices for virtually all paper grades but also to a much stronger competition in international raw material markets.

Existing technologies for treating recovered paper are well developed and able to provide acceptable recycled pulp quality. However, such quality is often at the expense of energy costs and the amount of waste generated. More research and development effort is required to improve energy saving, and to develop highly selective separation techniques and process lines.

Recyclability will become increasingly important to ensure the increasing use of recycled fibres. It will need to be considered throughout the manufacture and conversion of new paper products to ensure that the recycling potential of the product is maximised.

The demand for recovered paper is getting stronger all over the world (Grossmann, 2007). Canada has become a leading importer of recovered fibre, which for the most part is sourced from the USA (Moore, 2002, 2004). Recycling is an expensive option for Canada as the paper industry was developed and built around virgin fibre use and most of the mills are located far from sources of recovered paper. Its relatively low population and large land mass add to the expense. In Japan, the corrugated sector is the largest user. Shifts in Japan's environmental policies have resulted in the creation of a recycle-orientated society. Changes in existing packaging laws have resulted in collected volumes for recovered paper exceeding internal recycled-based capacity. The growth and expansion of the Chinese paper industry has fuelled a huge demand for recovered fibre. Such demand is being satisfied by imports. The primary factor driving this sustained growth in imports has been the enormous investment in large, modern paper machines that are using recovered paper as the primary furnish. Corrugated-based mills have shown the greatest growth in volume. Recovery of waste paper and board continues to attract political and social attention in North America and Western Europe. For example, the European Commission in conjunction with governments of the Member States has introduced a range of measures to implement such policies and set waste diversion and recycling targets that have to be met. Such moves will mean that recovery will continue to be ahead of utilisation, and the already large export surplus of North America is still forecast to expand. Some of the key countries within Western Europe will continue to develop and enlarge their export markets. China and other countries within Asia will have a soaring recovered fibre deficit, as local paper and board production continues to develop and expand. They will not be able to find sufficient domestic collection for their raw material needs. North American and European collection operations will further develop to service this market need. Imports of packaging grades are expected to grow the most, with mixed paper and board being sourced alongside old corrugated containers.

Competing utilisation and, in particular, energy production may use significant quantities of recovered paper. If it is used as a renewable fuel, recovered paper does not require costly sorting. Such competing uses could make the overall recycle loop very efficient (Moore, 2012). The best material goes for paper recycling and the rest is used for energy production or other uses. There is a need to undertake continued research and development on sorting systems and maximising the overall effectiveness of the paper recycling loop.

Attempts have been made to improve the collection of material from households, as the recycling targets grow. Local authorities are the main mechanism for collection of household waste. In the UK, there are three methods by which local authorities collect or receive recyclable materials: kerbside collection schemes, where material is collected directly from the householder; 'bring' schemes, to which the public can bring their recyclables to recycling banks; civic amenity sites provided by local authorities for the reception of excess household and garden waste. The use of kerbside collections is well established globally. They are for the most part still in their infancy in the UK. Although they are generally referred to as being multi-material collection schemes, paper tends to dominate the material collected. The 'bring' schemes have traditionally been the single most important route by which household waste in the UK is recycled. Paper, predominantly newspapers, is the major material recycled.

For the paper industry to be able to use any recovered paper arising from such collection routes effectively, then the material must not be mixed with any other household material and should be sorted at source by the householder. Such requirements require effective collection and segregation systems to be provided by local authorities. However, recovered paper processing is moving from a highly source-separated collection and manually separated system to one involving mixed incoming streams and state-of-the-art manufacturing techniques. To improve the economics of the operation, waste paper collectors are now picking up large quantities of mixed paper and delivering them to sorting facilities for processing into graded paper stock. To handle these new householder and commercial fibre streams, sorting facilities have to change. Progress has been made using mechanical separation equipment, such as special disc screens. However, such technologies are limited to the mechanical properties of paper such as size and weight. To meet this challenge, research and development effort has been directed at developing high-speed optical based systems. One such system, PaperSort, is now commercially available. The continuing increase in the use of recovered fibre in newsprint and other printing and writing grades has seen a steady stream of new deinking plants and rebuilds around the world. It has also prompted development and refinement of the deinking process. These have included dual-loop water circuits, high-speed disc dispersion and pre- and post-flotation. Much of this development could be described as evolution. Developments in printing technologies have meant the arrival of new inks and toners in the recycling stream. To counter some of the potentially adverse effects of such materials, and to ensure the effective optimisation and efficiency of the deinking process in the future, a revolution in deinking technology will be required.

There are still no widely accepted methods to calculate the carbon footprint of products. Some proposals being put forward may produce detrimental, through unintended, consequences for paper recycling. Moreover, carbon dioxide emissions created by transport may limit the transportation of recovered paper over long distances. This may create local limits to paper recycling.

Recycled fibre has been an integral part of papermaking for many years and, despite the impact of the issues raised above, will continue to be so. Currently, China is an eager recipient of the material to service its rapidly expanding papermaking industries and will remain so. Packaging, particularly old corrugated containers, is and will remain the highest traded material. Declines in the use of paper in many printing and writing applications will result in the reduced availability of some recovered paper grades, for example old newspapers. Most predictors believe such declines will be slow over the short to medium term. Issues of availability, however, may become more apparent in the long term.

Continued developments in printing technologies have meant the arrival of new inks and toners in the recycling stream. To counter some of the potentially adverse effects of such materials, research has enabled alternative resins to be developed that are more easily deinked. The variability of incoming raw material and changes arising from seasonal and climatic variations have all been identified as factors that can affect the efficiency of deinking operations. Manipulation of deinking conditions and chemistry can limit such effects. Further minimisation can be achieved through the design and flexibility of deinking hardware. New systems from Lamort and Metso are examples of developments in new deinking hardware. Smithers Pira predicts that deinking will be further developed and used for newsprint manufacture worldwide, particularly in Asia, and older deinking lines in Europe and elsewhere will be continually upgraded and expanded; new developments will enable further use of recovered fibre in mechanical fibre-based publication grades; tissue grades will continue to use deinked pulp, but the market requirements for softness, high absorbency, etc., may prevent the universal use of recovered fibres in all grades; the use of recovered fibre in wood-free printing and writing grades is not going to increase much above existing levels unless there is legislative intervention (Moore, 2005). The final end use and market acceptance of a product depends on its fitness for purpose and cost. The quality of recycled fibre pulps is a function of the recovered paper grades, the processing techniques applied and the operating mode of the processing plant. Depending on their end use, recycled fibre pulps are subject to certain strength requirements and may only contain controlled amounts of filler. Recycled furnishes need to be free of adhesive contaminants (stickies) and for graphic paper end uses need to have appropriate optical properties. Currently such criteria are met, and recovered fibre and the resultant recycled fibre-containing products have market acceptance in many grades. Such acceptance may however become increasingly difficult to meet as end-use demands become more stringent, incoming raw materials become more difficult to recycle, fibre properties deteriorate or a combination of two or more of such factors.

Recent years have seen a steady increase in the use of recovered fibre in newsprint and other printing and writing grades, as well as several new and rebuilt deinking plants around the world. They have also seen the development and refinement of the deinking process. This has included dual-loop water circuits, high-speed disc dispersion and pre- and post-flotation, but much of this development could be described as evolution. Developments in printing technologies have meant the arrival of new inks and toners in the recycling stream. However, a revolution in deinking technology will be required to counter some of the potentially adverse effects of these materials, and to ensure the effective optimisation and efficiency of future deinking.

The future trend is that recovered paper contents in paper production will be maximised. The rotation time for virgin pulp (from seedling to pulp) is counted in years; in Nordic countries in decades. By comparison, in the best circumstances the recycling time for recovered paper is only weeks or some months: this will give recovered paper a good cost advantage compared with wood as a raw material. These issues, combined with environmental perception, give recovered paper a clear advantage compared with virgin wood pulp. Ever-increasing utilisation of recovered paper is, however, limited by quality and availability issues of highest quality recovered paper grades. The best white grades are already effectively collected from printing and converting sources, and good quality old corrugated container sources like converting and retail waste (pre-consumer) are already fully exploited. A further challenge arises from the use of recovered paper as an energy source. It remains to be seen how much of the recovered paper potential will be used for energy from waste despite the European Waste Hierarchy favouring material recycling.

Future changes in recycling technologies are believed to be rather due to future changes in recovered paper components as a result of the development of new paper additives, and new printing and converting technologies (Stawicki & Barry, 2009). The increase in recycling rates, namely in graphic paper grades, progress in papermaking technology and the development of new approaches for use of recovered paper components as bio-energy sources should also affect future recycling technologies. Future changes in paper additives, including green chemicals, coating binders and new types of pigments and fillers, such as organic or hybrid pigments, as well as the development of new types of fibre raw materials, such as functionalised fibres and nano-fibres from celluloses, could lead to significant changes in the recycling process. New green coating binders or new organic pigments should behave much differently in the deinking process, namely at the flotation step, and may lead to the development of new deinking chemistry. The development of new paper grades with reinforced properties, for example using nano-fibrillated cellulose, and functional properties like water, vapour and air barrier properties, may lead to changes at the pulping and dispersing steps. Future changes in printing technologies may also significantly affect the deinking processes. For example the development of water-based inks in the newsprint sector, using the flexo printing process, is still a major threat for deinking, although no increase of flexo prints is predicted in that sector (Moore, 2008). Digital printing will develop and lead to changes in the deinking ability of recovered paper. Laser inks are, for example, more difficult to detach from fibres, but the problem can be solved with reinforced dispersing, but at the expense of higher energy consumption. On the other hand, the development of ink jet printing will be more difficult to handle, as the inks are soluble and tend to redeposit on fibres. The development of flexo print and other water-soluble inks may in general lead to large decreases in the deinking yield. Future changes in converting technologies should also lead to changes in recycling and deinking. For example, the development of more efficient pressure-sensitive adhesives has led to softer and thus smaller adhesive particles after pulping, specifically smaller macro-stickies and more micro-stickies. Small macro-stickies, also called mini-stickies (Galland, Fernandez de Grado, Lascar, Delagoutte, & Kumar, 2009), cannot be removed by fine slot screening, so that more efficient flotation conditions have to be implemented. Future development of green adhesives and glues may lead to more colloidal particles and soluble fractions. This means that, as for water-soluble inks, recycling and deinking yields could be reduced in the future, and more efforts will have to be devoted to process water and effluent cleaning processes. Finally, further development of recycling in high-quality graphic papers will lead to higher quality requirements for deinked pulp. Such developments have already started and led to changes in the deinking process in terms of reduced yield.

References

Bajpai, P. (2006). *Advances in recycling and deinking* (180 pp.). Leatherhead, Surrey, UK: Smithers Pira.

CEPI (2009). Key statistics 2008: European pulp and paper industry. June. <http://www.cepi.org/docshare/docs/1/CGNAAHIAAGGEBEKPJCFHAAOIBDVAV9VNBW9QQ7B19YBD/CEPI/docs/DLS/KeyStats08-final-20090625-00011-01-E.pdf> (Accessed on March 2012).

Galland, G., Fernandez de Grado, A., Lascar, A., Delagoutte, T., & Kumar, S. (2009). Mini-stickies or the challenge of small macro-stickies management. Characterisation, mill survey and removal strategy. *Progress in Paper Recycling*, *18*(4), 12–23.

Grossmann, H. (2007). The limits of paper recycling – An European approach to identify and extend the limits of paper recycling. *TAPPI eighth research forum on recycling*.

Moore, G. (2002). Making the most of the 'Urban Forest'. *Pulp and Paper International*, *44*(5), 34.

Moore G. (2004). Future forecast for the recycled paper industry. *Eighth Pira International conference on paper recycling technology*. Prague.

Moore, G. (2005). Where next for deinking? Evolution must give way to revolution. *Pulp and Paper International*, *47*(7), 24.

Moore, G. (2008). Assessment of the changing print market and its consequences for the deinking fibre. *COST E46, final conference*. Bordeaux, 22–23 October.

Moore, G. (2012). The future of paper recycling. *Paper360°, September/October*, 43–44.

Stawicki B., & Barry B. (2009). The future of paper recycling in Europe: Opportunities and limitations. <www.cost-e48.net/10_9_publications/thebook.pdf> (Accessed on March 2012).

Villanueva A., & Eder P. (2011). European Commission, Joint Research Centre Institute for Prospective Technological Studies, Scientific and Technical Reports. End-of-waste criteria for waste paper. <ftp://ftp.jrc.es/pub/EURdoc/JRC64346.pdf> (Accessed on March 2012).

Printed and bound by CPI Group (UK) Ltd, Croydon, CR0 4YY
08/05/2025
01864886-0001